Principles of Abrasive Water Jet Machining

Springer
*London
Berlin
Heidelberg
New York
Barcelona
Budapest
Hong Kong
Milan
Paris
Santa Clara
Singapore
Tokyo*

Andreas W. Momber and Radovan Kovacevic

Principles of Abrasive Water Jet Machining

With 306 Figures

 Springer

Dr. Andreas W. Momber
WOMA Apparatebau GmbH, PO BOX 141820, D-47208 Duisburg, Germany

Radovan Kovacevic
Southern Methodist University, PO BOX 750337, Dallas, TX 75275, USA

ISBN 3-540-76239-6 Springer-Verlag Berlin Heidelberg New York

British Library Cataloguing in Publication Data
Momber, Andreas W.
  Principles of abrasive waterjet machining
  1.Water jets
  I.Title II. Kovacevic, Radovan
  621.2'0422
  ISBN 3540762396

Library of Congress Cataloging-in-Publication Data
Momber, Andreas W., 1959-
    Principles of abrasive water jet machining / Andreas W. Momber and
  Radovan Kovacevic.
       p.   cm.
    Includes bibliographical references   (p.).
    ISBN 3-540-76239-6 (casebound : alk. paper)
    1. Water jet cutting. I. Kovacevic, Radovan, 1947-
  II. Title.
  TJ840.M58 1998   98-4667
  621.9'3- -dc21   CIP

Apart from any fair dealing for the purposes of research or private study, or criticism or review, as permitted under the Copyright, Designs and Patents Act 1988, this publication may only be reproduced, stored or transmitted, in any form or by any means, with the prior permission in writing of the publishers, or in the case of reprographic reproduction in accordance with the terms of licences issued by the Copyright Licensing Agency. Enquiries concerning reproduction outside those terms should be sent to the publishers.

© Springer-Verlag London Limited 1998
Printed in Great Britain

The use of registered names, trademarks, etc. in this publication does not imply, even in the absence of a specific statement, that such names are exempt from the relevant laws and regulations and therefore free for general use.

The publisher makes no representation, express or implied, with regard to the accuracy of the information contained in this book and cannot accept any legal responsibility or liability for any errors or omissions that may be made.

Typesetting: Camera ready by authors
Printed and bound at the Athenæum Press Ltd., Gateshead, Tyne & Wear
69/3830-543210 Printed on acid-free paper

Preface

This book addresses the state-of-the-art knowledge in abrasive water-jet machining for a high-level engineering reading. The book attempts to support researchers as well as practical engineers who are considering this non-traditional manufacturing method for application in their field.

Following Chapter 1 that gives a brief review of the water jet technology, the topic is developed following the 'history' of the abrasive water-jet machining. Chapters 2 through 4 discuss the properties of the 'tools' of the process: the abrasive particles, water jet and abrasive water jet. Chapter 5 reviews the material-removal process in the materials that are machined by the abrasive water-jet technique. Chapter 6 and Chapter 7 show how to model and optimize the manufacturing processes. Chapter 8 addresses aspects of the quality of abrasive water-jet cutting. Chapter 9 covers advanced applications of the technique, such as turning, milling, piercing and polishing. Finally, Chapter 10 reviews works that focus on the control of abrasive water-jet machining processes.

We wish to thank all colleagues and friends that actively contributed to the success of this book: Mr. P. DuBois, Arlington, USA, Dr. E. Geskin, Newark, USA, Drs. T.J. Kim and D.G. Taggard, Kingston, USA, Drs. H. Louis and J. Ohlsen, Hannover, Germany, Dr. R. Mohan, Tulsa, USA, Dr. K. Neusen, Madison, USA, Dr. C. Öjmertz, Goeteborg, Sweden, Dr. H. Peters, Sigless, Austria, Dr. M. Ramulu, Seattle, USA, Mr. T. Sausen, Duisburg, Germany, and Dr. S. Shimizu, Hiroshima, Japan.

This book project was heavily supported by the Alexander von Humboldt-Foundation, Bonn, Germany, through several fellowships for the first author. Since we both are Humboldt-fellows, special thank is addressed to the foundation.

We also would like to thank the Department of Mechanical Engineering of the Southern Methodist University, Dallas, TX, USA, the Department of Mechanical Engineering of the University of Kentucky, Lexington, KY, USA, and WOMA Apparatebau GmbH, Duisburg, Germany, for their active support.

Finally, we would like to thank Mrs. Lorie Carter who edited and proofread the manuscript.

Andreas W. Momber
Radovan Kovacevic

November 1997
Dallas, Texas

Nomenclature

A	collision number
A_C	step-area parameter
A_h	cutting rate
A_P	particle projection-area
A_W	wall projection-area
b	cut width
b_B	bottom cut-width
b_{IDZ}	initial-damage zone width
b_P	particle breadth
b_T	top cut-width
c_{Air}	sound velocity in air
c_{Col}	collision factor
c_D	particle friction-coefficient
c_f	fluid-jet friction coefficient
c_L	radial crack length
c_M	target-material sound velocity
C_N	contact number
c_P	abrasive-material sound velocity
d^*	particle-size distribution parameter
d_0	orifice diameter
d_{50}	median particle diameter
d_{Circle}	grain-projection diameter
d_F	focus diameter
d_T	final workpiece diameter after turning
d_{jet}	fluid-jet diameter
D_M	target-material average grain size
d_{max}	maximum projection diameter
d_{min}	minimum projection diameter
d_P	abrasive-particle diameter
d_P	geometric mean particle-diameter
d_{SH}	delivery-hose diameter
d_{ST}	statistical particle diameter
d_{Dst}	statistical wear-particle diameter
d_{WP}	workpiece diameter
e	lateral cutting distance
E_A	abrasive water-jet kinetic energy
E_{AbsD}	dynamic absorbed fracture energy
E_{Act}	activation energy
E_d	erosion-debris kinetic energy
E_D	energy dissipated by erosion-debris formation
E_{Diss}	energy dissipated in a workpiece
E_{Ex}	abrasive water-jet exit kinetic energy
E_F	particle-fragmentation energy
E_{Fr}	fracture energy

viii   Nomenclature

E_M	target-material Young's modulus
E_N	erosion number
E_P	abrasive-particle kinetic energy
E_{SP}	specific energy
E_{STR}	stress-wave energy
E_T	plastification energy
E_{thr}	material threshold energy
E_W	water-jet kinetic energy
F_0	particle circularity
F_A	abrasive water-jet impact force
f_{comp}	compressibility factor
f_D	interference frequency
F_D	virtual mass force
F_{DR}	drag force
F_{EX}	abrasive water-jet exit impact force
F_f	wall-friction force
F_G	pressure-gradient force
F_{jet}	fluid-jet impact force
f_P	particle impact-frequency
F_P	particle impact-force
F_{shape}	particle-shape factor
F_{UL}	uplift force
F_V	Basset force
f_W	stress-wave factor
F_W	water-jet theoretical impact force
F_{Weff}	water-jet effective impact force
Fr	Froude Number
g	gravity constant
h	depth of cut, workpiece thickness
h_1	geodetic height
h_{burr}	burr height
h_C	'cutting-wear' depth
h_{cr}	radial crack depth
h_d	'deformation-wear' depth
h_D	particle diving-depth
h_{max}	maximum depth of cut
H_F	focus-material hardness
h_F	particle impact-height
h_{Fcr}	critical particle impact-height
h_{IDZ}	initial-damage zone depth
H_{Knoop}	Knoop hardness
H_M	target-material hardness
H_{Mohs}	Mohs hardness
H_P	abrasive-material hardness
h_{RC}	rough-cutting zone depth
h_{ref}	reference cut depth

h_S	separation thickness
h_{SC}	smooth-cutting zone depth
h_T	turning depth
H_V	Vickers hardness
I_W	water jet impulse-flow
k_R	reaction velocity
K_A	acceleration coefficient
K_0	stiffness
K_{Ic}	fracture toughness
l	liquid-film thickness
l_{cr}	crack length
l_F	focus length
L_h	cut length
l_P	particle length
M	particle mass
M_F	modulus of fracture
m_i	absolute sieve-overflow mass
m_M	removed material mass
M_O	sieve-overflow percentage
m_P	abrasive-particle mass
M	one-dimensional material-removal rate
m_A	abrasive-mass flow rate
m_L	air-mass flow rate
m_M	erosion-debris mass flow rate
m_W	water-mass flow rate
n	particle-size distribution modulus
N_{corner}	number of corners on a particle
m	order of reaction
N_D	debris number
N_m	machinability number
N_{mP}	machinability number for piercing
n_P	number of passes
N_P	abrasive-particle number
N_T	machinability number for turning
n_V	acceleration ratio
n^*	particle-size distribution parameter
p	pump pressure
P	perimeter
P_{abr}	abrasive-evaluation parameter
p_{air}	air-flow pressure loss
p_{amb}	ambient pressure
p_{at}	atmospheric pressure
p_{cr}	transition pump pressure
P_E	erosion probability
p_{mix}	mixing-chamber pressure
p_{solid}	solid transport pressure-loss

x   Nomenclature

p_{thr}	material threshold pressure
p_V	pressure loss
P_W	water-jet power
PSD	peak of the power-spectrum density
q_H	heat flux
q_M	quality-level parameter
q_O	overlapping ratio
Q_A	abrasive-volume flow rate
Q_L	air-volume flow rate
Q_M	mixture-volume flow rate
Q_W	water-volume flow rate
r	radial direction
R	ratio m_A/m_W
R_a	roughness average
r_{corner}	particle corner-radius
Re_P	particle Reynolds number
r_E	particle elongation-ratio
r_F	particle flatness-ratio
R_F	force ratio
R_G	gas constant
r_{jet}	fluid-jet radius
R_{th}	thermal shock factor
R_E	erosion resistance
R_f	roundness parameter
R_{ku}	kurtosis
R_P	particle roundness
R_q	root-mean square roughness
R_{Sk}	skewness
R_W	striation (waviness) height
r_W	water jet radius
R_Y	peak-to-valley height
R^*	thermal shock-factor
S	particle-shape factor
s	cutting-front (arc) length
S_M	grain-sample surface
S_P	particle sphericity
S_R	particle roundness
t	exposure time
T	temperature
T_0	room temperature
t_D	dwelling time
t_{Dr}	drilling time
t_H	hole diameter
t_{JP}	jet-penetration period
t_{opt}	optimum exposure time
t_p	particle thickness

t_P	piercing time
T_R	cut taper
t_{TL}	focus lifetime
T_U	turbulence
t_W	water-jet impact period
v	traverse rate
v_0	effective water-jet velocity
v_{0cr}	threshold water-flow velocity
v_{0F}	water-jet velocity in the focus
v_{0th}	theoretical water-jet velocity
v_A	abrasive water-jet velocity
v_C	characteristic particle velocity
v_{cr}	threshold traverse rate
v_e	fluid-jet backflow velocity
v_{EX}	slurry exit velocity
V_{FR}	material removal by fracture
v_{jet}	jet velocity
v_M	erosion-debris velocity
V_M	removed material volume
v_L	air-flow velocity
v_{opt}	optimum traverse rate
v_P	abrasive-particle velocity
v_{P0}	abrasive initial velocity
v_{Pcr}	critical particle impact-velocity
V_P	particle volume
v_{pipe}	focus flow velocity
V_{PL}	material removal by plastic flow
v_{rel}	relative velocity
v_{rot}	rotational speed
$\overline{v}_P$	average abrasive-particle velocity
V_{ch}	characteristic volume-removal rate
V_D	material-displacement rate
V_M	material-volume removal rate
v_{thr}	threshold traverse rate
W_F	focus-wear rate
w_i	workability index
x	standoff distance, jet length
x_C	water-jet core length
x_L	traverse distance
x_{Tr}	water-jet transition zone length
α	impact angle
α_T	water-jet energy-transfer efficiency
χ_G	gas fraction
χ_i	energy-dissipation parameters
χ_M	material-removal efficiency parameter
Δp	pressure difference

Δp_V	vacuum pressure
ε	strain
ε_M	specific erosion energy
$\dot{\varepsilon}$	strain rate
φ	jet (particle) impact angle
φ_C	abrasive water-jet exit angle
φ_P	particle water interaction-angle
ϑ	particle deflection-angle
η_h	global energy-transfer parameter
η_T	momentum-transfer parameter
ϕ_D	particle disintegration-factor
Φ	relative depth of cut
γ_M	target material surface-energy
γ_l	energy loss parameter
Γ_M	work of fracture
Λ	stress-wave energy parameter
λ_S	striation wave-length
μ	water jet momentum-transfer parameter
μ_{Mix}	mixture viscosity
μ_{Sus}	momentum-transfer parameter
μ_W	water viscosity
ν_M	target material Poisson's ratio
ν_P	abrasive material Poisson's ratio
θ_A	abrasive inlet-device angle
θ_S	shock angle
ρ_A	slurry density
ρ_{jet}	fluid-jet density
ρ_L	air density
ρ_M	target-material density
ρ_{Mix}	mixing density
ρ_P	abrasive-material density
ρ_W	water density
σ_C	target-material compressive strength
σ_f	target-material flow stress
σ_{Ep}	abrasive-particle energy standard-deviation
σ_B	fracture strength
σ_S	target-material shear strength
σ_t	target-material tensile strength
σ_U	ultimate stress
σ_{Vp}	abrasive-particle velocity standard-deviation
σ_Y	yield stress
Ω	density ratio ρ_P/ρ_W
Ω_F	impact-force ratio
ξ_D	drilling parameter

ξ_{pipe}	pipe-flow friction coefficient
ζ	debris-particle acceleration coefficient

Table of Contents

Preface v
Nomenclature vii

1   Introduction 1

1.2 Classification of High-Speed Fluid Jets 1
1.3 State-of-the-Art Application of the Water-Jet Technique 2

2   Classification and Characterization of Abrasive Materials 5

2.1 Classification and Properties of Abrasive Materials 5
    2.1.1 General Classification of Abrasive Materials 5
    2.1.2 Global Abrasive-Evaluation Parameter 5
2.2 Abrasive-Material Structure and Hardness 7
    2.2.1 Structural Aspects of Abrasive Materials 7
    2.2.2 Hardness of Abrasive Materials 7
2.3 Abrasive-Particle Shape Parameters 10
    2.3.1 Relative Proportions of Abrasive Particles 10
    2.3.2 Geometrical Form of Particles 11
2.4 Abrasive-Particle Size Distribution and Abrasive-Particle Diameter 13
    2.4.1 Particle-Size Distribution 13
        2.4.1.1 General Definitions 13
        2.4.1.2 Sieve Analysis 14
        2.4.1.3 Particle-Size Distribution Models 15
    2.4.2 'Average' Particle Diameter 16
2.5 Number and Kinetic Energy of Abrasive Particles 17
    2.5.1 Abrasive-Particle Number and Frequency 17
    2.5.2 Kinetic Energy of Abrasive Particles 18

3   Generation of Abrasive Water Jets 20

3.1 Properties and Structure of High-Speed Water Jets 20
    3.1.1 Velocity of High-Speed Water Jets 20
        3.1.1.1 Integral Pressure Balance 20
        3.1.1.2 Momentum-Transfer Efficiency 20
    3.1.2 Kinetic Energy of High-Speed Water Jets 21
    3.1.3 Structure and Properties of High-Speed Water Jets 22
        3.1.3.1 Structure in Axial Direction 22
        3.1.3.2 Structure in Radial Direction 24
3.2 Abrasive Particle - Water Jet Mixing Principles in Injection Systems 27

3.1.3.2 Structure in Radial Direction 24
3.2 Abrasive Particle - Water Jet Mixing Principles in Injection Systems..........27
   3.2.1 General Design Principles... 27
   3.2.2 Internal Design Parameters... 28
      3.2.2.1 Distance Between Orifice Exit and Focus Entrance..............28
      3.2.2.2 Distance Between Abrasive Inlet and Focus Entrance...........28
      3.2.2.3 Alignment Between Orifice and Focus............................29
      3.2.2.4 Mixing-Chamber Length.. 31
   3.2.3 Alternative Injection-System Designs....................................31
      3.2.3.1 Annular Jet Systems.. 31
      3.2.3.2 Vortex-Flow System... 31
      3.2.3.3 Multiple Water-Jet System... 33
3.3 Abrasive Suction in Injection Systems..34
   3.3.1 Pressure Difference for Pneumatic Transport............................ 34
   3.3.2 Air-Flow Rate.. 35
   3.3.3 Abrasive-Particle Entry Velocity... 36
   3.3.4 Internal Focus Pressure-Profile..37
3.4 Abrasive-Particle Acceleration in Injection Systems............................. 38
   3.4.1 Simplified Momentum-Transfer Model.....................................38
      3.4.1.1 Integral Impulse Balance... 38
      3.4.1.2 Momentum-Transfer Efficiency................................... 39
   3.4.2 Improved Acceleration Model... 40
      3.4.2.1 Velocity Components...40
      3.4.2.2 Force Balance in Axial Direction................................. 41
      3.4.2.3 Friction Coefficient and Reynolds-Number...................... 41
      3.4.2.4 Force Balance in Radial Direction................................42
      3.4.2.5 Approximate Solution...43
      3.4.2.6 Rigorous Solution... 44
      3.4.2.7 Numerical Solutions in Axial Direction.......................... 45
      3.4.2.8 Numerical Solutions in Radial Solution.......................... 47
      3.4.2.9 Results of Steel-Ball Projection Experiments.....................47
   3.4.3 Regression Model.. 48
3.5 Abrasive-Particle Fragmentation in Injection Systems...........................49
   3.5.1 Solid-Particle Impact Comminution...................................... 49
      3.5.1.1 Impact Velocity and Impact Angle............................... 49
      3.5.1.2 Fracture Zones During Impact..................................... 51
      3.5.1.3 Size Effects... 51
      3.5.1.4 Other Material Properties.. 51
   3.5.2 Abrasive-Particle Size Reduction During Mixing and Acceleration... 52
      3.5.2.1 General Observations..52
      3.5.2.2 The 'Disintegration-Number'...................................... 52
      3.5.2.3 Influence of Abrasive-Particle Structure and Properties........ 54
      3.5.2.4 Energy Absorption During Abrasive-Particle Fragmentation...55
   3.5.3 Abrasive-Particle Shape Modification During Mixing and
      Acceleration... 55
3.6 Focus Wear in Injection Systems... 57

  3.6.1 General Features of Focus Wear.. 57
  3.6.2 Focus-Exit Diameter... 58
    3.6.2.1 Early Observations... 58
    3.6.2.2 Focus-Wear Rate.. 58
    3.6.2.3 Process-Parameter Influence..................................... 58
    3.6.2.4 Hardness Influence... 60
  3.6.3 Other Focus-Wear Features.. 61
    3.6.3.1 General Aspects... 61
    3.6.3.2 Focus-Mass Loss and Focus-Wear Pattern................. 61
    3.6.3.3 'Selective' Focus Wear... 63
    3.6.3.4 Eccentricity of Focus-Exit Wear................................ 63
  3.6.4 Modeling the Focus-Wear Process... 63
    3.6.4.1 Phenomenological Focus-Wear Model....................... 63
    3.6.4.2 'Two-Material' Focus Concept.................................. 65
    3.6.4.3 Lifetime-Estimation Model.. 65
3.7 Generation of Suspension-Abrasive Water Jets.................................. 66
  3.7.1 General System Features.. 66
    3.7.1.1 System Components.. 66
    3.7.1.2 Bypass-Systems.. 66
    3.7.1.3 Direct-Pumping Systems... 69
  3.7.2 Abrasive-Particle Acceleration.. 70
    3.7.2.1 Acceleration-Nozzle Design...................................... 70
    3.7.2.2 Simple Momentum-Transfer Model............................ 70
    3.7.2.3 Numerical Simulations.. 72
    3.7.2.4 Finite-Element Modeling.. 74
    3.7.2.5 Acceleration-Nozzle Wear.. 76

4 Structure and Hydrodynamics of Abrasive Water Jets....................77

4.1 General Structure of Injection-Abrasive Water Jets............................ 77
  4.1.1 General Structural Features... 77
  4.1.2 Optical Examinations.. 77
4.2 Phase Distributions in Injection-Abrasive Water Jets......................... 79
  4.2.1 Average Abrasive-Density Distribution.................................. 79
  4.2.2 Radial-Zone Model.. 79
  4.2.3 Phase Estimation by X-Ray Densitometer.............................. 81
    4.2.3.1 Water-Phase Distribution... 81
    4.2.3.2 Abrasive-Phase Distribution..................................... 82
    4.2.3.3 Air Content.. 82
4.3 Abrasive-Particle Velocity Distribution in Injection-Abrasive Water Jets..... 83
  4.3.1 Radial Velocity-Profile.. 83
  4.3.2 Turbulence Profile.. 85
  4.3.3 Statistical Abrasive-Particle Velocity Distribution................. 86
4.4 Structure of Suspension-Abrasive Water Jets..................................... 87

5 Material-Removal Mechanisms in Abrasive Water-Jet Machining 89

5.1 Erosion by Single Solid-Particle Impact 89
    5.1.1 General Aspects of Solid-Particle Impact 89
    5.1.2 Erosion of Ductile-Behaving Materials 90
        5.1.2.1 Generalized Erosion Equation 90
        5.1.2.2 'Micro-Cutting' Model 91
        5.1.2.3 'Extended 'Cutting-Deformation' Model 91
        5.1.2.4 'Ploughing-Deformation' Model 92
        5.1.2.5 Low-Cycle Fatigue and Thermal Effects 93
        5.1.2.6 Comparison of Models for Ductile-Behaving Materials 93
    5.1.3 Erosion of Brittle-Behaving Materials 93
        5.1.3.1 Generalized Erosion Equation 93
        5.1.3.2 Elastic Model 94
        5.1.3.3 Elastic-Plastic Model 94
        5.1.3.4 Grain-Ejection Model 94
        5.1.3.5 Comparison of Models for Brittle-Behaving Materials 94
5.2 Micro-Mechanisms of Abrasive-Particle Material-Removal in Abrasive Water-Jet Machining 95
    5.2.1 Observations on Ductile-Behaving Materials 95
        5.2.1.1 SEM-Observations 95
        5.2.1.2 Stress Measurements 99
    5.2.2 Observations on Composite Materials 99
        5.2.2.1 SEM-Observations on Metal-Matrix Composites ... 99
        5.2.2.2 SEM-Observations on Fiber Reinforced Composites 100
    5.2.3 Observations on Brittle-Behaving Materials 101
        5.2.3.1 SEM-Observations on Polycrystalline Ceramics 101
        5.2.3.2 SEM-Observations on Refractory Ceramics 102
        5.2.3.3 Acoustic-Emission Measurements on Brittle-Behaving Materials 102
        5.2.3.4 Photoelasticity Investigations on Brittle-Behaving Materials 105
        5.2.3.5 Microboiling in Ceramics and Metal-Matrix Composites 105
        5.2.3.6 Observations on Glass 107
5.3 Material Removal by the High-Speed Water Flow 108
    5.3.1 General Observations 108
    5.3.2 Observations in Pre-Cracked Materials 108
        5.3.2.1 Effect of 'Water Wedging' 108
        5.3.2.2 'Transition-Velocity' Concept 109
        5.3.2.3 Pocket Formation in Soft Materials 110
5.4 Macro-Mechanisms of Abrasive Water-Jet Material Removal 111
    5.4.1 Some Observations of the Surface Topography 111

		5.4.1.1 General Statement... 111
		5.4.1.2 Surface-Profile Inspections... 112
		5.4.1.3 Wavelength Decomposition... 114
	5.4.2	Two-Dimensional Model of the Integral Material Removal............ 115
		5.4.2.1 Traverse-Direction Stages... 115
		5.4.2.2 Penetration-Direction Stages...................................... 116
		5.4.2.3 Further Development of the Model............................... 116
		5.4.2.4 Step Formation on the Cutting Front............................. 117
	5.4.3	Three-Dimensional Model of the Integral Material Removal.......... 118
		5.4.3.1 Three-Dimensional Step Formation............................... 118
		5.4.3.2 Influence of Machine Vibrations................................. 120
5.4.4	Alternative Models of the Integral Material Removal..................... 122	
		5.4.1.1 General Comments... 122
		5.4.1.2 Two-Stage Impact Zone Model................................... 122
		5.4.1.3 'Three-Zone' Cutting Front Model.............................. 122
		5.4.1.4 Energetic Cutting Model... 123
		5.4.1.5 Numerical Simulation of the Cutting Front 124
5.5	Energy Balance of Abrasive Water-Jet Material Removal........................ 125	
	5.5.1	General Energy Situation.. 125
		5.5.1.1 Dissipated Energy... 125
		5.5.1.2 Energy-Dissipation Function..................................... 126
	5.5.2	Geometrical Energy-Dissipation Model................................... 127
		5.5.2.1 Special Solutions of the Energy-Dissipation Function........... 127
		5.5.2.2 Basics for a General Solution..................................... 128
		5.5.2.3 Striation Geometry.. 128
		5.5.2.4 General Solution of the Energy-Dissipation Function........... 130
		5.5.2.5 Solution for the Relative Depth of Cut.......................... 132
		5.5.2.6 Local Energy-Dissipation Intensity.............................. 132
5.6	Erosion-Debris Generation and Acceleration.. 133	
	5.6.1	Properties of Generated Erosion Debris.................................... 133
		5.6.1.1 Structure, Size and Shape of Erosion Debris..................... 133
		5.6.1.2 Contact-Number Estimation...................................... 135
		5.6.1.3 Erosion-Debris Size Distribution Function..................... 136
	5.6.2	Efficiency of Erosion-Debris Generation................................... 137
		5.6.2.1 Surface-Based Efficiency Estimation-Model..................... 137
		5.6.2.2 Fracture-Based Efficiency Estimation-Model.................... 139
		5.6.2.3 Parameter Influence on the Efficiency........................... 139
	5.6.3	Erosion-Debris Acceleration.. 140
5.7	Damping Effects in Abrasive Water-Jet Material Removal....................... 142	
	5.7.1	Damping During Single Particle-Impact................................... 142
		5.7.1.1 Observations in Solid-Particle Erosion........................... 142
		5.7.1.2 Damping of Free-Falling Objects................................. 142
		5.7.1.3 Critical Particle Velocities for Damping......................... 143
	5.7.2	Damping During Abrasive Water-Jet Penetration........................ 145
		5.7.2.1 Concept of Force Measurements for Damping Estimation...... 145
		5.7.2.2 Results of Force Measurements................................... 145

5.7.2.3 Efficiency Losses due to Damping.................................. 146
5.8 Heat Generation During Abrasive Water-Jet Material Removal................ 146
   5.8.1 Sources of Heat Generation.. 146
   5.8.2 Results from Thermocouple Measurements............................ 147
      5.8.2.1 General Results.. 147
      5.8.2.2 Process-Parameter Influence............................... 147
      5.8.2.3 Local Temperature Distribution........................... 148
   5.8.3 Results from Infrared-Thermography Measurements............... 149
      5.8.3.1 General Remarks... 149
      5.8.3.2 Process-Parameter Influence on Linescans........... 149
      5.8.3.3 Material Isotherms.. 150
   5.8.4 Comparison Between Thermocouple and Infrared-Thermography..... 150
   5.8.5 Modeling of the Heat-Generation Process.............................. 152
      5.8.5.1 Basic Equations.. 152
      5.8.5.2 Results of the Modeling....................................... 153
5.9 Target-Material Property Influence on Material Removal..................... 154
   5.9.1 Hardness and Modulus of Fracture... 154
      5.9.1.1 General Observations.. 154
      5.9.1.2 'Two-Stage' Resistance Approach........................ 156
   5.9.2 Concepts of Material Machinability.. 156
      5.9.2.1 The 'Machinability-Number'............................... 156
      5.9.2.2 Other Machinability Concepts............................. 158
   5.9.3 Properties of Pre-Cracked Materials....................................... 159
      5.9.3.1 Stress-Strain Behavior... 159
      5.9.3.2 Relations to Conventional Testing Procedures..... 160
   5.9.4 Other Material Properties... 161
      5.9.4.1 Material Porosity... 161
      5.9.4.2 Thermal-Shock Factor... 162

6 Modeling of Abrasive Water Jet Cutting Processes........................... 163

6.1 Introduction.. 163
6.2 Volume-Displacement Models.. 164
   6.2.1 Volume-Displacement Model for Ductile Materials................ 164
   6.2.3 Volume-Displacement Model for Brittle Materials................. 169
   6.2.2 Generalized Volume-Displacement Model............................. 171
6.3 Energy-Conservation Models.. 174
   6.3.1 Two-Parameter Energy-Conservation Model......................... 174
   6.3.2 Regression Energy-Conservation Model................................ 175
   6.3.3 Semi-Empirical Energy-Conservation Model......................... 176
   6.3.4 Elasto-Plastic Energy-Conservation Model............................ 178
   6.3.5 Energy-Conservation Models for Pre-Cracked Materials........ 179
6.4 Regression Models... 181
   6.4.1 Multi-Factorial Regression Models.. 181
   6.4.2 Further Regression Models... 183
   6.4.3 Regression Model for Cutting with Suspension-Abrasive Water Jets... 184

6.5 Kinetic Model of the Abrasive Water-Jet Cutting Process........................ 184
6.6 Fuzzy Rule-Based Model of the Abrasive Water-Jet Cutting Process 189
6.7 Numerical Models.. 191
    6.7.1 Numerical Simulations... 191
    6.7.2 Numerical Process Model... 193

7 Process Parameter Optimization... 195

7.1 Definition of Process and Target Parameters....................................... 195
    7.1.1 Process Parameters.. 195
    7.1.2 Target Parameters...196
7.2 Influence of Hydraulic Process Parameters... 197
    7.2.1 Influence of Pump Pressure.. 197
        7.2.1.1 General Trendss...197
        7.2.1.2 Incubation Stage and Threshold Pressure.................. 199
        7.2.1.3 Linear Stage and Decreasing Stage.......................... 200
        7.2.1.4 Optimization Aspects... 201
    7.2.2 Influence of Water-Orifice Diameter................................... 202
        7.2.2.1 General Trends... 202
        7.2.2.2 Threshold Orifice Diameter...................................204
        7.2.2.3 Optimization Aspects... 204
7.3 Influence of Cutting Parameters... 204
    7.3.1 Influence of Traverse Rate... 204
        7.3.1.1 General Trends... 204
        7.3.1.2 Threshold Traverse Rate....................................... 205
        7.3.1.3 Exposure Time... 206
        7.3.1.4 Particle-Impact Frequency and Damping Effects.................206
        7.3.1.5 Influence on the Cutting Rate................................. 207
    7.3.2 Influence of Number of Passes.. 208
        7.3.2.1 General Trends... 208
        7.3.2.2 Multipass Cutting.. 209
    7.3.3 Influence of Standoff Distance.. 210
        7.3.3.1 General Trends... 210
        7.3.3.2 Special Observations.. 211
    7.3.4 Influence of Impact Angle.. 211
        7.3.4.1 Influence on Ductile-Behaving Materials...........................211
        7.3.4.2 Influence on Brittle-Behaving Materials..................... 211
7.4 Influence of Mixing Parameters... 212
    7.4.1 Influence of Focus Diameter... 212
        7.4.1.1 General Trends... 212
        7.4.1.2 Optimum Focus Diameter..................................... 213
    4.4.2 Influence of Focus Length.. 214
        7.4.2.1 General Trend.. 214
        7.4.2.2 Optimum Focus Length.. 215
7.5 Influence of Abrasive Parameters...216
    7.5.1 Influence of Abrasive-Mass Flow Rate................................ 216

7.5.1.1 General Trends... 216
7.5.1.2 Optimization Aspects.. 217
7.5.1.3 Influence on Cutting Rate..218
7.5.2 *Influence of Abrasive-Particle Diameter*................................... 219
7.5.2.1 General Trends... 219
7.5.2.2 Optimization Aspects.. 220
7.5.3 *Influence of Abrasive-Particle Size Distribution*......................... 221
7.5.4 *Influence of Abrasive-Particle Shape*..221
7.5.4.1 General Trends... 221
7.5.4.2 Influence on Ductile-Behaving Materials..........................222
7.5.4.3 Influence on Brittle-Behaving Materials.......................... 222
7.5.5 *Influence of Abrasive-Material Hardness*.................................. 223
7.5.5.1 General Trends... 223
7.5.5.2 Observations in Abrasive Water-Jet Cutting...................... 223
7.5.6 *Recycling Capacity of Abrasives*..224
7.5.6.1 Early Observations... 224
7.5.6.2 Parameter Influence on Disintegration............................. 225
7.5.6.3 Particle-Shape Modification..226
7.5.6.4 Suspension Abrasive Water Jets..227
7.5.6.5 Modelling of Recycling Processes.....................................228

8 Geometry, Topography and Integrity of Abrasive Water-Jet Machined Parts... 230

8.1 Cut Geometry and Structure... 230
8.1.1 *Definition of Cut Geometry Parameters*..................................... 230
8.1.2 *Width on Top of the Cut*.. 233
8.1.2.1 Ductile-Behaving Materials...233
8.1.2.2 Brittle-Behaving Composite Materials.............................. 233
8.1.2.3 Ceramics, Glass and Metal-Matrix Compounds................... 235
8.1.2.4 Models for Top-Width Estimation....................................... 236
8.1.3 *Width on the Bottom of the Cut*.. 237
8.1.3.1 Ductile-Behaving Materials...237
8.1.3.2 Brittle-Behaving Composite Materials............................... 237
8.1.3.3 Ceramics and Glass..237
8.1.4 *Taper of the Cut and Flank Angle*... 239
8.1.4.1 Ductile-Behaving Materials...239
8.1.4.2 Brittle-Behaving Composite Materials............................... 242
8.1.4.3 Ceramics, Glass and Metal-Matrix Compounds................... 245
8.1.4.4 Models for Taper Estimation..248
8.1.5 *General Cut Profile*... 248
8.1.5.1 Experimental Results...248
8.1.5.2 General Cut-Geometry Model...248
8.1.6 *Initial-Damage Geometry*..250
8.1.6.1 General Relations... 250
8.1.6.2 Ductile-Behaving Materials...251

　　　　　　　8.1.6.3 Brittle-Behaving Composite Materials............................251
　　　　　　　8.1.6.4 Model for Initial Damage Zone Geometry.......................252
　　8.2 Topography of Abrasive Water-Jet Generated Surfaces.........................253
　　　　8.2.1 General Characterization...253
　　　　　　　8.2.1.1 Introductional Aspects..253
　　　　　　　8.2.1.2 Static Characterization...253
　　　　　　　8.2.1.3 Dynamic Characterization...255
　　　　　　　8.2.1.4 Wavelength Decomposition..256
　　　　8.2.2 Surface Roughness..257
　　　　　　　8.2.2.1 General Relations..257
　　　　　　　8.2.2.2 Influence of Hydraulic Parameters 257
　　　　　　　8.2.2.3 Influence of Cutting Parameters....................................258
　　　　　　　8.2.2.4 Influence of Mixing Parameters....................................258
　　　　　　　8.2.2.5 Influence of Abrasive Parameters..................................260
　　　　　　　8.2.2.6 Influence of Target-Material Structure...........................261
　　　　　　　8.2.2.7 Models for Roughness Estimation.................................262
　　　　8.2.3 Surface Waviness..264
　　　　　　　8.2.3.1 General Relations..264
　　　　　　　8.2.3.2 Influence of Process Parameters....................................266
　　　　　　　8.2.3.3 Models for Waviness Estimation...................................267
　　8.3 Integrity of Abrasive Water-Jet Generated Surfaces............................271
　　　　8.3.1 Fatigue Life...271
　　　　8.3.2 Surface Hardening..273
　　　　　　　8.3.2.1 Hardness Measurements...273
　　　　　　　8.3.2.2 Stress Measurements...273
　　　　8.3.3 Micro-Structural Aspects..275
　　　　　　　8.3.3.1 General Aspects of Alteration......................................275
　　　　　　　8.3.3.2 Surface Cracking in Brittle-Behaving Materials.................275
　　　　　　　8.3.3.3 Phase Modifications in Ceramics..................................277
　　　　8.3.4 Abrasive-Particle Fragment Embedding................................. 278
　　　　8.3.5 Delamination in Composite Materials....................................279
　　　　8.3.6 Burr Formation.. 281

9　Alternative Machining Operations with Abrasive Water Jet...............284

9.1 Capability of Abrasive Water Jets for Alternative Machining...................284
9.2 Milling with Abrasive Water Jets..284
　　9.2.1 Concepts of Abrasive Water-Jet Milling 284
　　9.2.2 Parameter Optimization in Abrasive Water-Jet Milling..................288
　　9.2.3 Quality of Abrasive Water-Jet Milling.......................................292
　　9.2.4 Modeling of Abrasive Water-Jet Milling..................................... 295
　　　　　9.2.4.1 General Milling Model..295
　　　　　9.2.4.2 Milling Model for Fiber-Reinforced Plastics......................298
　　　　　9.2.4.3 Model for Discrete Milling..299
　　　　　9.2.4.4 Numerical Milling Model... 301
9.3 Turning with Abrasive Water Jets.. 301

    9.3.1 Macromechanism of Abrasive Water-Jet Turning........................ 301
    9.3.2 Parameter Optimization in Abrasive Water-Jet Turning................ 302
    9.3.3 Quality of Abrasive Water-Jet Turning...................................... 307
    9.3.4 Modeling of Abrasive Water-Jet Turning................................... 308
        9.3.4.1 Analytical Turning Model.. 308
        9.3.4.2 Regression Turning Model.. 310
9.4 Piercing with Abrasive Water Jets... 312
    9.4.1 Macromechanism of Abrasive Water-Jet Piercing....................... 312
    9.4.2 Parameter Optimization in Abrasive Water-Jet Piercing................ 314
    9.4.3 Geometry and Quality of Abrasive Water-Jet Pierced Holes........... 315
        9.4.3.1 Hole Geometry... 315
        9.4.3.2 Hole Quality... 317
    9.4.4 Modeling of Abrasive Water-Jet Piercing................................... 319
        9.4.4.1 Phenomenological Piercing Model................................ 319
        9.4.4.2 Analytical Piercing Model.. 321
        9.4.4.3 Regression Piercing Model... 323
        9.4.4.4 Simulation Model for Piercing...................................... 323
9.5 Hole Trepanning and Deep-Hole Drilling with Abrasive Water Jets........... 325
    9.5.1 Hole Trepanning with Abrasive Water Jets................................ 325
    9.5.2 Deep-Hole Drilling with Abrasive Water Jets............................. 326
9.6 Polishing with Abrasive Water Jet... 328
    9.6.1 Abrasive Water-Jet Polishing Concepts..................................... 328
    9.6.2 Quality Aspects of Abrasive Water-Jet Polishing........................ 329
9.7 Screw-Thread Machining with Abrasive Water Jets............................... 331

10 Control and Supervision of Abrasive Water-Jet Machining Processes.. 333

10.1 General Aspects of Process Control... 333
10.2 Control of the Abrasive-Particle Suction Process................................ 334
    10.2.1 General Demands.. 334
    10.2.2 Acoustic Sensing... 334
    10.2.3 Workpiece Reaction-Force Measurement............................... 335
    10.2.4 Vacuum Sensor.. 335
    10.2.5 Actual Abrasive-Mass Flow Rate... 335
10.3 Control of Water-Orifice Condition and Wear.................................... 336
    10.3.1 Optical Jet Inspection... 336
    10.3.2 Vacuum-Pressure Measurement.. 336
10.4 Control of Focus Condition and Wear.. 337
    10.4.1 General Comments.. 337
    10.4.2 Direct Tracking.. 338
    10.4.3 Jet-Structure Monitoring.. 339
    10.4.4 Air-Flow Measurements... 340
    10.4.5 Infrared Thermography.. 341
    10.4.6 Acoustic Sensing... 343
    10.4.7 Workpiece Reaction-Force Measurement............................... 344

  10.4.8 Off-Line Focus-Diameter Measurement.................................... 346
10.5 Measurement and Control of Abrasive Water-Jet Velocity.....................346
  10.5.1 Inductive Methods... 346
  10.5.2 Measurement by Impact Crater Counting................................347
  10.5.3 Laser-Based Methods.. 349
    10.5.3.1 Laser-2-Focus-Velocimeter...................................... 349
    10.5.3.2 Laser-Transit-Velocimeter...350
    10.5.3.3 Laser-Doppler-Velocimeter...................................... 351
    10.5.3.4 Laser-Light-Section Procedure Technique...................... 352
  10.5.4 Other Optical Methods... 352
    10.5.4.1 Schlieren-Photography... 352
    10.5.4.2 High-Speed-Photography... 353
  10.5.5 Jet Impact-Force Measurements.. 354
10.6 Measurement and Control of Abrasive Water-Jet Structure. 356
  10.6.1 Scanning-X-Ray Densitometry... 356
  10.6.2 Flow-Separation Technique... 357
10.7 Control of Material-Removal Processes...358
  10.7.1 Acoustic-Emission Technique... 358
    10.7.1.1 Material-Removal Visualization.................................358
    10.7.1.2 Cutting-Process Visualization and Cutting-Through
    Control.. 359
    10.7.1.3 Cutting-Efficiency Control.. 362
  10.7.2 Control by Infrared Thermography.. 362
10.8 Control of Depth of Penetration...,.... 364
  10.8.1 Acoustic Sensing.. 364
  10.8.2 Acoustic-Emission Technique.. 365
  10.8.3 Workpiece Reaction-Force...366
  10.8.4 Supervision and Copntrol of Piercing Processes........................ 366
    10.8.4.1 Monitoring by Pressure Sensors................................. 366
    10.8.4.2 Monitoring by Acoustic-Emission Technique.................. 367
10.9 Control of the Generated Surface Topography.................................... 368
  10.9.1 Roughness Control by Static Workpiece Reaction-Force............. 368
  10.9.2 Roughness Control by Dynamic Workpiece Reaction-Force..........371
  10.9.3 Surface Quality Monitoring by Acoustic-Emission Technique........ 372
10.10 Expert Systems for Abrasive Water-Jet Machining............................. 374

References... 376

1. Introduction

1.1 Classification of High-Speed Fluid Jets

The term 'high-speed water jet' includes several modifications of jets. Momber [1] subdivides high-speed water jets as shown in Figure 1.1.

It is technically difficult to define a critical pump pressure that separates a low-pressure jet and a high-pressure jet. Louis [2] suggests that jets generated by plunger pumps should be defined as 'low-pressure jet'; whereas, jets generated by hydraulically-driven intensifiers are called 'high-pressure jet'. Figure 1.2 shows a state-of-the-art intensifier for abrasive water-jet applications. These definitions create several problems because recently developed commercial plunger pumps are able to generate water pressures up to p=270 MPa which is in the range of hydraulic pressure intensifiers [3].

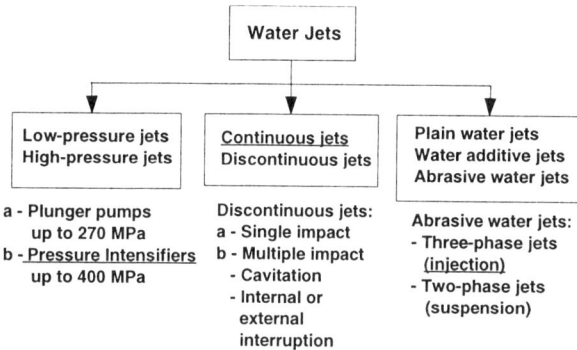

Figure 1.1 *Subdivision of water jets [1]*

In relation to the loading regime, there can generally be distinguished between continuous jets and discontinuous jets. Wiedemeier [4] defines a jet as discontinuous if it generates a discontinuous load on the impact site. But as pointed out by Momber [4], every water jet generates discontinuous phases during the impact because of pressure fluctuations and droplet formation. He suggests that 'discontinuous jets' are artificially broken up by external mechanisms; whereas, 'continuous jets' are not influenced by external mechanisms. Momber [1], Labus [5], and Vijay [6] give reviews about the formation, action and application of discontinuous water jets. Figure 1.1 presents a systematization of discontinuous water jets.

According to the fluid medium, there can generally be distinguished between plain water jets, water jets with additives (soluble), and abrasive (non-soluble) water jets.

Hollinger et al. [7] make an attempt to add a mixture of additives and abrasives to water jets. Abrasive water jets devide further according to their generation and their phase composition. Figure 1.1 shows that there exists suspension-abrasive water jets and injection-abrasive water jet. An injection-abrasive water jet consists of water, abrasives and air, and is considered to be a three-phase jet. In contrast, a suspension-abrasive water jet does not contain air and, therefore, is a two-phase jet. This book mainly focuses on injection-abrasive water jets.

1.2 State-of-the Art Applications of the Water-Jet Technique

Water jets are applied in nearly all areas of modern industry, such as automotive industry, airspace industry, construction engineering, environmental technology, chemical process engineering, and industrial maintenance. Table 1.1 gives a review about industrial applications of the water jet-technique. Generally, water jets are used for:

- industrial cleaning [8]
- surface preparation [9, 10]
- paint, enamel and coating stripping [11, 12]
- concrete hydrodemolition [1, 13]
- rock fragmentation [8]
- assisted mining operations [8]
- rock and soil drilling [8]
- soil stabilization [8, 14, 15]
- decontamination [16]
- demolition [1, 17]
- material recycling [18, 19, 20]
- *manufacturing operations*

Figure 1.3 shows a commercial state-of-the-art abrasive water-jet cutting system for manufacturing applications. In the area of manufacturing, the water jet-technique is used for:

- material cutting by plain water jets (e.g., plastics, thin metal sheets, textiles, foam; Figure 1.4a)
- deburring by plain water jets [21]
- surface peening by plain water jets [22]
- conventional machining with water-jet assistance [23, 24, 25]
- cutting of difficult-to-machine materials by abrasive water jets (Figure 1.4b)
- milling and 3-D-shaping by abrasive water jets
- turning by abrasive water jets
- piercing and drilling by abrasive water jets
- polishing by abrasive water jets

This book focuses on the last five applications mentioned above.

1. Introduction   3

Figure 1.2 *Hydraulically-driven high-pressure water jet intensifier (Innoweld, Sigless)*

Figure 1.3 *State-of-the-art 2-dimensional abrasive water-jet cutting system 3 x 2 m, 60 kW (WOMA Apparatebau GmbH, Duisburg)*

Table 1.1 *Industrial applications of high-pressure water jets (WOMA Apparatebau GmbH, Duisburg)*

Industrial area	Application
Civil engineering	Concrete hydrodemolition
	Surface cleaning
	Vibration-free demolition
	Soil stabilization
	Soil decontamination
	Water jet supported pile driving
	Joint cleaning
Chemical process engineering	Pipeline cleaning and decoating
	Tube bundle cleaning
	Vessel, container and autoclave cleaning
Maintenance and corrosion prevention	Coating removal
	Emission-free surface preparation
	Selective paint stripping
Municipal engineering	Sewer cleaning
Automotive engineering	Lacquer stripping
	Engine reconditioning
	Deburring
Environmental engineering	Material recycling
	Emission-free decontamination

Figure 1.4 *Cutting applications with high-pressure water jets (Innoweld, Sigless)*
left: a – plastics, plain water jet   right: b – marble, abrasive water jet

2 Classification and Characterization of Abrasive Materials

2.1 Classification and Properties of Abrasive Materials

2.1.1 General Classification of Abrasive Materials

A large number of different types of abrasive materials are used in the abrasive water-jet technique. A survey in 1995 [26] shows that most of the abrasive water-jet shops use garnet (90 %), followed by olivine (15 %), slag (15 %), aluminum-oxide (11 %), and silica-sand (11 %). Martinec [27] distinguishes between two major groups of abrasive materials: oxides and silicates. Table 2.1 lists subgroups of both of these types.

Table 2.1 *Classification of abrasive materials [27]*

Oxide	Silicate	
	Garnet	Other silicate
Magnetide	Almandine	Zircone
Ilmenide	Spessartine	Topas
Corundum	Porype	Olivine
Rutile	Grossularite	Staurelite
Quartz	Andradite	Olivine

The evaluation of an abrasive material for abrasive water-jet processes includes the following important parameters:

- material structure
- material hardness
- mechanical behavior
- grain shape
- grain-size distribution
- average grain size

2.1.2 Global Abrasive-Evaluation Parameter

Agus et al. [28, 29] introduce a parameter to evaluate abrasive material,

$$P_{Abr} = H_P^{a_1} \cdot S^{a_2} \cdot \rho_P^{a_3} \cdot d_P^{a_4} \cdot \dot{m}_A^{a_5}, \qquad (2.1)$$

that includes the following parameters:

- abrasive-material hardness
- abrasive-particle shape
- abrasive-material density
- abrasive-particle diameter
- abrasive-mass flow rate

In Eq. (2.1), H_P is the Knoop hardness, and S is a particle-shape factor. The evaluation parameter P_{Abr} can be directly related to the specific erosion capability of injection-abrasive water jets as well as of suspension-abrasive water jets (Figure 2.1).

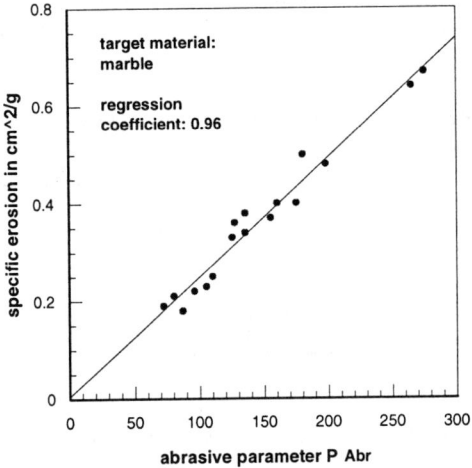

Figure 2.1 *Relation between the abrasive-material evaluation parameter and erosion capability of an injection-abrasive water jet [28]*

Table 2.2 lists typical parameters for the power exponents in Eq. (2.1). These cases evidence that abrasive-material hardness and abrasive-particle shape show the most remarkable influence on the cutting process. For hard rock (granite and porphyry) cutting, the abrasive-material hardness is predominant; whereas, the particle shape is most important for cutting softer materials (marble).

Table 2.2 *Abrasive-material characterization exponents for Eq. (2.1) [29]*

Target material	Power exponent				
	a_1	a_2	a_3	a_4	a_5
Granite	1.4	0.2	-0.4	0.1	-0.5
Porphyry	1.5	-0.1	-0.2	0.1	-0.5
Basalt	1.2	0.7	-0.2	0.1	-0.5
Marble	0.7	2.0	-0.2	0.1	-0.5

2.2 Abrasive-Material Structure and Hardness

2.2.1 Structural Aspects of Abrasive Materials

Structural aspects of abrasive materials include the following features [27]:

- lattice parameters
- crystallographical group and symmetry
- chemical composition
- crystallochemical formular
- cleavage
- inclusions (water-gas inclusion, mineral inclusion)

Table 2.3 lists typical values for some abrasive materials. Table 2.4 displays a commercial technical data and physical characteristics sheet for a typical garnet abrasive.

Table 2.3 *Structural properties of abrasive materials [30]*

Material	Damaged grains (%)	Lattice constant [A]	Cell volume [A^3]
Almandine	5 - 60	11.522 (0.006)	1529.62
Spessartine	-	11.613 (0.005)	1566.15
Pyrope	-	11.457 (0.005)	1503.88
Grossulare	30	11.867 (0.005)	1671.18
Andradite	80 - 90	12.091 (0.009)	1767.61

2.2.2 Hardness of Abrasive Materials

The hardness of brittle materials, such as the abrasive materials discussed here, are usually estimated by two types of tests: a scratching test that gives the *Mohs hardness*, and an indentation test that gives the *Knoop hardness*.

The Mohs hardness-test is based on a scale of ten minerals such that each mineral scratches the mineral on the scale below it, but does not scratch the mineral above it. The Knoop hardness-test is a test with an indenter producing an elongated indentation.

Figure 2.2 shows that Knoop hardness-values can be related to hardness values estimated by the Vickers hardness measurement-method [31]. In Figure 2.3, H_{Knoop} is in g·100, and $H_{Vickers}$ is in kg/mm^2 (load 100 g and 1,000 g).

Table 2.4 *Technical data and physical characteristics for Barton-garnet abrasive grains and powders (Barton Mines Corp., North Creek)*

Feature	Comments
General Description	- Combination of almandite and pyrope - Homogeneous mineral - No free chemicals - Oxides and dioxides are combine chemically as follows: $Fe_3Al_2(SiO_4)_3$ - Iron and aluminum ions are partially replaceable by calcium, magnesium, and manganese
Chemical Analysis	- Silicon Dioxide (SiO_2) 41.34 % - Ferrous Oxide (FeO) 9.72 % - Ferric Oxide (Fe_2O_3) 12.55 % - Aluminum Oxide (Al_2O_3) 20.36 % - Calcium Oxide (CaO) 2.97 % - Magnesium Oxide (MgO) 12.35 % - Manganese Oxid (MnO) 0.85 %
Hardness	Between 8 and 9 on Mohs scale
Strength	Friable to tough
Particle shape	Sharp, angular, irregular
Cleavage	Pronounced laminations, irregular cleavage planes
Colour	Red to pink
Streaks	White
Transparency	Translucent
Lustre	Vitreous
Specific gravity	3.9 g/cm^3 to 4.1 g/cm^3
Mean refractive index	1.83
Facet angles	37 °C and 42 °P
Crystallization	Cubic (isometric) system as rhombic dodecahedrons or tetragonal trisoctahedrons (trapezohedrons) or in combinations of the two
Melting Point	1,315 °C (2,300 °F)
Magnetism	Slightly magnetic (volume susceptibility = 9.000375)
Electrostatic properties	- Mineral conductivity: 18,000 volts - Non-reversible
Moisture absorption	Non-hygroscopic, inert
Dispersion	Self-dispersing
Uniformity	Garnet mineral in the deposit was formed simultaneously under identical natural conditions. It has been proven uniform throughout during over 100 years of use in technical abrasive applications.
Pathological effects	None
Harmful free silica content	None (silicosis free)

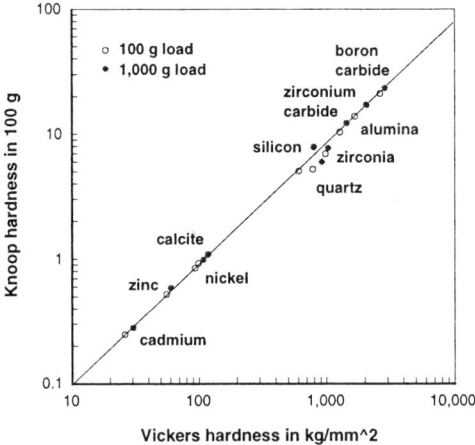

Figure 2.2 *Relation between the Mohs hardness and Knoop hardness [31]*

Bowden and Tabor [32] find that Knoop hardness and Mohs hardness can be related to each other; exceptions are diamond and corundum. Table 2.5 lists the hardness values of several minerals that are frequently used as abrasive materials for abrasive water-jet cutting.

Section 7.5.5 discusses the influence of the abrasive-material hardness on the process of abrasive water-jet machining.

Table 2.5 *Hardness values of abrasive materials*

Material	Hardness	
	Mohs	Knoop
Aluminum oxide	8 - 9	2,100
Copper slag	-	1,050
Garnet	7.5	1,350
Glass bead	5.5	400 - 600
Hard rock garnet	8+	-
Olivine	5.5	1,100
Silicone carbide	9.15	2,500
Silica sand	-	700
Steel grit	-	400 - 800
Zirconium	-	1,300

2.3 Abrasive-Particle Shape Parameters

2.3.1 Relative Proportions of Particles

Shape factors characterize the shape of single particles. Wadell [33] and Heywood [32] give rigorous analyses of shape factors. Heywood [34] considers the shape of a particle to have two distinct characteristics: the relative proportions of length, breadth, and thickness, and the geometrical form.

Figure 2.3 illiustrates the relative proportions: the elongation ratio

$$r_E = \frac{l_P}{b_P} \qquad (2.2)$$

and the *flatness ratio*

$$r_F = \frac{b_P}{t_P}. \qquad (2.3)$$

Bahadur and Badruddin [35] use Eq. (2.2) to investigate the influence of the abrasive-particle shape on solid-particle impact erosion processes. They find relations between the abrasive type, abrasive-grain diameter, and abrasive-grain shape. Silica-carbide particles become more elongated and less circular with an increase in the particle size while the opposite is the case with silica-oxide particles. The general variation of aluminum-oxide is similar to that of silica-carbide particles, though not as systematic (Figure 2.4).

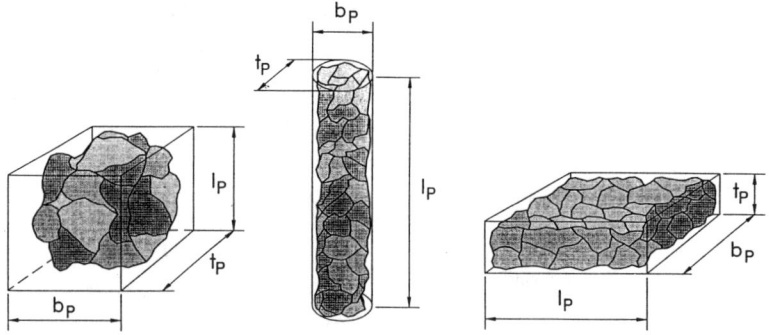

Figure 2.3 *Elongation ratio and flatness ratio for the shape characterization of abrasive-particles [36]*

2.3 Abrasive-Particle Shape Parameters 11

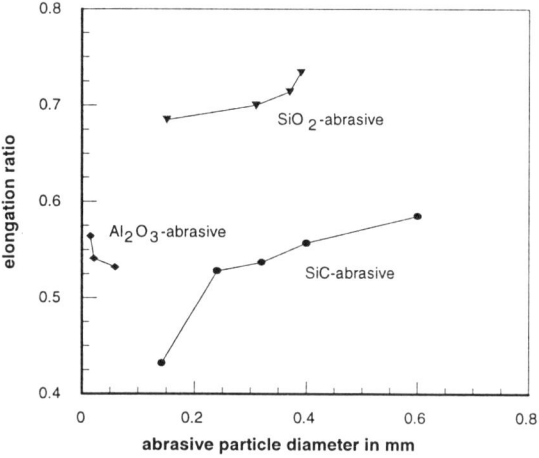

Figure 2.4 *Relation between the abrasive-particle size and abrasive-particle shape [35]*

2.3.2 Geometrical Form of Particles

The geometrical form is a volumetric shape factor representing the degree to which a particle approximates an ideal geometric form (cube, sphere, tetrahedron). One of these shape factors is the sphericity (Figure 2.5a) given by Wadell [33]

$$S_P = \frac{\sqrt{\frac{4}{\pi} \cdot b_P \cdot l_P}}{d_{circle}}. \quad (2.4)$$

In two dimensions, Eq. (2.5) is related to the projection area of the sphere yielding the roundness

$$S_R = \frac{\sum \left(\frac{2 \cdot r_{corner}}{d_P}\right)}{N_{corner}} \quad (2.5)$$

as illustrated in Figure 2.5b. Spherity as well as roundness ranges from 0 for very angular particles to 1 for ideal round particles. Several references use roundness-spherity diagrams as presented in Figure 2.6a to characterize the shape of single abrasive particles for abrasive water jet cutting-processes [37, 38].

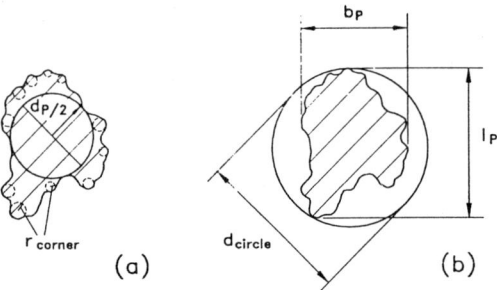

Figure 2.5 *Geometric form characteristics of abrasive particles*
a - sphericity b – roundness

Vasek et al. [30] and Martinec [39] suggest a circularity factor that is originally developed by Cox [40] to characterize abrasive particles

$$F_0 = \frac{4 \cdot \pi \cdot A_P}{P^2} \tag{2.6}$$

$$P = P_X + P_Y + \sqrt{2} \cdot P_{XY}$$

and a *shape factor*

$$F_{shape} = \frac{d_{min}}{d_{max}}. \tag{2.7}$$

Figure 2.6b illustrates both parameters. For circles, $F_0=1$ and $F_{shape}=1$. Table 2.6 gives some typical values for these parameters for different abrasive-material types.

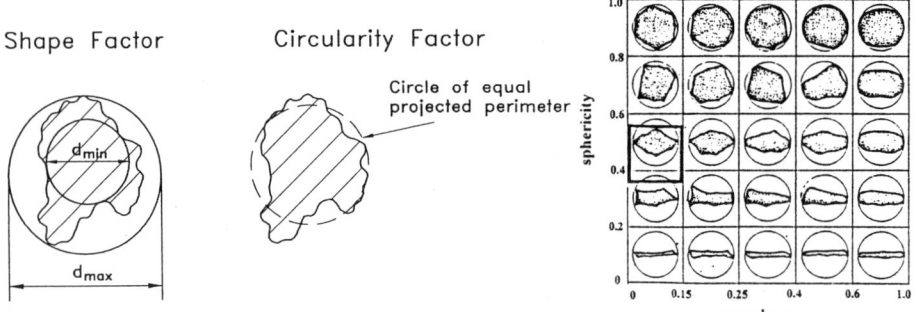

Figure 2.6 *Shape characterizations for abrasive particles*
a - Roundness-sphericity diagram for abrasive-particles [37]
b - Circularity and shape factor for abrasive-particles

In Figure 2.7, Vasek et al. [30] define seven different types of individual grain shapes for garnet abrasives.

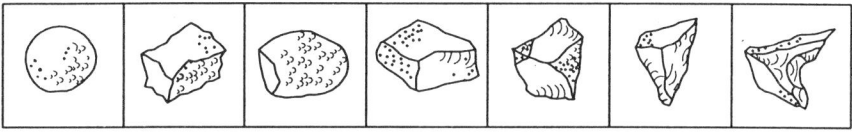

Figure 2.7 *Typical shapes of garnet abrasive used for abrasive water-jet machining [30]*

Table 2.6 shows the proportion of individual grain-shape factors on different abrasive materials. Section 7.5.4 discusses the influence of the abrasive-particle shape on the abrasive water-jet machining process.

Table 2.6 *Shape factors and shape characteristics of garnet abrasive-materials [30]*

Mineral	Subtype	Shape parameter	
		F_0	F_{shape}
Almandite	B	0.66	0.65
	M	0.69	0.67
	K	0.68	0.66
	G	0.66	0.64
Gossulare	-	0.71	0.70
Andradite	V-A	0.67	0.65
	V-B	0.68	0.68
	V-C	0.65	0.68

2.4 Abrasive-Particle Size Distribution and Abrasive-Particle Diameter

2.4.1 Particle-Size Distributions

2.4.1.1 General Definitions

In general, the term 'diameter' is specified for any equidimensional particle. By conversion, particle sizes are expressed in different units depending on the size ranges involved. Coarse grains are measured in inches or millimeters, fine particles in terms of screen size, very fine particles are measured in micrometer or manometer.

2. Classification and Characterization of Abrasive Materials

A number of 'diameter' definitions are known. The diameter is defined either in terms of some real property of the particle, such as its volume or surface area, or in terms of behavior of the particle in some specific circumstances, such as settling in water under defined conditions [41].

In the area of abrasive water-jet machining, the particle size is usually given in mesh designation according to the Tyler-Standard-Screen sieve-series that bareley mentions the related particle-size distributions or the shape of the used particles. A regression study made to link the Tyler-sieve-series to the corresponding average particle diameter gives

$$d_p = 17,479 \cdot \text{mesh}^{-1.0315} \tag{2.8}$$

with a regression coefficient of $R^2=0.998$. Most of the users of abrasive water jets (86 %) prefer the mesh designations # 50, # 60, # 80, and # 100 [26].

2.4.1.2 Sieve Analysis

Because it is impracticable to individually estimate each particle, size analysis is carried out by dividing the particles into a number of suitably narrow size ranges. Table 2.7 presents a sieve analysis typical for an abrasive-particle sample used in abrasive water-jet cutting.

Table 2.7 *Typical sieve analysis for two garnet abrasive-materials (Source: Barton Mines Corp., North Creek)*

Sieve Opening [μm]	# 36 CG Average retained [%]	# 36 CG Standard deviation	# 50 HP Average retained [%]	# 50 HP Standard deviation
246			6.1	2.5
295	0.1	0.0	20.9	4.7
351	2.6	0.2	28.0	2.9
417	18.1	1.0	24.6	3.3
495	38.0	1.1	14.2	4.4
590	37.1	1.6	4.3	2.1
700	4.0	0.6		
Rest	0.1	0.0	2.0	1.0

Graphically, data are conventionally presented by plotting the particle size horizontally and the measured quantity of property vertically. Two approaches are used to present the quantity: the first plots the absolute amount in each size fraction

(Figure 2.8a), and the second plots the cumulative amount above or below a certain size (Figure 2.8b).

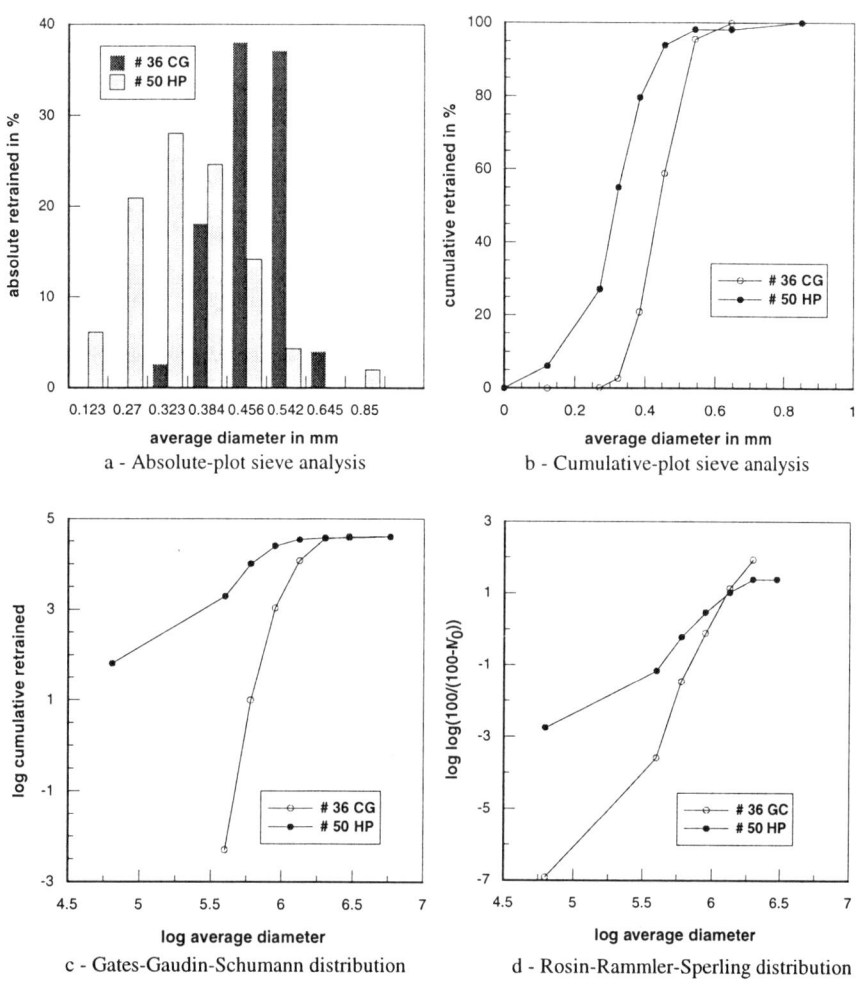

Figure 2.8 *Sieve-analysis functions and grain-size distributions of two abrasive materials (values from **Table 2.7**)*

2.4.1.3 Particle-Size Distribution Models

A number of models are developed to mathematically describe the size distributions of fine-grained comminution products, that includes abrasive particles. These models are empirical relationships, which to a greater or lesser extent are found capable of describing comminution size-distributions. Table 2.8 lists the most frequently used models. These equations are all of the general form

$$M_0(d_P) = f\left(\frac{d_P}{d^*}\right)^n. \qquad (2.9)$$

The size modulus, d^*, is an indication of the average particle diameter. When the equation has an upper size limit, d^* is in fact the maximum particle size in the distribution (Table 2.8). Eq. (2.9) also includes a second parameter, n, that is frequently called the distribution modulus, since it is a measure of the spread of particle sizes. The higher the n, the more homogeneous is the grain-size structure of the sample. For n→∞, the sample consists of grains with identical diameters.

Momber et al. [42] apply Eq. (2.9b) in Table 2.8 to characterize the grain-size distribution of abrasive mixtures for abrasive water-jet cutting (section 7.5.3). Figures 2.8c,d show fits for the sieve analysis from Table 2.7 by two common particle-size distribution functions.

Table 2.8 *Particle-size distribution functions [41, 43]*

Name	Formula $M_0(d_P)$	Significance of d^*	Equation
Logarithmic Probability	$\mathrm{erf}\left[\dfrac{\ln\frac{d_P}{d^*}}{\sigma}\right]$	medium particle diameter	2.9a
Rosin-Rammler-Sperling (RRSB)	$1 - \exp\left[\left(\dfrac{d_P}{d^*}\right)^n\right]$	particle diameter at M_0=63.2 %	2.9b
Gates-Gaudin-Schumann (GGS)	$\left(\dfrac{d_P}{d^*}\right)^n$		2.9c
Gaudin-Meloy	$1 - \left[1 - \dfrac{d_P}{d^*}\right]^2$	maximum particle diameter	2.9d
Broadbent-Callcott	$\dfrac{1 - \exp\left[-\dfrac{d_P}{d^*}\right]}{1 - \exp(-1)}$	-	2.9e

2.4.2 'Average' Particle Diameter

If the particle-size distribution is known from the sieve analysis, several 'average' diameter values of the particle sample can be estimated.

The median diameter, d_{50}, is the 50% point on any cumulative distribution-curve (Figure 2.9b). For the examples presented in Table 2.7 and Figure 2.8, this diameter is d_{P50}=475µm (# 36 CG) and d_{P50}=340 µm (# 50 HP), respectively.

The geometric mean-diameter, d_{PG}, is based on the assumption of an even graduation in size from maximum to minimum and assumes an equal number of particles in each size average:

$$d_{PG} = \frac{d_{P\,min} - d_{P\,max}}{2}. \qquad (2.10)$$

In the examples given in Table 2.7 and Figure 2.8, this diameter is d_{PG}=497 µm (# 36 CG) and d_{PG}=418 µm (# 50 HP), respectively.

A third possibility is the definition of a statistical diameter, d_{PSt}, that follows the equation

$$d_{PSt} = \frac{\sum_{i=1}^{n}(m_i \cdot d_{Pi})}{100}. \qquad (2.11)$$

For the examples in Table 2.7 and Figure 2.8, the statistical diameter is d_{PSt}=520 µm (# 36 CG) and d_{PSt}=373 µm (# 50 HP), respectively.

Sections 7.5.2 and 7.5.3 discuss the influence of the abrasive-particle diameter and abrasive-particle size distribution on the abrasive water-jet machining.

2.5 Number and Kinetic Energy of Abrasive Particles

2.5.1 Abrasive-Particle Number and Frequency

The number of particles contained in an abrasive sample is approximated by

$$N_P = \frac{\dot{m}_A}{m_P} \cdot t \qquad (2.12)$$

with

$$m_P = \frac{\pi}{6} \cdot d_P^3 \cdot \rho_P \qquad (2.13)$$

for spherical particles. If, for example, the statistical particle diameter d_{PSt} is used, as recommended for abrasive-particle samples by Guo et al. [44], the number of particles in a transversal (x-direction) moving abrasive water jet is

$$N_P = \frac{6 \cdot 100^3 \cdot \dot{m}_A \cdot x}{\pi \cdot \rho_P \cdot \left[\sum_{i=1}^{n}(m_i \cdot d_{Pi})\right]^3 \cdot v}. \quad (2.14)$$

Therefore, for a given traverse distance, the higher the abrasive-mass flow rate, the higher the number of abrasive particles. The higher abrasive-material density and average abrasive-grain diameter, the lower the number of abrasive particles. The abrasive-particle impact frequency is

$$f_P = \frac{N_P}{t} = \frac{\dot{m}_A}{m_P}. \quad (2.15)$$

For a given exposure time, the impact frequency increases with an increase in the abrasive-mass flow rate and with a decrease in the average-particle diameter. As the abrasive-material density increases, the impact frequency decreases.

2.5.2 Kinetic Energy of Abrasive Particles

The kinetic energy of a spherical abrasive particle is

$$E_P = \frac{\pi}{12} \cdot d_P^3 \cdot \rho_A \cdot v_P^2. \quad (2.16)$$

During the abrasive mixing, the kinetic energy of a high-speed water jet as given by Eq. (3.6) is partially absorbed by accelerating the abrasive particles. For a constant abrasive-mass flow rate, a given traverse rate, a given traverse distance, and a given abrasive material, the kinetic energy of an abrasive particle is after mixing

$$E_P = \alpha_T \cdot \frac{E_W}{N_P}. \quad (2.17)$$

In this equation, α_T is an energy-transfer coefficient with values between $\alpha_T=0.075$ and $\alpha_T=0.09$ [45]. The energy-transfer parameter increases as the abrasive-mass flow increases. The influence of the particle size and shape on the acceleration behavior of a particle is neglected (chapter 3). From this simple analysis, the kinetic

2.5 Number and Kinetic Energy of Abrasive Particles   19

energy of an abrasive particle decreases as the abrasive-mass flow rate increases and as the average particle diameter decreases.

Eq. (2.17) illustrates the relation between the impact energy of an abrasive particle and the impact frequency of an abrasive-grain sample. From Eq. (2.15) and (2.17),

$$E_P = C \cdot \frac{1}{f_P}. \tag{2.18}$$

In this equation, C considers the energy content of the high-speed water jet and the geometrical and mechanical properties of the abrasive material. Figure 2.9 illustrates Eq. (2.18) for a selected parameter combination.

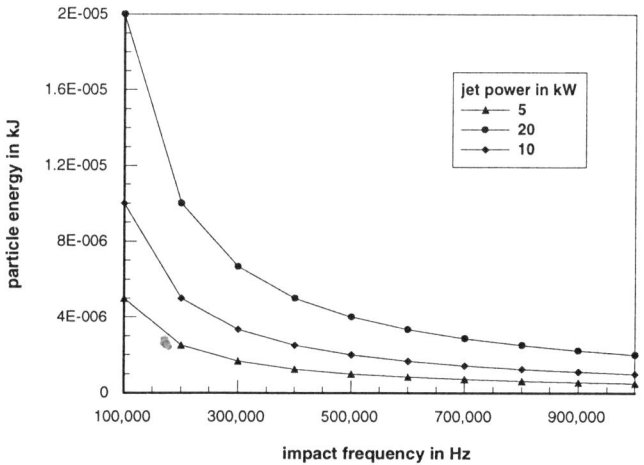

Figure 2.9 *Relation between the abrasive-particle impact energy and impact frequency for different values of the water-jet power*

3 Generation of Abrasive Water Jets

3.1 Properties and Structure of High-Speed Water Jets

3.1.1 Velocity of High-Speed Water Jets

3.1.1.1 Integral Pressure Balance

The acceleration of a certain volume of pressurized water in an orifice generates high-speed water jets. In this case, Bernoulli's law gives

$$p_{at} + \frac{\rho_w}{2} \cdot v_0^2 + \rho_w \cdot g \cdot h_1 = p + \frac{\rho_w}{2} \cdot v_{Pipe}^2 \cdot \rho_w \cdot g \cdot h_2. \qquad (3.1)$$

With $h_1=h_2$, $p_{at}\ll p$, and $v_0 \gg v_{pipe}$, the approximate velocity of the exit water-jet is

$$v_{0th} = \sqrt{\frac{2 \cdot p}{\rho_w}}. \qquad (3.2a)$$

In practice,

$$v_0 = \mu \cdot v_{0th} = \mu \cdot \sqrt{\frac{2 \cdot p}{\rho_w}}. \qquad (3.2b)$$

In Eq. (3.2b), μ is an efficiency coefficient that characterizes momentum losses due to wall friction, fluid-flow disturbances, and the compressibility of the water.

3.1.1.2 Momentum-Transfer Efficiency

As $\mu=v_0/v_{0th}$, values for μ can be obtained by measuring real water-jet velocities. Section 10.4 presents methods and results of water-jet velocity measurements. From these measurements, typical values for μ are $\mu=0.85$ [46], $0.88<\mu<0.95$ [47], and $\mu=0.98$ [48]. The certain value depends, among other factors, on the pump pressure and orifice geometry (Figure 3.1a).

Another way to estimate the velocity of a high-speed water jet as well as the efficiency coefficient is the measurement of the jet impact-force (section 10.4.9). The theoretical impulse flow of a fluid jet is

$$\dot{I}_w = \dot{m}_w \cdot v_{0th} = F_w. \qquad (3.3)$$

3.1 Properties and Structure of High-Speed Water Jets

Therefore, if the jet diameter is measured independently,

$$\mu = \frac{v_0}{v_{0th}} = \sqrt{\frac{2 \cdot F_{Weff}}{\pi \cdot p \cdot d_{jet}^2}} \,. \tag{3.4}$$

In Eq. (3.4), F_{Weff} is the measured water-jet impact force (Figure 10.22). Typical values for μ based on jet-force measurements are $0.83 < \mu < 0.93$, which agree well with those estimated by optical methods [49].

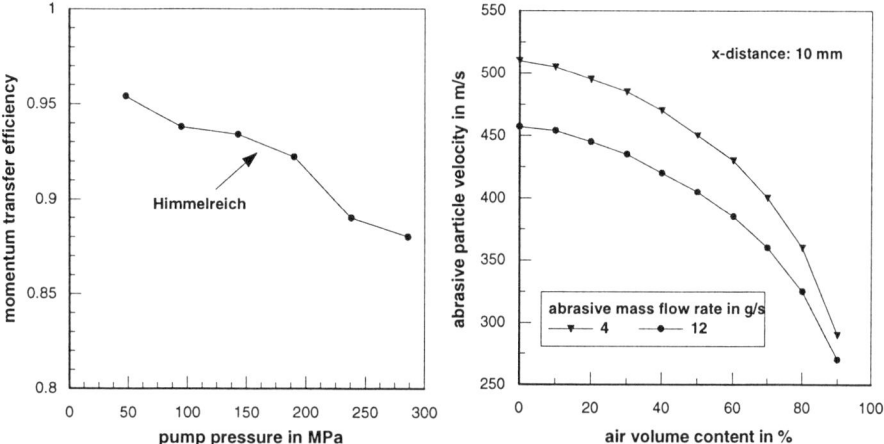

Figure 3.1 *Momentum transfer values [47]*   **Figure 3.2** *Influence of sucked air [50]*

Neusen et al. [48] and Tazibt et al. [50] show that the water-jet velocity substantially reduces if air is sucked into the mixing chamber (Figure 3.2).

3.1.2 Kinetic Energy of High-Speed Water Jets

As the high-speed water jet leaves the orifice, the kinetic energy of the jet is

$$E_W = \frac{1}{2} \cdot \dot{m}_W \cdot v_0^2 \cdot t \,. \tag{3.5}$$

With $\dot{m}_W = \frac{\pi}{4} \cdot d_0^2 \cdot v_0 \cdot \rho_W$ and Eq. (3.2b),

$$E_W = \alpha \cdot \frac{\pi}{4} \cdot d_0^2 \cdot \mu^3 \cdot \left[\sqrt{\frac{2 \cdot p}{\rho_W}}\right]^3 \cdot \rho_W \cdot t = \frac{\alpha \cdot \pi \cdot \mu^3}{\sqrt{2 \cdot \rho_W}} \cdot d_0^2 \cdot p^{1.5} \cdot t \,. \tag{3.6}$$

In Eq. (3.6), α is a non-dimensional number that considers the reduction in the water-mass flow rate due to the sudden changes in the fluid-mechanic conditions on the orifice outlet as well as the reduced jet velocity due to orifice-wall friction. Estimate values for α by measuring the real water-mass flow rate, and relate this rate to the theoretical water-mass flow rate. Typical values for sharp-edged sapphire orifices are $0.6<\alpha<0.8$ (Figure 3.3).

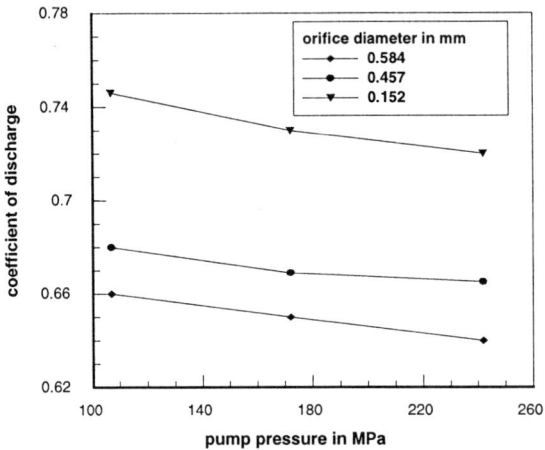

Figure 3.3 *Orifice outflow-coefficients [45]*

3.1.3 Structure of High-Speed Water Jets

3.1.3.1 Structure in Axial Direction

Shavlovsky [51] and Yanaida and Ohashi [52, 53] systematically investigate the structure of a high-speed water jet for pump pressures up to p=30 MPa. According to Whiting et al. [54], these results may be generally valid even for higher pump pressures (p=340 MPa), which are more typical for abrasive water-jet machining processes.

Yanaida [55] looks at the geometrical structure of a water jet escaping from an orifice in the air, as shown in Figure 3.4.

In the axial (x-) direction, the jet divides into three zones: a core zone, a transition zone, and a final zone.

3.1 Properties and Structure of High-Speed Water Jets

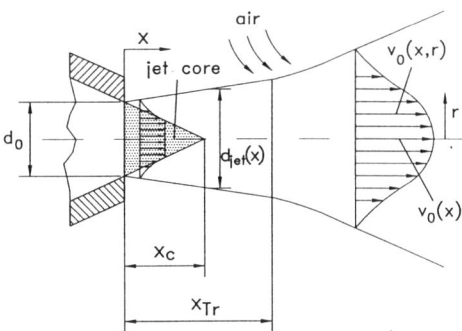

Figure 3.4 *Structure of a high-speed water jet [55]*

In the conical-shaped core zone, the flow properties, especially the stagnation pressure, are constant along the jet axis as shown in Figure 3.5a. Usually, the length of this zone relates to the orifice diameter,

$$\frac{x_C}{d_0} = A \ . \tag{3.7}$$

In Eq. (3.7), A depends on the Reynolds-number of the jet flow, and on the orifice geometry and quality. Several values estimated from measurements of the stagnation pressure in a water jet are published, such as 73<A<135 [55], A=90 [51], and 20<A<150 [56]. Nikonov et al. [57] show that A is independent on the Reynolds-number for Re>450·10^3. Beyond this value, A is a constant value for a given orifice configuration. Neusen et al. [58] publish values for A that are based on water-jet velocity measurements. As Figure 3.5b shows, values for A are sensitive to the pump pressure and jet velocity, respectively. The length of the core zone is between x_C=50·d_0 and x_C=125·d_0.

Based on high-speed motion picture inspections, Whiting et al. [54] develop an empirical relation between the pump pressure and the length of the water-jet core zone,

$$x_C = -3.545 \cdot 10^{-11} \cdot p + 2.535 \cdot 10^{-2} \ . \tag{3.8}$$

Eq. (3.8) holds for pump pressures between p=200 MPa and p=340 MPa, and for orifice diameters between d_0=0.1 mm and d_0=0.3 mm. In the water-jet core region, the velocity profile is almost rectangular, $v_0(r)$=constant.

In contrast, the water velocity is a function of the jet radius for x>x_C, which is in the transition zone. Also, the axial water velocity (axial stagnation pressure, respectively) drops if the transition zone is reached (Figure 3.5).

24   3. Generation of Abrasive Water Jets

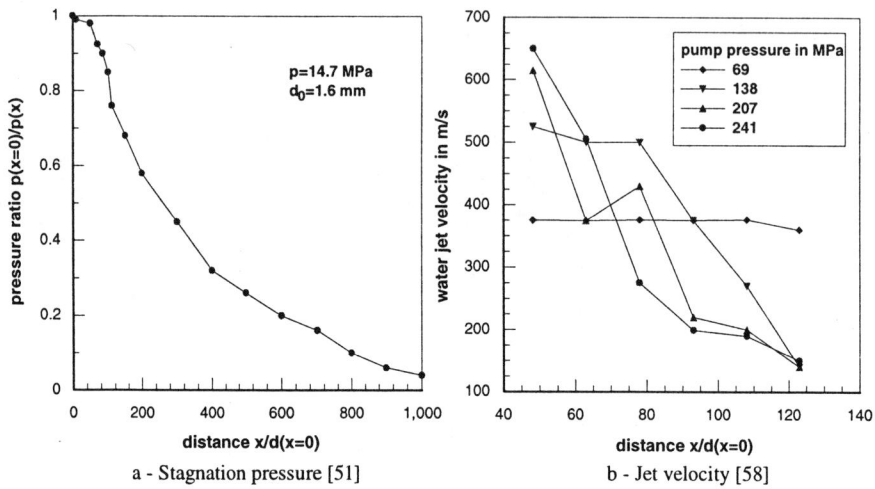

a - Stagnation pressure [51]   b - Jet velocity [58]

Figure 3.5 *Performance of a high-speed water jet in axial direction*

Thus,

$$v_0(x) < v_0(x \leq x_C), \tag{3.9}$$

and,

$$v_0 = f(r). \tag{3.10}$$

The length of the transition zone also relates to the orifice diameter,

$$\frac{x_T}{d_0} = B. \tag{3.11}$$

Based on the stagnation-pressure measurements, one finds $90 < B < 600$ [51], and $x_T \cong 5.33 \cdot x_C$ [55] respectively.

3.1.3.2 Structure in Radial Direction

The radial profile, $v_0 = f(r)$, of the water velocity has a typical bell shape that exponential functions mathematically describe (Table 3.1).

In the transition zone, the influence of the ambient air that enters the water jet becomes more pronounced. Figure 3.6a illustrates how this air entrainment influences the jet density.

3.1 Properties and Structure of High-Speed Water Jets   25

Table 3.1 *Radial stagnation-pressure distribution functions*

Function $\dfrac{p(r)}{p(r=0)}$	Reference
$\left[1-\left(\dfrac{r}{r_{jet}}\right)^{1.5}\right]^2$	Yanaida [55] Davies and Jackson [59]
$\exp\left[-a_1 \cdot \left(\dfrac{r}{r_{jet}}\right)^{a_2}\right]$	Shavlovsky [51]
$1 - 3 \cdot \left(\dfrac{r}{r_{jet}}\right)^2 + 2 \cdot \left(\dfrac{r}{r_{jet}}\right)^3$	Leach and Walker [60]
$\exp\left[\dfrac{-r}{2} \cdot a_3^2\right]$	Yahiro and Yoshida [61]
$\exp\left[-a_4^2 \cdot r^2\right]$	Rehbinder [62]

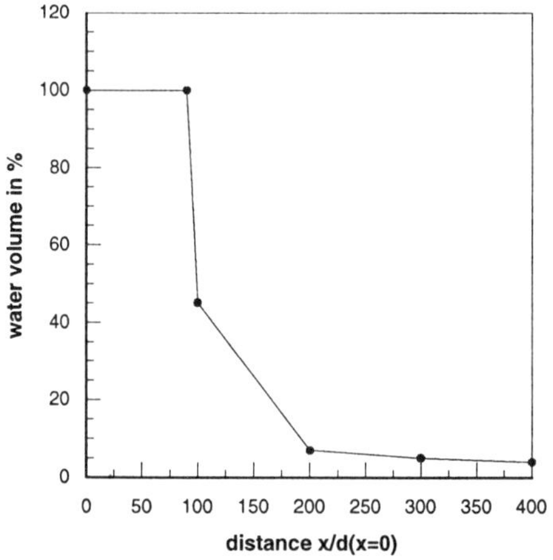

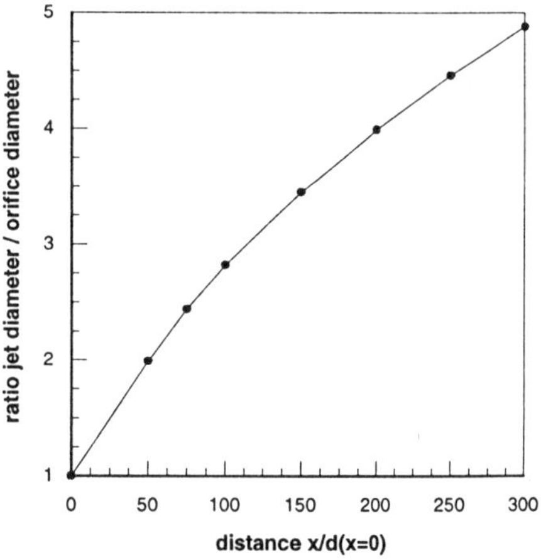

Figure 3.6 *Air content and structure of a high-speed water jet in radial direction*

a - Air content [59]

b - Jet diameter [55]

Shavlovsky [51] gives an empirical relation for the change in the jet density with an increase in the jet length. This relation is

$$\rho_{jet}(x) = \left[1 - a_1 \cdot \dfrac{x}{d_0}\right] \cdot \rho_w \qquad (3.12a)$$

for the water-jet core zone, and

$$\rho_{jet}(x) = \left[\frac{x}{d_0 \cdot \left(2.7 \cdot \frac{x}{d_0} - 20\right)} + \frac{8 \cdot d_0}{x}\right] \cdot \rho_w \qquad (3.12b)$$

for the water-jet transition zone. The parameter a_1 in Eq. (3.12a) considers the gas content in the jet and is tabulated in Table 3.2.

Table 3.2 *Tabulation of the parameter a_1 [51]*

x/d_0	5	10	15	20	30	40	50	60
a_1	0.028	0.024	0.02	0.017	0.014	0.011	0.009	0.008

A further very important aspect is the radial increase in the water jet with an increase in the jet length,

$$d_{jet} = f(x). \qquad (3.13)$$

For x=0, $d_{jet}=d_0$. Several investigators attempt to solve Eq. (3.13). Yanaida and Ohashi [53] propose

$$d_{jet}(x) = 0.24 \cdot d_0 \cdot \sqrt{x}. \qquad (3.13a)$$

Nienhaus [63] suggests

$$d_{jet}(x) \propto \operatorname{arccos} h(x). \qquad (3.13b)$$

Both equations lead to a decreasing progress in the jet diameter with an increase in the jet length (Figure 3.6b).

A problem is that there does not yet exist a definition of the jet diameter, d_{jet}, which is a fuzzy parameter because the transition from the 'water jet' to the ambient air is not steady. In fact, there exists a transition layer. Yanaida and Ohashi [53] define the jet edge as the location where first drop impact signals can be measured. In contrast, Wulf [64] defines the jet edge as the location where the jet impact-force is about 5 % of the maximum measured impact force.

3.2 Abrasive Particle - Water Jet Mixing Principles in Injection Systems

3.2.1 General Design Principles

In injection-abrasive water-jet systems, abrasive particles as characterized in chapter 2, a high-speed water jet as described in section 3.1, and air enter a cutting head from different entries. In the cutting head, these phases are mixed, and the abrasive particles and the air are accelerated. As a result, an abrasive water jet is formed in the cutting head.

There are several demands that have to be covered by a generation system for abrasive water jet, such as

- optimum abrasive-particle acceleration
- high energy density of the generated abrasive water jet
- low wear of the parts of the system, especially of the focus
- reliable performance
- simple function

Figure 3.7 illustrates the general structure of a system for the generation of an injection-abrasive water jet.

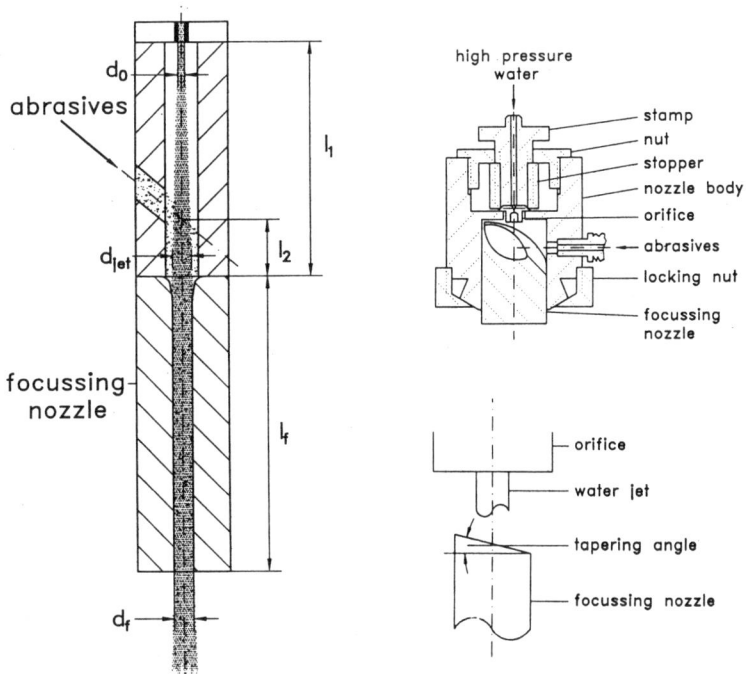

Figure 3.7 *General structure of a cutting-head for an injection-abrasive water jet (NJIT)*

Pneumatic transport drags the abrasive particles in the mixing chamber of the cutting head. Section 3.3 discusses this process. To overcome the cohesion forces, transport very fine abrasive particles in a suspension.

Regarding the location of the abrasive input, two principles distinguish between mixing prior to the nozzle and mixing after the nozzle. The first principle is usually referred as to the injection-abrasive water jet, that is discussed in the next sections; whereas, the second principle is called the suspension-abrasive water jet. Section 3.7 briefly discusses the generation of suspension-abrasive water jets.

A mixing-and-acceleration system for injection-abrasive water jets mainly consists of three features (Figure 3.7). First, there are entries for the high-pressure water and for the abrasive particles. Second, there is the generation of a suction pressure to suck in the abrasive particles and air. Third, the abrasives are accelerated, and the three-phase jet is focused.

To input the water jet, the classical design is the central entry on top of the cutting head. Abrasives and air circumferencely enter the mixing chamber, usually through a single-side input port. Most of all cutting heads for injection-abrasive water jets are based on this solution.

3.2.2 Internal Design Parameters

3.2.2.1 Distance Between Orifice Exit and Focus Entrance

Figure 3.8 shows several design parameters to optimize.

Li et al. [65] who use the depth of the generated cut as an optimization criterion, and Galecki and Mazurkiewicz [66] who define the pressure transmission ratio $p/p(l_1)$ as an evaluation parameter, investigate the influence of the distance between the water jet orifice-exit and focus entrance (l_1) on the cutting capability of a cutting head. As Figure 3.8a shows, there exists an optimum distance with a corresponding maximum depth of cut. The cutting capability of the abrasive water jet improves up to 80 % by selecting the proper distance between the orifice and focus. The optimum distance is independent of the taper of the focus entry [65] as well as on the orifice diameter [66]

3.2.2.2 Distance Between Abrasive Inlet and Focus Entrance

Mazurkiewicz et al. [67] carry out piercing experiments to optimize the distance between the abrasive inlet-level and the focus entrance (l_2). As Figure 3.8b shows, the performance of an abrasive water jet increases as this distance increases. This effect is very pronounced for large focus diameters (d_F=3.175 mm). In this case, the piercing time reduces up to 30 %. Small-diameter focuses are less sensitive to changes in the distance between the abrasive inlet and the focus entrance. For the smallest focus (d_F=0.75 mm), the piercing time is nearly unaffected.

3.2 Abrasive Particle – Water Jet Mixing Principles in Injection Systems

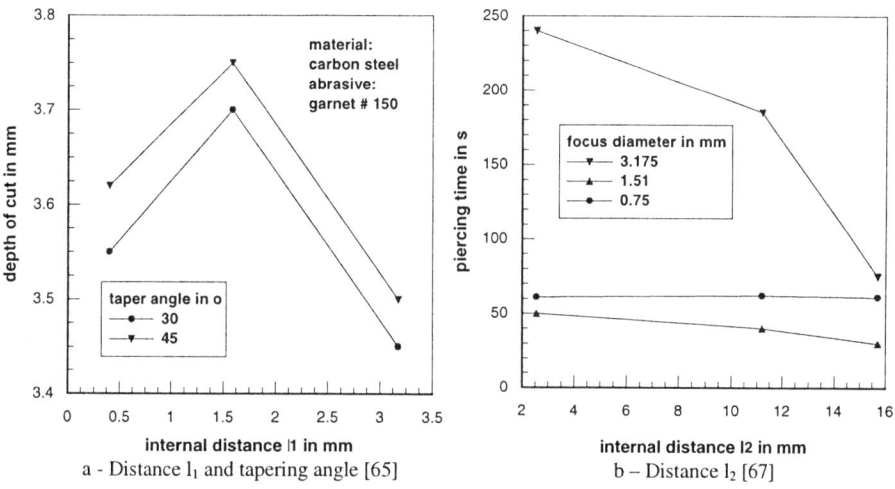

a - Distance l_1 and tapering angle [65]   b – Distance l_2 [67]

Figure 3.8 *Optimization of the internal geometry of a cutting head*

Sections 3.4.2, 7.2.2, and 7.4 discuss the influences of the orifice and focus dimensions on the process of abrasive acceleration and material cutting.

3.2.2.3 Alignment Between Orifice and Focus

An essential feature of the efficiency of standard cutting heads for injection-abrasive water jets is the alignment between the orifice and focus. Under ideal conditions, the focus and orifice are perfectly aligned, i.e., bore axes for both lie along the same line. But in practice, there are two types of misalignment, the linear misalignment (Figure 3.9a), and the angular misalignment (Figure 3.9b).

In linear misalignment, the two axes are parallel but not colinear. Galecki and Mazurkiewicz [66] show that a linear misalignment influences the pressure transmission between the water-jet orifice and focus significantly. This type of misalignment can be controlled by specifying close tolerances on the orifice and focus holders. Such tolerances that result in a maximum linear misalignment of $\Delta y = 51$ µm, are generally stated in current equipment [68]. Angular misalignment needs not be very large to seriously affect the efficiency of a cutting head as well as of the cutting process. For a typical focus with $l_F = 51$ mm, even a 1°-misalignment of the orifice forces the jet to impact the focus wall first before exiting. This misalignment results in an accelerated, non-symmetric focus wear, an uneven distribution of the abrasive particles, and a gradually deteriorating cutting performance.

30   3. Generation of Abrasive Water Jets

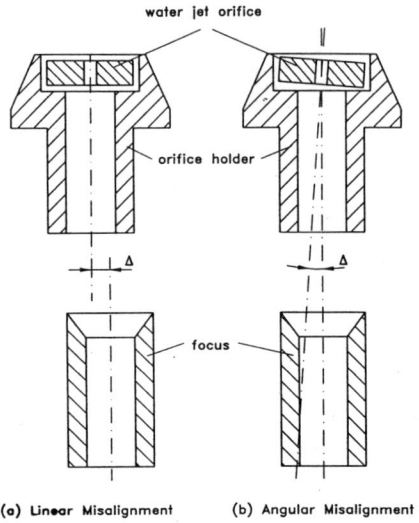

Figure 3.9 *Misalignment between the orifice and focus in a cutting-head for injection-abrasive water jets [68]*

Therefore, several cutting heads are designed to control and to minimize the misalignment. Blickwedel [69] introduce a concept of a high-alignment cutting head that consists of two independent parts that are connected by a ball-and-socket joint. Singh and Munoz [68] who also give a review of alignment-related problems, develop a similar concept.

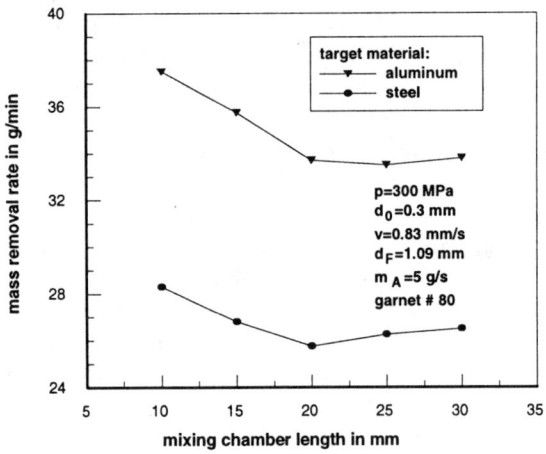

Figure 3.10 *Influence of the mixing-chamber length [70]*

3.2 Abrasive Particle - Water Jet Mixing Principles in Injection Systems

3.2.2.4 Mixing-Chamber Length

Figure 3.10 shows the influence of the mixing-chamber length on the mass-removal rate in two materials. The mass-removal rate decreases with an increase in the mixing-chamber length up to a length of about 20 mm. As the chamber exceeds this length, the material-removal rate remains on a stable level. The material-removal rate increases up to 15 % if a suitable mixing-chamber length is applied [70].

3.2.3 Alternative Injection-System Designs

3.2.3.1 Annular-Jet Systems

Figure 3.11 illustrates several alternative developments of cutting-heads for injection-systems in addition to the standard designs for injection-heads.

Figure 3.11a shows a nozzle that is designed with an annular slit connected to a conical cylinder. The slit supplies the high-speed water that passes through the conical cylinder and deforms into a spiral flow. An inlet on top of the nozzle feeds the abrasives. The water jet focuses well and the abrasive particles concentrate in the central axis of the water jet. Also, turbulence and focus wear are reduced [71]. Nevertheless, the highest reported water-jet velocity is about $v_0=35$ m/s.

Figure 3.11b illustrates a similar principle. In this case, the abrasives mix into an annular air jet through an inner steel pipe. The high-speed water jet enters the mixing chamber through a side entry accelerates the mixture. Visualization experiments show that the abrasives mix very homogeneously. But, as in case of the system discussed earlier, this system can be run just at low pump pressures of about $p=14$ MPa [72]

3.2.3.2 Vortex-Flow System

Figure 3.11c illustrates a further alternative mixing principle. The water flow that centrally enters the mixing chamber is directly turned into a vortex flow that flows through the nozzle and forms a vortex water-jet. The rotated movement of the water jet generates improved abrasive suction-capability and mixing efficiency [73]. This system is limited to pump pressures of about $p=10$ MPa, and requires a large orifice ($d_0=3$ mm) and focus ($d_F=7$ mm).

32   3. Generation of Abrasive Water Jets

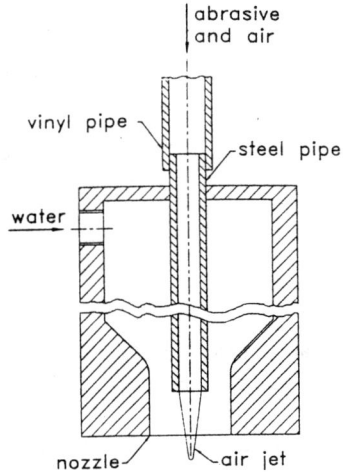

Figure 3.11a *Central annular water-jet [71]*

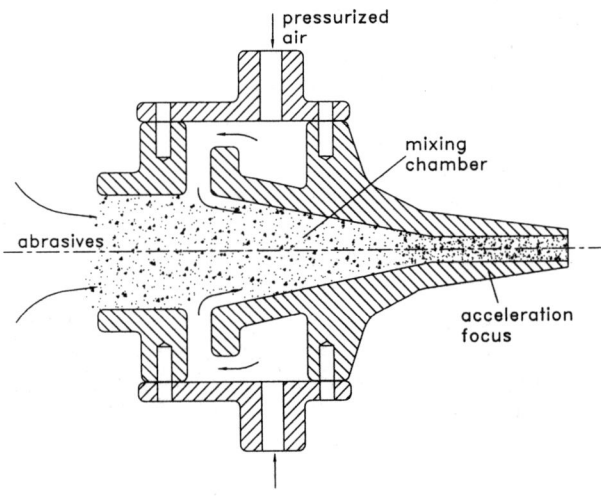

Figure 3.11b *Central annular air-jet [72]*

3.2 Abrasive Particle - Water Jet Mixing Principles in Injection Systems 33

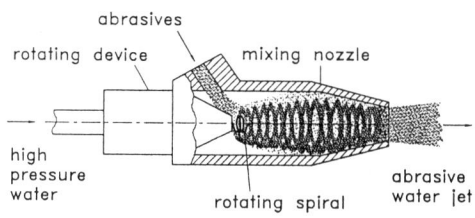

Figure 3.11c *Central rotated water-jet [73]*

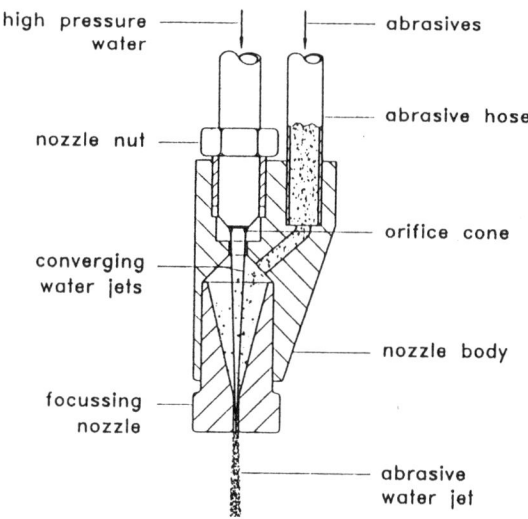

Figure 3.11d *Multiple water-jet head [74]*

3.2.3.3 Multiple Water-Jet System

Yie [74] reports on a cutting-head for an injection-abrasive water jet that is designed to issue multiple, parallel water jets arranged in a circular pattern. The water jets axially converge to form a single stream in order to create favourable conditions for entraining abrasive particles (Figure 3.11d). In this design, the water jets space apart in the region near the orifice cone such that the abrasive particles can easily pass through two adjecent water jets and enter into them. This cutting head works at moderate pump pressures of about p=100 MPa.

3.3 Abrasive Suction in Injection Systems

3.3.1 Pressure Difference for Pneumatic Transport

In cutting heads for injection-abrasive water jets, an air-volume stream carries the abrasive particles. This air stream results from a pressure difference between the pressure in the mixing chamber and the ambient air pressure,

$$\Delta p = p_{amb} - p_{mix}. \tag{3.14}$$

In pneumatic-transport systems, this pressure difference is identical to the summary of the pressure loss of the air flow and the pressure loss due to the solid-particle transport [75]. Thus,

$$\Delta p = p_{air} + p_{solid}. \tag{3.15}$$

For a fixed diameter of the suction hose [47],

$$\Delta p = a_1 \cdot \dot{Q}_L^2 + a_2 \cdot \dot{m}_A \cdot \dot{Q}_L. \tag{3.16}$$

Fig. 3.12 shows this relation in comparison with experimentally-estimated pressure differences. The vacuum pressure in the mixing chamber increases as the pump pressure increases and mixing-chamber length decreases [70].

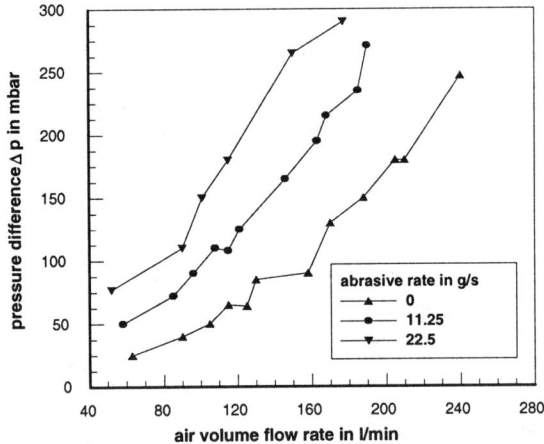

Figure 3.12 *Relation between the air-volume flow rate, abrasive-mass flow rate, and pressure difference [47]*

3.3.2 Air-Flow Rate

The air-volume flow rate depends on several process parameters, including the pump pressure, abrasive-mass flow rate, mixing-chamber design, focus diameter, and suction-hose diameter (Figure 3.13). Figure 3.13a illustrates the influence of the pump pressure and focus diameter. Both Himmelreich [47] and Tazibt et al. [50] and this figure measure how the air-volume flow rate almost linearly increases with the square root of the pump pressure. This result occurs with abrasives as well as without abrasives. Therefore,

$$\dot{Q}_L = a_3 \cdot \sqrt{p} \, . \tag{3.17}$$

Figure 3.13b shows that the sucked air-volume flow rate significantly decreases as the abrasive-mass flow rate increases and the focus diameter decreases. Even for small abrasive-mass flow rates (3.2 g/s), the air volume reduces up to 50 %. The air volume-flow rate linearly increases with an increase in the mixing-chamber length [70].

Assuming that pressure difference and air-volume flow rate expresses the 'pump characteristics' of a water jet, Figure 3.14 partially characterizes the suction process. This figure confirms Eq. (3.17) for fixed suction-hose diameters.

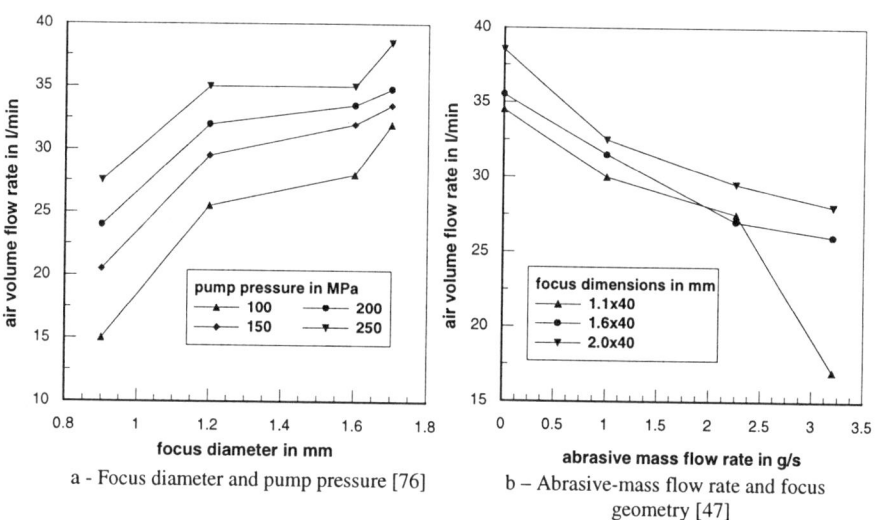

a - Focus diameter and pump pressure [76]
b – Abrasive-mass flow rate and focus geometry [47]

Figure 3.13 *Parameter influence on the air-volume flow rate*

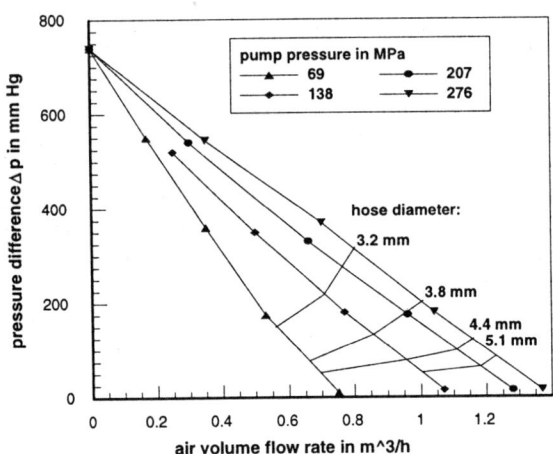

Figure 3.14 *'Pump characteristics' of high-speed water jets [45]*

3.3.3 Abrasive-Particle Entry Velocity

As the abrasive particles enter the mixing chamber, assume the velocity of the abrasive particles to be identical to the transport velocity due to pneumatic transport in the air stream. Using relations for the pneumatic transport of solid particles [75], the velocity is

$$v_{P0} = \frac{v_L}{1 + C^* \cdot \left[\frac{\rho_P}{\rho_L} - 1\right]^{2/3} \cdot \left[\frac{d_P}{d_{SH}}\right]^{2/3} \cdot \left[1 + \frac{200}{Fr - Fr_0}\right]}. \quad (3.18)$$

In this equation, v_L is the air velocity that is obtained from the law of continuity,

$$v_L = \frac{4 \cdot \dot{Q}_L}{\pi \cdot d_{SH}^2}. \quad (3.19)$$

Fr is the Froude-Number that is given by definition,

$$Fr = \frac{v_L^2}{d_{SH} \cdot g}. \quad (3.20)$$

The parameters C^* and Fr_0 are tabulated [75]. For quartz-grains with diameters between $d_P = 500$ μm and $d_P = 700$ μm, C^* is 0.09 and Fr_0 is 100. For standard conditions (ρ_L, ρ_P, d_P), approximate Eq. (3.18),

$$v_{PO} = a_4 \cdot \cos\theta_A \cdot \frac{4 \cdot \dot{Q}_L}{\pi \cdot d_{SH}^2}. \tag{3.21}$$

3.3.4 Internal Focus Pressure-Profile

The pressure inside the mixing-and-acceleration device is another criterion for the focus performance. This pressure characterizes the air velocity during the mixing process. If the pressure drops below the atmospheric pressure ($p_{mix}<p_{at}$), the air accelerates. In case the pressure exceeds the atmospheric pressure ($p_{mix}>p_{at}$), the air compresses and does not accelerate. Figure 3.15 illustrates both cases.

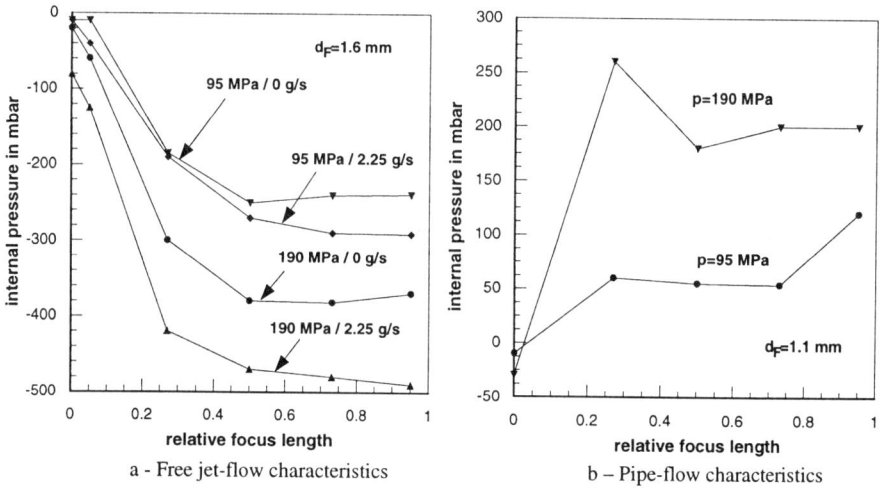

Figure 3.15 *Static pressure distribution over the focus length [47]*

In Figure 3.15a that illuminates the behavior of a comparatively large-diameter focus (d_F=1.6 mm), the air effectively accelerates due to the suction pressure. For plain air (without abrasives), the pressure difference is not very sensitive to changes in the pump pressure. The addition of abrasive grains changes the situation. The air accelerates much more, and the pump pressure significantly influences the acceleration process. Himmelreich [47] explains the higher air velocity that results from the addition of abrasives: as results of interactions between the abrasives and water jet, the water jet breakup and drop formation increases. The generated drops and water slugs act as 'pistons' that suck air into the mixing head. The behavior of a comparatively small-diameter focus (d_F=1.1 mm) as illustrated in Figure 3.15b, is very different: the pressure inside the focus exceeds the atmospheric pressure over the entire focus length. Therefore, the air must be compressed. Obviously, the square area of the focus is too small to allow an unrestrained expansion of the air. Some air particles decelerate as a result of friction between the air and the focus.

Himmelreich [47] concludes from pressure measurements that the character of the air flow is determined by the ratio between the focus length to focus diameter, $D^* = l_F/d_F$. For $D^* \cong 2.75$, one can expect a typical pipe-flow (Figure 3.15b); whereas, $D^* \cong 4$ allows the conservation of the free-jet flow (Figure 3.15a).

3.4 Abrasive-Particle Acceleration in Injection Systems

3.4.1 Simplified Momentum-Transfer Model

3.4.1.1 Integral Impulse Balance

The process of the abrasive-particle acceleration in an injection-abrasive water jet is a momentum transfer from the high-velocity water jet to the abrasive particles that are injected at comparatively low velocities, and the air sucked with the abrasive material (Figure 3.16). An impulse balance gives

$$\sum F = \text{const.} \tag{3.22}$$

and leads to

$$(\dot{m}_A \cdot v_{P0} + \dot{m}_W \cdot v_0 + \dot{m}_L \cdot v_L) = (\dot{m}_A + \dot{m}_L + \dot{m}_W) \cdot v_P . \tag{3.23}$$

The amount of the air on the entire mass of the abrasive water jet is about 3% (Figure 4.9). For simplification, consider this amount of air neglected. Another simplification is to omit the amount of neglect of a slip between the velocities of the water and of the abrasive particles if they leave the focus exit. Himmelreich [47] experimentally estimates the slip to be about 10 % Also, neglect the input velocity of the abrasive particle since $v_{P0} \ll v_P$.

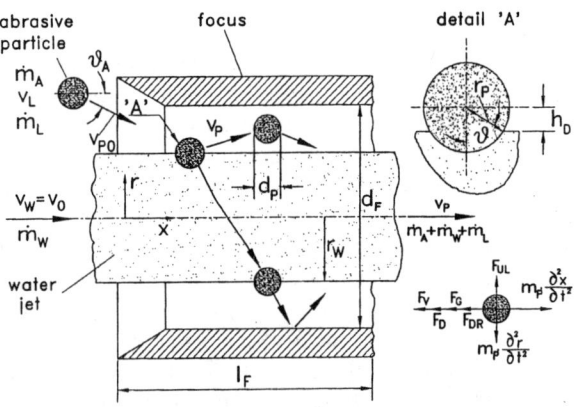

Figure 3.16 *Momentum (force) balance for the acceleration of abrasive particles*

Losses during the acceleration process are characterized by a momentum-transfer parameter η_T. Thus,

$$v_P = \eta_T \cdot \frac{v_0}{1 + \frac{\dot{m}_A}{\dot{m}_W}}. \tag{3.24}$$

3.4.1.2 Momentum-Transfer Efficiency

The momentum-transfer parameter is a function of several process parameters including the pump pressure, abrasive-mass flow rate, abrasive-particle characterization, and orifice and focus geometry. The momentum-transfer parameter also covers the simplifications and assumptions that are made in order to derive Eq. (3.24).

The momentum-transfer parameter is experimentally estimated by at least two methods. First, solve Eq. (3.24) for η_T,

$$\eta_T = \frac{v_P}{v_0} \cdot [1 + R], \quad R = \dot{m}_A / \dot{m}_W. \tag{3.25}$$

Measure the abrasive-particle velocity and the water-jet velocity by methods outlined in section 10.5. Based on velocity measurements that Himmelreich [47] carries out, the efficiency of the abrasive-particle acceleration is between $\eta_T=0.65$ and $\eta_T=0.85$. Applying the results of Isobe et al. [77] who use extremely high abrasive-mass flow rates (67 g/s), gives comparatively low values of $\eta_T=0.5$ that are not typical for abrasive water-jet cutting processes.

The second experimental method is based on force measurements of plain and abrasive-water jets as described in section 10.5.5. Use Eq. (3.24) and neglect the abrasive-particle input velocity as well as the air-mass flow rate to give

$$\eta_T = \frac{(\dot{m}_W + \dot{m}_A) \cdot v_P}{\dot{m}_W \cdot v_0} = \frac{F_A}{F_W}. \tag{3.26}$$

Therefore, estimate the momentum-transfer parameter by measuring the impact forces of a plain water jet and an abrasive water jet, respectively, under identical process conditions. Hashish [45] and Momber and Kovacevic [49] exploit this method. These authors obtain efficiency values between $\eta_T = 0.73$ and $\eta_T = 0.94$ that are slightly higher than the values estimated from the optical abrasive-particle velocity measurements.

Nevertheless, both experimental methods give comparative results for the momentum-transfer parameter.

In regard to the influence of other process parameters on the momentum-transfer efficiency, velocity measurements [47] as well as force measurements [45] indicate that an optimum momentum transfer exists in the range of medium pump pressures. Figure 3.17a shows that higher pump pressures lead to a deteriorated acceleration process.

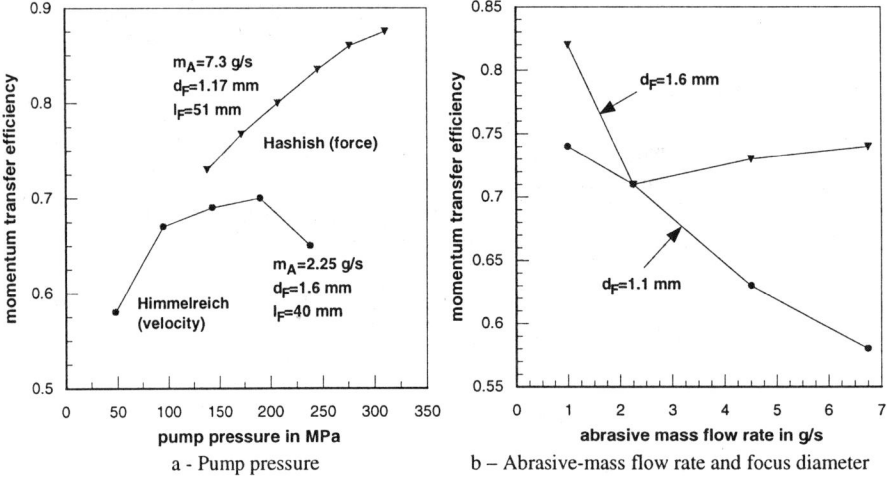

Figure 3.17 *Parameter influence on the momentum-transfer efficiency [47]*

Figure 3.17b shows the relation between the abrasive-mass flow rate and the momentum-transfer efficiency. The acceleration in the range of commonly used abrasive-mass flow rates improves by using small abrasive-mass flow rates. The results demonstrate that a saturation value is achieved for very high abrasive-mass flow rates. As Figure 3.17b exemplifies, large focus diameters improve the momentum transfer.

3.4.2 Improved Acceleration Model

3.4.2.1 Velocity Components

The velocity of the accelerated abrasive particles, v_P, is divided into an axial component, v_{Px}, and a radial component, v_{Pr}. Thus,

$$v_p = \sqrt{v_{Px}^2 + v_{Pr}^2} \ . \tag{3.27}$$

3.4.2.2 Force Balance in Axial Direction

For an abrasive particle in a fluid flow, the force equilibrium in axial (x-) direction is (Figure 3.16) gives

$$m_p \cdot \frac{\partial^2 x}{\partial t^2} = F_{DR} + F_G + F_D + F_V . \tag{3.28}$$

In Eq. (3.28), F_{DR} is the drag force. F_G is the force due to the pressure gradient in the fluid surrounding the particle, F_D is the force to accelerate the apparent mass of the particle relative to the fluid, and F_V, due to Basset, takes into account the effect of the deviation in the flow pattern from a steady state. Neglect these terms since they become important only if the density of the fluid has a similar or higher order of magnitude than that of the solid particle.

The force of the friction between a solid-particle surface and flowing water is [78]

$$F_{DR} = c_D \cdot A_P \cdot \frac{1}{2} \cdot \rho_{jet} \cdot |v_{rel}| \cdot v_{rel} . \tag{3.29}$$

The relative velocity between the fluid and the abrasive particle is

$$v_{rel} = v_0 - v_{P0} . \tag{3.30}$$

Assume pneumatic transport of the abrasive particles by air flow due to the pressure difference between the abrasive hopper and mixing chamber. Thus, the initial velocity of the sucked abrasive particles, v_{P0}, is given by Eqs. (3.18) and (3.21). The components of the initial abrasive-particle velocity are

$$v_{P0x} = \cos\theta_A \cdot v_{P0} , \tag{3.31a}$$
$$v_{P0r} = \sin\theta_A \cdot v_{P0} . \tag{3.31b}$$

In the equations, θ_A is the angle of the mounted abrasive-particle inlet device.

3.4.2.3 Friction Coefficient and Reynolds-Number

The friction coefficient in Eq. (3.29) depends on the Reynolds-number of the particle flow that is given by

$$Re_P = \frac{d_P \cdot \rho_{jet} \cdot v_{rel}}{\mu_{jet}} . \tag{3.32}$$

An analytical solution for the function $c_D = f(Re_P)$ for high Reynolds-numbers does not exist but experimental results cited by Brauer [75] indicate that the coefficient of friction is constant for high Reynolds-numbers as they appear in high-pressure water jet flow ($10^5 < Re < 7 \cdot 10^5$). For completely submerged spheres,

$$c_D = \text{constant} = 0.44. \tag{3.33a}$$

If the sphere is just partially under water (Figure 5.16, detail 'A'), the friction coefficient is [77]

$$c_D(r) = 0.225 \cdot \frac{\cos(3 \cdot \vartheta) - 9 \cdot \cos \vartheta + 8}{24}. \tag{3.33b}$$

The area A_P in Eq. (3.29) is the area of the abrasive particle subjected to the water flow. This area depends on the diving depth, h_D, of the particle. For spheres,

$$A_P(r) = \frac{1}{2} \cdot \left[r_w - r + \frac{d_P}{2} \right] \cdot h_D \cdot \pi, \qquad r > r_w - \frac{d_P}{2}. \tag{3.34a}$$

$$A_P(r) = \frac{1}{2} \cdot d_P \cdot \pi, \qquad r < r_w - \frac{d_P}{2}. \tag{3.34b}$$

Consider the parameters ρ_{jet} and μ_{jet} in Eq. (3.32) as parameters of a two-phase flow (water and air) that depend on the values of x and r, respectively. Thus,

$$\rho_w = \rho_{mix}(x, r), \tag{3.35a}$$
$$\mu_w = \mu_{mix}(x, r). \tag{3.35b}$$

For the solution of Eq. (3.35a) in x-direction, use Eqs. (3.12a,b). Miller and Archibald [79] suggest

$$\rho_{mix}(x) = \chi_G(x) \cdot \rho_L + [1 - \chi_G(x)] \cdot \rho_w, \qquad \text{and} \tag{3.36a}$$
$$\mu_{mix} = \chi_G(x) \cdot \mu_L + [1 - \chi_G(x)] \cdot \mu_w. \tag{3.36b}$$

In these equations, χ_G is the gas fraction in the water jet. Consider the gas fraction as air, and use experimental results from Davies and Jackson [59] to estimate the air content along the x-axis (Figure 3.6a).

3.4.2.4 Force Balance in Radial Direction

For an abrasive particle in a fluid flow, the force equilibrium in radial (r-) direction gives (Figure 3.16)

$$m_P \cdot \frac{\partial^2 r}{\partial t^2} = F_{UL}. \tag{3.37}$$

Ramsauer et al. [80] derive the uplift force, F_{UL}, that an impacting sphere experiences in water,

$$F_{UL} = \frac{\partial \varphi_P}{\partial x} \cdot m_P \cdot v_P^2. \qquad (3.38)$$

Ramsauer et al. [80] experimentally estimate the function $\partial \varphi_P/\partial x$ as shown in Figure 3.18. The function does not depend on the particle velocity but on the diving depth, h_D, of the particle. The function $\partial \varphi_P/\partial x = f(h_D)$ fits in a high-order polynomial [77] (reference [77] uses wrong values for $\partial \varphi_P/\partial x$) or in an exponential regression [47], respectively. The function shows a typical maximum at a diving depth of $h_D = d_P/2$.

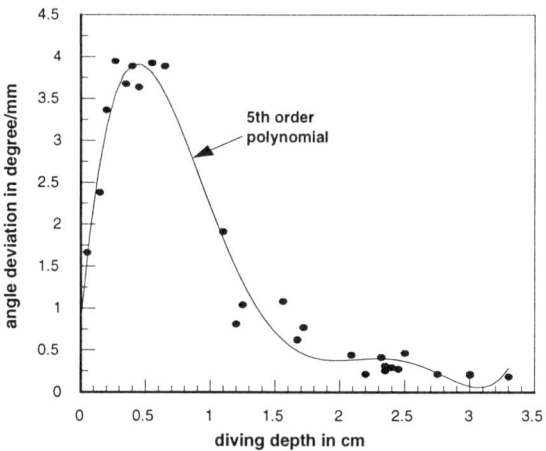

Figure 3.18 *Relation between the trajectory and diving depth of a spherical solid particle [80]*

3.4.2.5 Approximate Solution

Abudaka and Crofton [81] obtain an approximate solution of Eq. (3.28) through simplifying the process (constant water-jet velocity, linearization). Nevertheless, they calculate acceleration distances of just 5 mm to 10 mm that are in contrast to the reality. Earlier, You et al. [82] try to estimate the axial abrasive-particle velocity in consideration of: the decrease in the water-flow velocity with increasing distance from the orifice, the radial water-flow velocity distribution.

3.4.2.6 Rigorous Solution

Blickwedel [69] rigorously solves Eq. (3.28) to estimate an acceleration distance, L_A, required for a spherical particle to obtain a velocity v_P. The final relation is

$$L_A = \frac{1}{K_A(1+R)^2} \cdot \left[\frac{1}{1-n_V \cdot (1+R)} - 1 + \ln[1 - n_V \cdot (1+R)] \right]. \quad (3.39)$$

$$n_V = \frac{v_P}{v_0}, \quad R = \frac{\dot{m}_A}{\dot{m}_W}$$

In the equation, K_A is an acceleration coefficient that augments with an increase in the acceleration efficiency,

$$K_A = c_D \cdot \frac{\rho_{mix}}{\rho_P} \cdot \frac{3}{4 \cdot d_P}. \quad (3.40)$$

Eq. (3.40) shows that the acceleration process improves by increasing the coefficient of friction, increasing the carrier-fluid density, reducing the abrasive-material density, and decreasing the abrasive-particle size.

Figure 3.19a contains a graphical expression of Eq. (3.39) showing the influence of the carrier-fluid density and abrasive-particle diameter on the velocity ratio, n_V. Figure 3.19b plots the relation between n_V and the mass-flow ratio, R. These figures are based on spherical garnet abrasives and an abrasive-material density of $\rho_P = 4.150$ kg/m^3. Eq. (3.33a) gives the coefficient of friction.

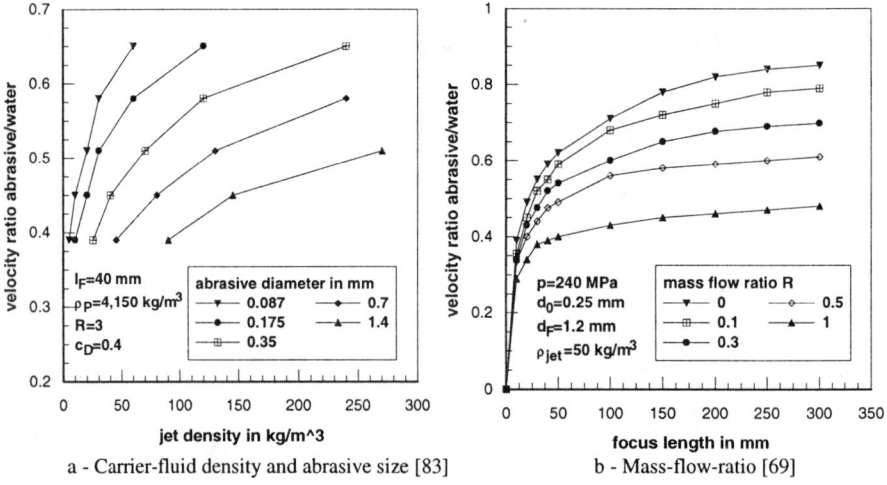

a - Carrier-fluid density and abrasive size [83]   b - Mass-flow-ratio [69]

Figure 3.19 *Parameter influence on the velocity ratio for the abrasive-particle acceleration*

Figure 3.19a illuminates the major influence of the jet density on the acceleration process. The higher the jet density, the higher the final velocity of the particle. The maximum possible velocity ratio ($n_V=1$) is reached with a plain continuous water jet, $\rho_{mix}=\rho_W$ which is an idealized condition. Estimate real jet densities using Eqs. (3.12a,b). The water jet accelerates small abrasive grains very quickly even if the jet density is low. To obtain identical velocity ratios, larger abrasive particles require a higher carrier-fluid density, which could be realized by using smaller focus diameters. Figure 3.19b shows that the abrasive-mass flow rate has a significant impact on the acceleration process; high abrasive-mass flow rates yield lower abrasive-particle velocities. This figure also illustrates also that a minimum acceleration distance (focus length) is necessary to introduce the acceleration process. For the condition in Figure 3.19b, this length is between $L_A=20$ mm and $L_A=40$ mm. Beyond an acceleration distance of $L_A=100$ mm, no substantial improvement in the acceleration process occurs.

Tazibt et al. [50] present both trends, that of the air content as well as that of the abrasive-mass flow rate, in a numerical model.

3.4.2.7 Numerical Solutions in Axial Direction

In order to estimate the acceleration of single abrasive particles in axial as well as in radial direction, Isobe et al. [77] and Himmelreich [47] solve Eqs. (3.28) and (3.38) by numerical integration. Figure 3.20 gives some results of the calculations for the axial movement. Figure 3.20a shows the acceleration process for different abrasive-mass flow rates. The trends are identical to those shown in Figure 3.19b. The exit abrasive-particle velocity decreases almost linearly with an increase in the abrasive-mass flow rate. The velocity slip between the abrasive particle and the water flow is about $(v_0-v_P)=75$ m/s for all conditions, which is about 10 % of the water velocity.

Figure 3.20b illustrates the influence of the abrasive-particle entry velocity on the acceleration process. Even small changes in the entry velocity ($\Delta v_{P0}=\pm 1.0$ m/s) generate very different acceleration conditions. Also, there is no general trend between the entry velocity and abrasive-particle exit velocity. The highest abrasive-particle entry velocity gives the highest abrasive-particle exit velocity. The steps that are observed at certain acceleration distances characterize areas where the abrasive particle moves between the water jet and the focus wall. Figure 3.20c illuminates the impact of the abrasive-material density on the acceleration process. Generally, the abrasive-particle exit velocity increases with a decrease in the abrasive-material density. Lightweight abrasive materials (silica sand) accelerate very effective using small focus lengths (in the case shown, about $L_A=10$ mm). The differences in the acceleration behavior of the two high-dense abrasive materials are not pronounced over the entire acceleration range. But these materials require significantly longer focuses to obtain the same final velocity as the silica abrasive. The numerical simulations also show that the abrasive-particle velocity increases by using small diameter focuses and small abrasive-grain diameters [47], which is in agreement with Figure 3.19a.

46   3. Generation of Abrasive Water Jets

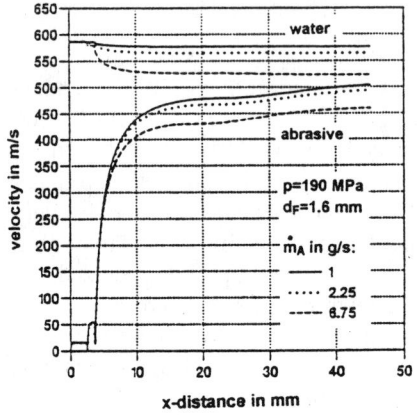

a – Abrasive-mass flow rate

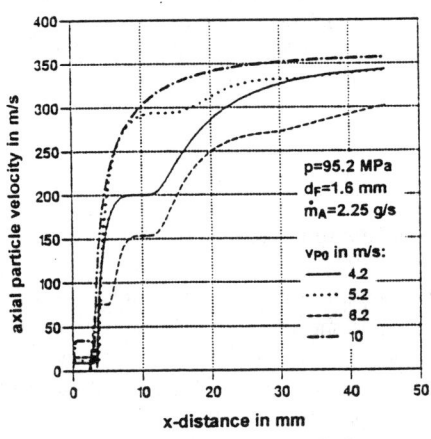

b – Abrasive-particle entry velocity

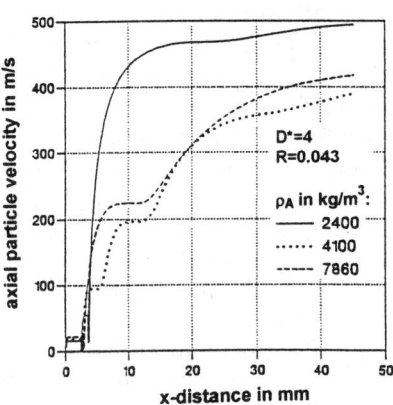

c – Abrasive-material density

Figure 3.20 *Numerical solutions for the axial abrasive-particle velocity [47]*

3.4.2.8 Numerical Solutions in Radial Direction

Figure 3.21 shows some numerical solutions of Eq. (3.38) for the radial abrasive-particle movement. More than Figure 3.20b, this figure illustrates the importance of the abrasive-particle entry velocity on the particle movement in the focus. A change in the entry velocity changes the radial movement catastrophically. The particle entrained at a high velocity of v_{P0}=5.2 m/s behaves extremely unsteadily during the acceleration process. The particle hits the focus wall as well as the water jet several times and shows the highest radial velocity after acceleration. Nevertheless, as shown in Figure 3.20b, this particle can reach a high axial velocity. Also, the steps in the axial velocity detected in Figure 3.20b are distinguished as ranges with increased radial components in Figure 3.21. Every tough between the water jet and the abrasive particle accelerates the particle in the axial direction as well as in radial direction. The radial velocity increases by using small abrasive particles [47].

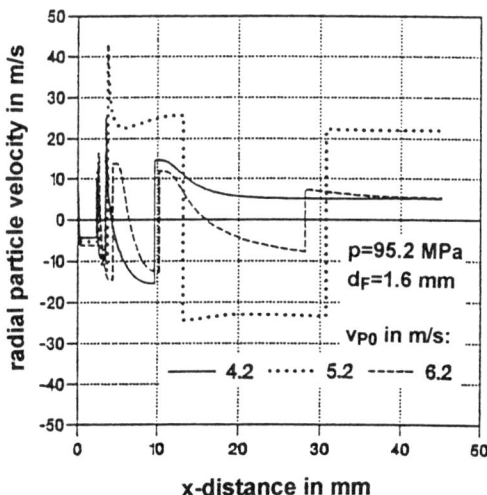

Figure 3.21 *Numerical solutions for the radial abrasive-particle velocity [47]*

3.4.2.9 Results of Steel-Ball Projection Experiments

Isobe et al. [77] investigate in more detail the interaction between water jet and abrasive particle based on steel-ball projection tests. Figure 3.22 plots some results. Figure 3.22a shows the relation between the entry velocity of a particle and the angle of deflection. In this figure, angles of $\vartheta_0<0°$ mean that the particle crosses the water jet, as is the case for high abrasive-particle entry velocities. Isobe et al. [77] also show that the highest particle-deflection velocity is achieved at deflection angles of $\vartheta_0=0°$, which means that the particle remains in the water jet, and the drag force can act over the maximum particle surface.

48   3. Generation of Abrasive Water Jets

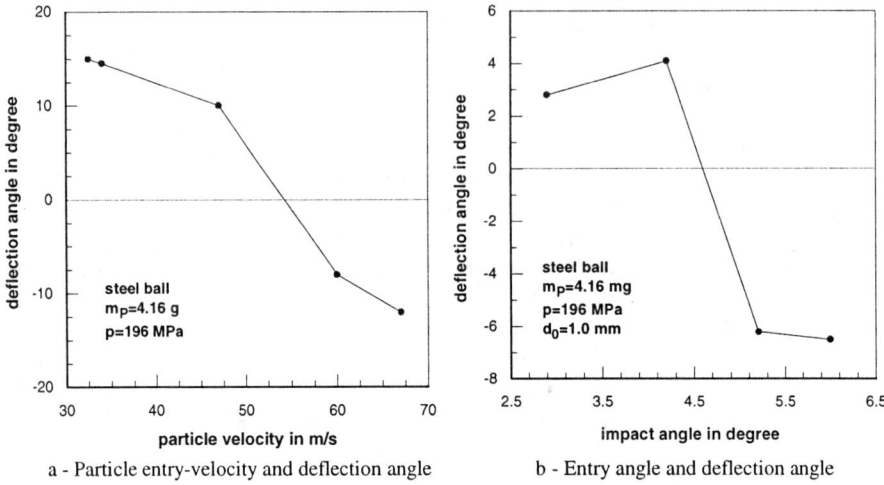

a - Particle entry-velocity and deflection angle   b - Entry angle and deflection angle

Figure 3.22 *Interaction between the solid particle and water jet [77]*

Therefore, at medium particle entry velocities, maximum acceleration is observed. This result is not completely in agreement with Figure 3.20b, but there is also some evidence that medium injection velocities contribute very effectively to the acceleration process. Figure 3.22b shows that small deflection angles are obtained by selecting an entry angle of about $\theta_A = 5°$. This result indicates that the commercially used abrasive-mixing units work with too large entry angles. The numerical results agree very well with the presented experimental results [77] that confirm the application of the uplift-force model for the estimation of the radial abrasive movement.

3.4.3 Regression Model

Chen and Geskin [84] develop a regression model that fits abrasive-particle velocities that are estimated by measurements with the laser-transit-velocimeter (section 10.5.3). The regression analysis contains the influence of several process parameters including the water-jet velocity, abrasive-mass flow rate, focus diameter, and orifice diameter. The analysis gives

$$\frac{v_{0F} - v_P}{v_0} = a_1 \cdot \left[\left(\frac{\dot{Q}_A}{\dot{Q}_W} \right)^{\left(\frac{d_0}{d_F} \right)^2} \right]^{a_2} . \qquad (3.41)$$

Figure 3.23 shows this relation. For ratios $0.11 \leq d_0/d_F \leq 0.41$, the regression constants are $a_1 = 0.627$ and $a_2 = 2.557$. The left term in Eq. (3.41) is a non-dimensional number that characterizes the relation between the abrasive-particle velocity, the water-jet velocity, and the velocity of the water jet passing the focus. The velocity v_{0F} is introduced to include the effect of the alignment between the water-jet orifice and the acceleration focus (section 3.2.2.3).

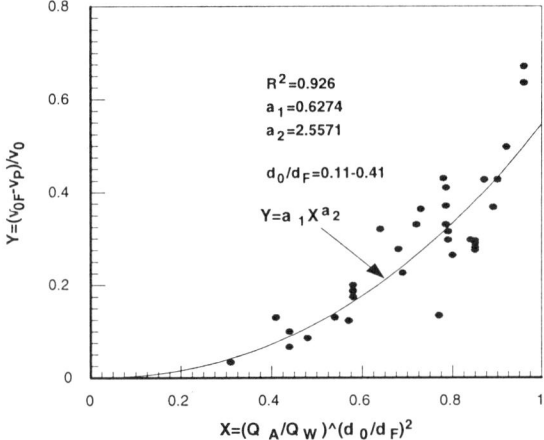

Figure 3.23 *Correlations between measured abrasive-particle velocities [84]*

The velocity difference, v_{0F}-v_P, is a measure of the momentum transfer between the water jet and the entrained abrasive particles. The higher this value, the lower the abrasive particle velocity, and the worse the acceleration process.

3.5 Abrasive-Particle Fragmentation in Injection Systems

3.5.1 Solid-Particle Impact Comminution

3.5.1.1 Impact Velocity and Impact Angle

One of the most pronounced features of the abrasive water-jet mixing-process is the fragmentation and comminution of the abrasive particles.

The elementary process of abrasive-particle fragmentation is assumed to be impact comminution. The major kinematic parameter in impact comminution is the impact velocity, that is identical to the velocity of the accelerated abrasive particle, v_P. The grain hits a solid or liquid surface creating stresses in the grain. Figure 3.24 illustrates the geometrical situation. A certain probability exists that the grain fractures during the impact. This fracture probability strongly depends on the impact velocity, impact angle, and abrasive-particle diameter.

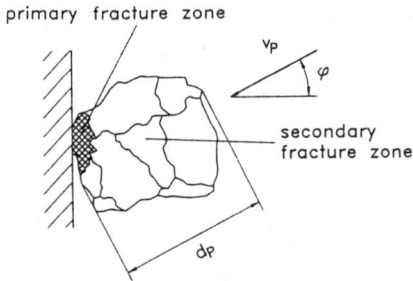

Figure 3.24 *Fracture structure of a single grain subjected to impact comminution [85]*

Figure 3.25a shows the influence of the impact velocity and impact angle for non-regular quartz particles. The comminution probability increases with an increase in the impact velocity and an increase in the impact angle. These relations indicate some influence of the pump pressure and mixing-chamber geometry on the abrasive-particle comminution. Eqs. (3.2b) and (3.24) relate the pump pressure to the abrasive impact-velocity. Changes in the mixing-chamber design affect the abrasive-particle velocity as well as the impact angle. Figure 3.25b gives relations between the fracture probability, particle impact-velocity, and particle diameter for two materials. These materials are not typically used as abrasives but are considered to represent a typical abrasive material behavior. As Figure 3.25b shows, an increase in the impact velocity fractures smaller grains. If the impact velocity is not high enough the grain is not destroyed. Hutchings [86] uses a very similar approach for the construction of erosion maps in brittle materials. He, too, suggests that beyond a critical particle size and a critical particle velocity, the fragmentation of the particle is introduced.

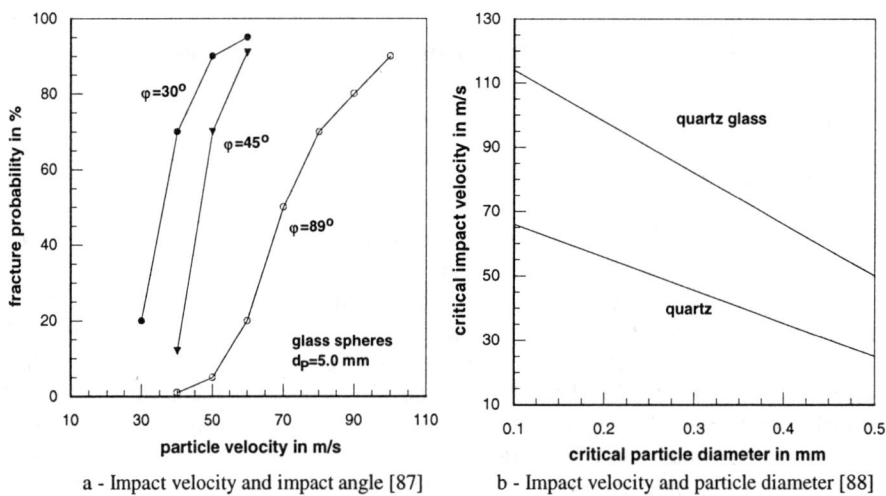

a - Impact velocity and impact angle [87]   b - Impact velocity and particle diameter [88]

Figure 3.25 *Parameter influence on the fracture of impacting solid particles*

3.5.1.2 Fracture Zones During Impact

Figure 3.24 shows a simplification of the fracture structure in a non-regular grain subjected to impact. In the figure, two fracture zones are distinguished. The 'primary zone' is a result of high-velocity stress waves generated during the impact. On their way through the grain, the waves decelerate and reflect on the rear side of the grain. The reflected waves hit the fracture front. When the reflected waves partially stop the fracture front, coarser fragments form on the rear side of the grain that is defined as the 'secondary zone'. Reiners [89] observes these processes among others by using a high-speed photography technique. These relations explain the high amount of very fine abrasive debris after the mixing and acceleration in the focus that Simpson [90] reports.

3.5.1.3 Size Effects

The fracture probability of abrasive grains strongly depends on their structure, in particular on the number and distribution of non-regularities, such as microcracks, grain boundaries, and dislocations. The number of flaws depends on the volume of the abrasive particle in question [91]. Experimental results by Martinec [39] show the validity of this concept for the abrasive comminution in mixing nozzles. He finds that garnet particles with very small diameters are extremely homogeneous, and that they show good stability of size and shape during the mixing-and-acceleration-process. Reiners [89] analyzes debris generated by the high-speed impact of glass spheres. For relatively low impact velocities (v_P=100 m/s), he finds large portions of sharp-shaped particles in all grain-size classes. At higher impact velocities (v_P=650 m/s), the debris consists mainly of cubic and rectangular grains. This result suggests a relation between the abrasive-particle velocity and the shape of the abrasive particles impacting the target material during the abrasive water-jet cutting.

3.5.1.4 Other Material Properties

Clever et al. [92] suggest a reduced impact fragmentation of single grains with lower hardness, lower density, larger fracture toughness, and lower grain diameter. Murugesh et al. [93] find similar relations. Larsen-Basse [94] observes a reduced fracture stress for single abrasive grains with an increase in the atmospheric humidity and explains this result due to the crack sharpening that results from the moisture attack at the crack tip. This effect plays a role considering the relatively high humidity in a closed machining-chamber for abrasive water jets.

3.5.2 Abrasive-Particle Size Reduction During Mixing and Acceleration

3.5.2.1 General Observations

Several authors observe that most of the abrasive particles fracture during the mixing process. Galecki and Mazurkiewicz [66] and Galecki et al. [95] are the first to note this aspect. These authors measure that about 70% to 80% of all particles are subjected to fragmentation and find that this number depends on the original abrasive-grain size, pump pressure, and focus diameter. They also show that changes in the focus length do not affect the particle-size distribution, but changes in the mixing-chamber design do affect the fragmentation behavior. Foldyna and Martinec [38] divide abrasive materials into two groups: materials that are highly sensitive to changes in the mixing-chamber geometry (garnet, ilmenite, staurolite), and materials that are less sensitive to the mixing-chamber design (quartz, silica carbide).

Labus et al. [96], Simpson [90] and Ohlsen [97] carry out more systematic investigations.

3.5.2.2 The 'Disintegration Number'

To quantitatively evaluate the process of abrasive-particle disintegration, Ohlsen [97] introduces a 'disintegration number',

$$\phi_D = 1 - \frac{\overline{d}_{Pout}}{\overline{d}_{Pin}}, \tag{3.42}$$

with $0 < \phi_D < 1$. If $\phi_D = 0$, no disintegration occurs. Typical values for garnet are between $\phi_D = 0.15$ and $\phi_D = 0.70$.

Figure 3.26 illustrates the influence of several process parameters on the disintegration number. Figure 3.26a shows that the disintegration number linearly increases with the pump pressure; the slop decreases at very high pump pressures. Simpson [90] observes the same trend and also notices a critical 'fragmentation pressure' at about p=40 MPa.

Figure 3.26b plots the influence of the focus length. The disintegration number decreases with an increase in the focus length. Although the increase in the focus length is 400 %, the increase in the disintagration number is about 10 %. Thus, particle impacts on the focus wall do not significantly contribute to the fragmentation.

More pronounced is the focus diameter influence as shown in Figure 3.26c. As the focus diameter decreases, the disintegration increases from $\phi_D = 0.22$ to $\phi_D = 0.46$. This result contributes to the quadratic decrease in the focus area.

3.5 Abrasive-Particle Fragmentation in Injection Systems 53

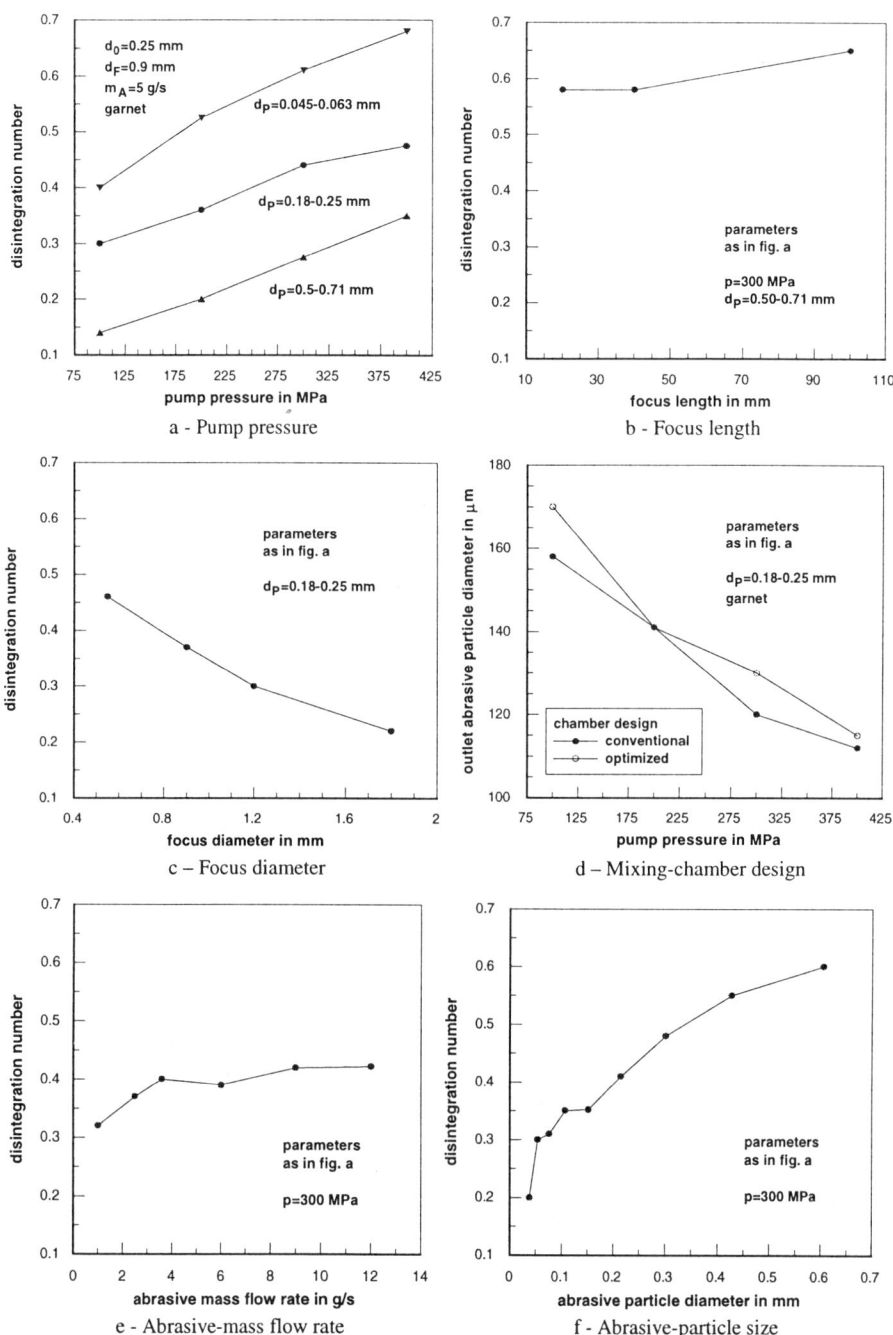

Figure 3.26 *Parameter influence on the abrasive-particle disintegration* [97]

Figure 3.26d shows the influence of the mixing-chamber geometry on the abrasive disintegration. The figure shows that the mixing-chamber design does not have a major impact on the disintegration number.

Figure 3.26e illustrates that, as the abrasive-mass flow rate exceeds a certain value (4 g/s), this parameter does not significantly influence the particle disintegration. In contrast, Labus et al. [96] find that the number of intact abrasive particles decreases with an increase in the abrasive-mass flow rate. This result may be due to the higher impact probability for the larger number of abrasive particles.

Figure 3.26f illustrates the influence of the particle diameter; the disintegration number almost linearly increases as the particle diameter increases. Simpson [90] observes that the total percentage of material that breaks down is greater for larger particles and explains this effect that large abrasive grains are not readily entrained in the inner core of the water jet. An additional plausible explanation can be deduced from Figure 3.25.

3.5.2.3 Influence of Abrasive-Particle Structure and Properties

The statistical fracture theory proves that the amount of weak zones, such as cracks, flaws, and interfaces, in a material depends on the size of the particle. The larger the grain, the more weak zones exist. Martinec [39], who uses garnet, finds that abrasive particles with diameters smaller than $d_P=100$ µm are extremely homogeneous and practically free of natural or gas-fluid inclusions. Consequently, their physical properties are closed to the properties of the ideal garnet monocrystals. For comparable particle diameters, the flaw density depends on the material type. Martinec [39] shows that garnet material (almandite B) with a low flaw density (ca. 5%) shows a comparatively high Vickers hardness as well as a relatively high disintegration number ($\phi_D=0.49$). In a comparative study, Foldyna and Martinec [38] find that Bohemia garnet has a high resistance against particle fragmentation. Interestingly, this material exhibits the highest Vickers hardness (HV=12,020 to 14,530) of the investigated abrasive materials.

Table 3.3 *Models for the impact fragmentation of solid-particles*

Reference	Model
Buhlmann [85]	$\dfrac{1}{\overline{d}_{Pout}} = \dfrac{f_C \cdot \rho_P \cdot v_P^2 \cdot F_{shape}}{2 \cdot \gamma_P \cdot F_{surf}} + \dfrac{1}{\overline{d}_{Pin}}$
Grady [100]	$\overline{d}_{Pout} = \left[\dfrac{\sqrt{20} \cdot K_{Ic}}{\rho_P \cdot c_P \cdot \dot{\varepsilon}} \right]^{\frac{2}{3}}$
Glenn et al. [101]	$\overline{d}_{Pout} = 2 \cdot \left[\dfrac{5 \cdot K_{Ic}^2}{(\rho_P \cdot c_P \cdot \dot{\varepsilon})^2} \right]^{\frac{1}{3}}$

Kiesskalt and Dahlhoff [98] and Dahlhoff [99] find a significant relation between the wave velocity in minerals and their resistance against the impact comminution. The higher the wave velocity, the higher the resistance against the comminution.

Table 3.3 lists references that develop analytical models for the estimation of debris diameters after the impact comminution. Buhlmann [85], Devaswithin et al. [102], and Grady [103] analytically derive distribution functions of the resulting fragment sizes.

3.5.2.4 Energy Absorption During Abrasive-Particle Fragmentation

The fragmentation of the abrasive particles absorbs a certain amount of the abrasive-particles' kinetic energy. Knowing the grain-size distributions of the particle mixture before it enters the mixing chamber and after it exits the acceleration focus, Bond's comminution formula [104] can be used to calculate the energy involved in this fragmentation process. The absorbed energy is

$$E_F = w_i \cdot \frac{\sqrt{d_{Pin}} - \sqrt{d_{Pout}}}{\sqrt{d_{Pout}}} \cdot \sqrt{\frac{100}{d_{Pin}}} \,. \qquad (3.43)$$

In Eq. (3.43), w_i is the index of workability that is estimated by comminution tests. Table 3.4 gives the tabulated values of some materials.

Table 3.4 *Workability indices of minerals [105]*

Material	Workability index [kWh/t]
Ferro-chromite	7.64
Ferro-magnesite	8.30
Glass	12.31
Gravel	16.06
Quartzite	13.57
Slag	9.39
Silica carbide	25.87
Silica sand	14.10

Mazurkiewicz and Galecki [106] use Eq. (3.43) to estimate the energy that is absorbed by the fragmentation of the abrasive particles (Figure 3.27). An average value is about 5% of the water-jet input energy.

3.5.3 Abrasive-Particle Shape Modification During Mixing and Acceleration

Another important aspect of the abrasive fragmentation is the *change in the shape* of the individual grains that may significantly influence the mechanism for material removal.

56   3. Generation of Abrasive Water Jets

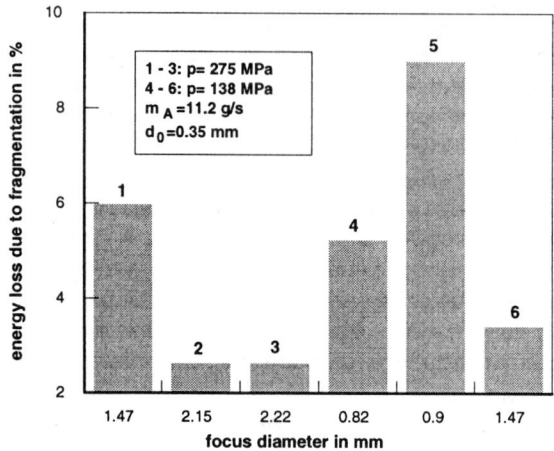

Figure 3.27 *Energy absorption due to abrasive fragmentation [106]*

The modification of the particle shape is expressed by changes in the shape parameters defined in section 2.2. Table 3.5 that shows results from different garnet modifications, illustrates that the shape factor in tendency increases due to the mixing process. This result indicates that sharp edges and corners are removed from the abrasive particles. The results also show that the shape changes are more pronounced for the larger abrasive particles. Martinec [39] observes that garnet particles with diameters smaller than $d_P=100$ µm exhibit a stable chip-like shape, regardless of the nature of the process of disintegration. The generation of these chip-like shapes agrees with the mode of fragmentation of cubic minerals.

Figure 3.28 shows the influence of the particle comminution on the abrasive shape in a roundness-sphericity diagram. The roundness as well as the sphericity reduce as a result of an intense particle fracture. This result is partially in opposition to Table 3.5 at least for the garnet material.

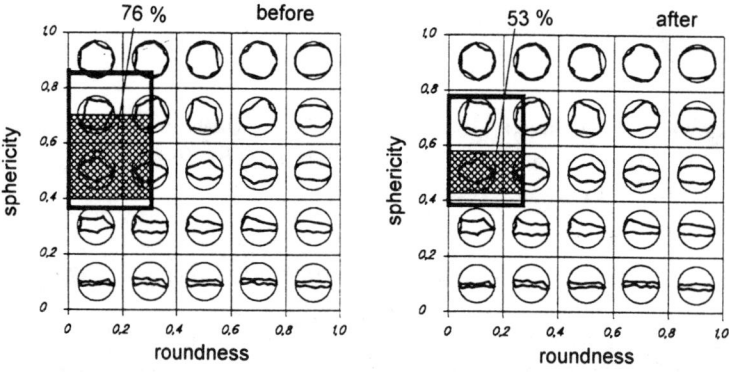

Figure 3.28 *Modification of the abrasive shape during the mixing and acceleration [38]*

Table 3.5 *Abrasive-particle shape modification during the mixing-and-acceleration process [107]*

Particle-size	Shape factor F_{shape}	
fraction [µm]	before mixing	after mixing
Almandite B		
> 200	0.65	0.73
100 - 125	0.65	0.68
< 63	0.65	0.67
Almandite K		
> 200	0.66	0.75
100 - 150	0.66	0.70
< 63	0.66	0.70
Grossular Z		
> 200	0.70	0.75
100 - 150	0.70	0.70
< 63	0.70	0.70
Andradite VC		
> 200	0.68	0.75
100 - 150	0.68	0.73
< 63	0.68	0.71

3.6 Focus Wear in Injection Systems

3.6.1 General Features of Focus Wear

The most important reasons for rejecting focusing nozzles are that they are worn (85 %) or broken (13 %). This criteria is valid for high-resistant focus materials as well as for tungsten-carbide nozzles [26]. Thus, the wear of the focusing nozzle is a significant feature of the abrasive water-jet machining. Also, as shown among others by Hulsley et al. [108], the age of a focus influences the cut geometry as well as the surface quality.

The term 'focus wear', discussed in this section, considers several phenomena, such as

- an increase in the focus exit-diameter,
- a mass loss of the focus,
- a generation of wear patterns of the inner focus surface,
- a change in the geometry of the focus.

The wear of a focusing tube can be tolerated inside certain ranges. These ranges depend on the quantitative and qualitative demands of the machining process, such as depth of cut, mass-removal rate, cut geometry, and cut-surface quality.

3.6.2 Focus-Exit Diameter

3.6.2.1 Early Observations

The most common method to estimate the focus wear is by measuring the focus outlet-diameter over a given period of time. In an early investigation in the focus wear, Nakaya et al. [109] find that the diameter ratio $d_F/d_F(t)$ depends on the combination of focus material and abrasive material. They observe that ceramic nozzles are more worn out by steel-grit abrasives compared to aluminum-oxide abrasives despite the much lower hardness of the steel grit. For the first time, this result indicates that the complex character of the focus-wear process is not a simple hardness relation between the focus material and the abrasive material.

3.6.2.2 Focus-Wear Rate

Figure 3.29 presents typical plots between the working time and focus diameter. Several references [69, 110-113] observe the general linear relationships shown in this figure for comparatively hard focus materials for a wide range of process conditions. Therefore,

$$d_F = W_F \cdot t. \qquad (3.44)$$

The progress of the function, W_F, is the focus-wear rate,

$$W_F = \frac{\Delta d_F}{\Delta t}, \qquad (3.45)$$

that depends on several material and process parameters (Figure 3.29).

3.6.2.3 Process-Parameter Influence

Hashish [45] and Kovacevic and Beardsley [114] investigate the influence of the pump pressure on the focus wear. Figure 3.29a shows that the wear increases with an increase in the pump pressure as a result of the increasing abrasive-particle velocity.

Figure 3.29b exhibits the relation between the wear and focus length. The wear rate reduces due to longer nozzles with a decreasing progress. Here, the explanation is that the abrasive particles travel at trajectories almost parallel to the focus wall far enough from the mixing point. Abudaka and Crofton [81] report that shorter mixing chambers yield lower wear rates of the focusing nozzles.

Hashish [110] address the influence of the orifice diameter on the focus wear. Interestingly, the trend between the focus diameter and time is non-linear, which means $W_F=f(t)$. This result is due to the fact that a commercial tool steel is used as a focus material. This relatively soft material is very sensitive to the location of the wear process. After a certain period of time, the soft focus material generates wear patterns that follow the water and abrasive-particle trajectories. Figure 3.30 that is an x-ray photograph of a worn focus made from low-resistant material, illustrates this patterns. If stable geometrical conditions are established, the wear rate drops. Hashish [115] observes a similar behavior for the wear of intermediate focus materials, such as soft-grade tungsten-carbide.

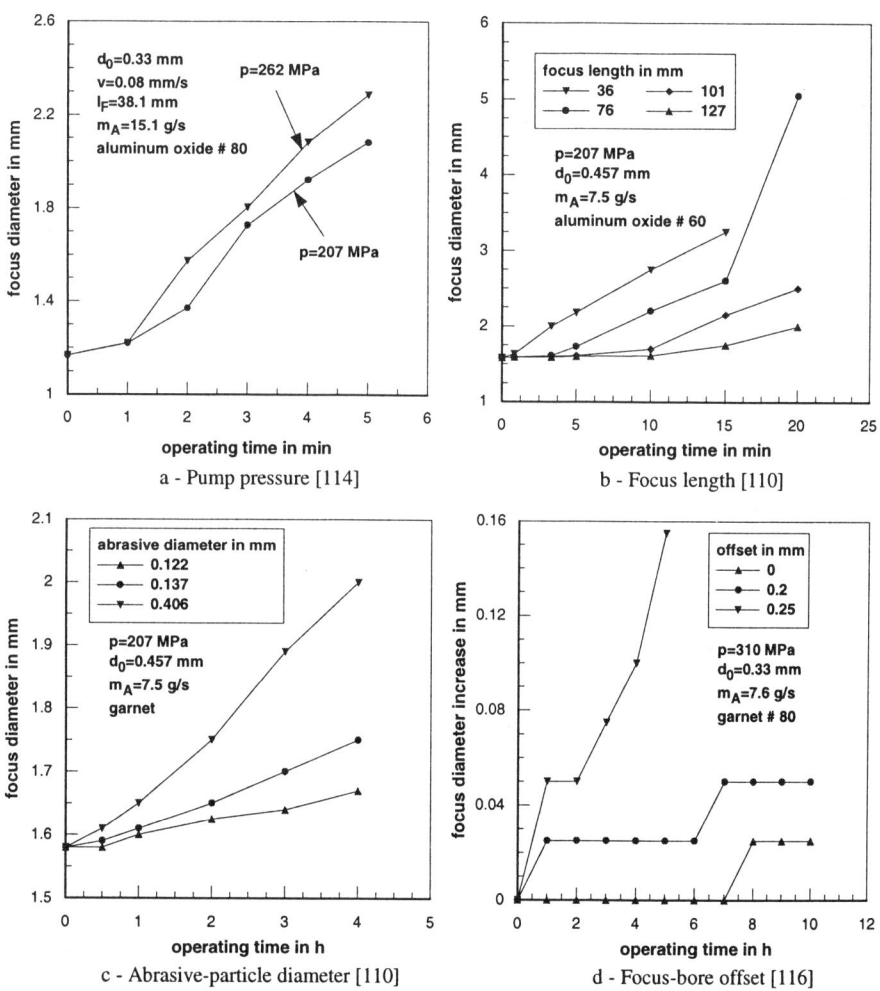

Figure 3.29 *Parameter influence on the focus exit-diameter*

The results given in Figure 3.29c illustrate the influence of the abrasive-particle diameter on the focus wear. Hashish [110] shows that a soft tool-steel focus is not sensitive to the particle size; whereas, a hard tungsten-carbide focus exhibits a strong dependence. In that case, the wear rate is significantly reduced with a

60   3. Generation of Abrasive Water Jets

decrease in the particle diameter. These results support the idea that soft focus materials are more sensitive to the flow conditions of the water jet.

Kovacevic and Beardsley [114] find that the wear rate increases with an increase in the abrasive-mass flow rate. This result is simply due to the higher number of impacting abrasive particles with an increase in the abrasive-mass flow rate.

Nanduri et al. [116] find that the offset in the focus bores that results from the manufacturing of the focuses, is uncritical as far as it is not larger than $l_{OB}=0.2$ mm. A focus with $l_{OB}<0.2$ mm performs as well as a focus with no offset (Figure 3.29d).

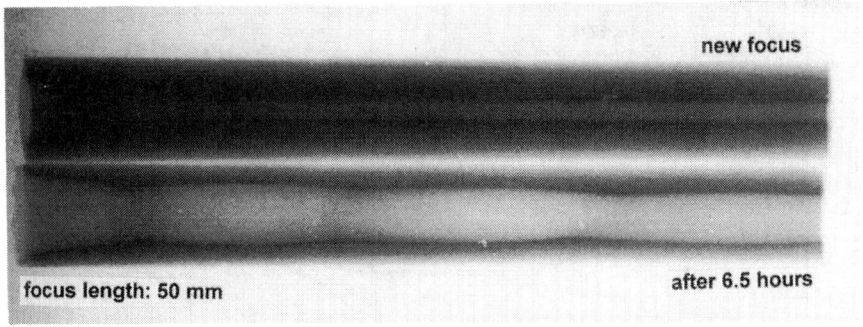

Figure 3.30 *X-ray photograph of a worn hard-metal focus (Univ. Hannover, IW)*

3.6.2.4 Hardness Influence

Neuss et al. [112] present a discussion into the influence of the hardness of both the abrasive material and the focus material. The results, as plotted in Figure 3.31, fit reasonably into

$$\log d_F = C_1 \cdot \log\left(\frac{H_P}{H_F}\right) - C_2. \tag{3.46}$$

In this equation, the hardness values are Vickers hardness values. The constants C_1 and C_2 consider the effect of other parameters. Fig. 3.31 shows that the values for the boron-carbide focus do not fit the regression. This discrepancy is a result of the high hardness of this material compared to the other focus materials. Therefore, the validity of Eq. (3.46) is restricted for hardness ratios between $H_P/H_F=0.7$-1.2 [112].

This assumption is in agreement with the measurements of the solid-particle erosion of ceramics and ultra-hard materials. These materials show a strong relation between the erosion rate and hardness if erodent hardness and target hardness is similar [117]. Srinivasan and Scattergood [118] also find that the erosion resistance of brittle-behaving materials goes through a marked transition as the target-to-particle hardness approaches unity.

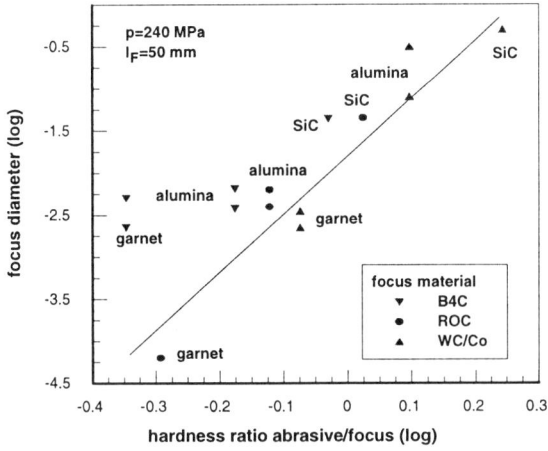

Figure 3.31 *Influence of the focus/abrasive hardness-ratio on focus wear [112]*

3.6.3 Other Focus-Wear Features

3.6.3.1 General Aspects

The wear rate based on the focus exit-diameter, and so on Eq. (3.46), is not a definite characterization of the condition of a worn focus. Abudaka and Crofton [81] for example find that materials with similar wear rates can show very different wear patterns. Neuss et al. [112] also observe that even when the exit diameter does not show any change after some minutes, the cutting efficiency drops. These facts indicate that material wear happens on other focus locations, too.

3.6.3.2 Focus-Mass Loss and Focus-Wear Pattern

The beginning focus wear is illustrated by the focus-mass loss due to the particle erosion rather than by the exit-diameter increase. Estimate the mass loss by weighing the focus at certain time steps. Figure 3.32a shows a typical plot between the exposure time and focus-mass loss. The figure shows that the mass loss linearly increases with the time. Figure 3.30 supports that this mass loss non-uniformly distributes over the focus length. Neuss et al. [112] detect a decrease in the wear intensity with an increase in the distance between the focus entry and the point of measurement. Figure 3.33 supports these observations. Figure 3.33a exhibits the inside wall profiles of worn focuses. Maximum wear is observed in the bore entry-section for both types of nozzles. The figure also shows that the wear patterns are more significant in the softer wall material, illuminating again the sensitivity of hard-metal focus materials against the fluid-flow conditions.

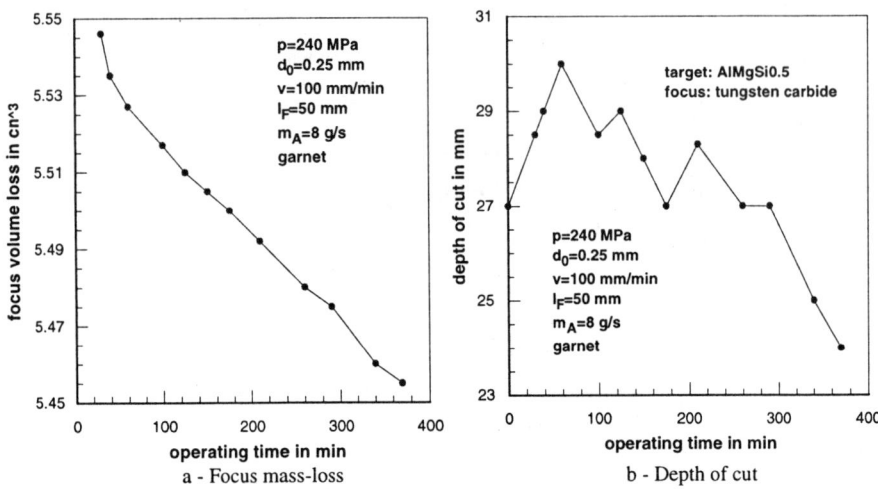

Figure 3.32 *Relation between the exposure time and focus performance [69]*

Figure 3.33b that illustrates the fast growth of the entry diameter in the beginning of the application period, presents more experimental evidence for an unsteady wear process. After a certain time (about t=40 minutes), the entry diameter remains more or less constant. This result indicates that the fast-erosion phase is a result of entry 'smoothing'.

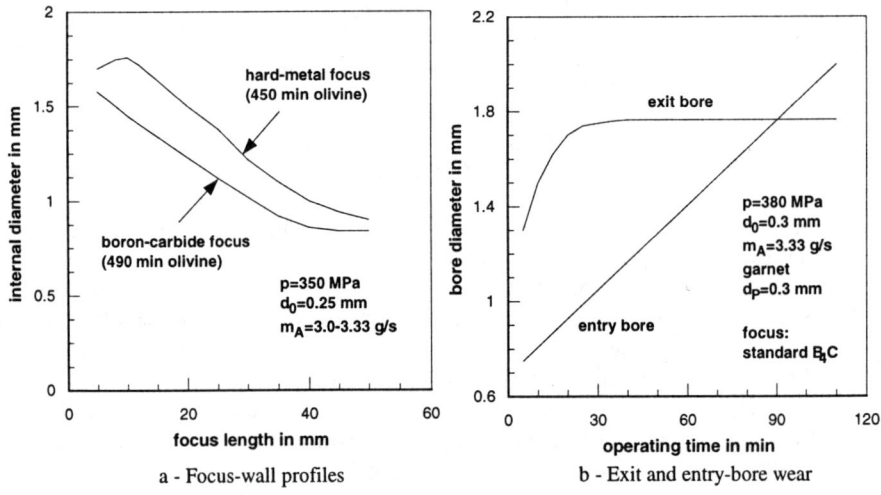

Figure 3.33 *Wear patterns in focus materials [113]*

3.6.3.3 'Selective' Focus Wear

Blickwedel [69] finds that the depth of cut in a material increases during the early focus-wear stage despite the measured focus wear (Figure 3.32b). Werner [119] also observes this behavior at continuous sapphire water-orifices. In the latter case, Werner [119] finds due to SEM-observations that the sharp-edged entry zone of the orifice is angled by the water-jet attack until optimum fluid-flow conditions are established. The same mechanism is assumed for the conditions given in Figure 3.32b. Some kind of 'selective wear' generates optimum conditions for the abrasive-water flow.

Hashish [110] shows that the inlet diameters of tungsten-carbide focuses are very sensitive to the abrasive-particle size. The larger the abrasive particles, the more non-symmetric performs the wear process.

3.6.3.4 Eccentricity of Focus-Exit Wear

Another aspect of focus wear is the eccentricity of the focus-exit wear. For example, worn nozzle outlets show elliptical shapes more than circles [120]. The eccentricity is defined as the ratio between the smallest and the largest dimension of the focus-exit area. Wightman and Dixon [121] and Singh and Munoz [68] prove that the focus wear is more consistent with proper alignment between the focus nozzle and water-jet orifice. Also, avoiding misalignments increases the focus life time up to 40% [121].

3.6.4 Modeling the Focus-Wear Process

3.6.4.1 Phenomenological Focus-Wear Model

Figure 3.34 shows a general model for the wear modes in abrasive water jet nozzles as presented by Hashish [115] and Nanduri et al. [116]. At the entrance, abrasive particles come in and impact the wall at different relatively large angles. When the focus is of sufficient length, the abrasives tend to eventually travel parallel to the wall. Consider the wear regime acting in the tube as shallow-impact erosion. Bell and Rogers [122] suggest that fracture toughness is the predominant property during large-angle erosive wear of brittle materials; whereas, at low-impact angles, the hardness is the most important property. The wear of ceramic focus-materials is assumed as a combination of ductile and brittle behavior. The dominating material-removal mechanisms are micro-ploughing and micro-cracking.

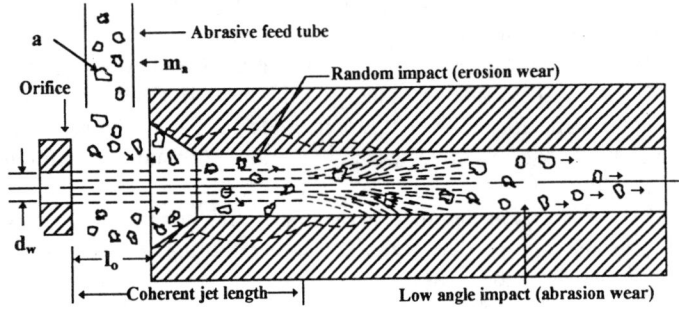

Figure 3.34 *Phenomenological model of the focus-wear [116]*

Wang et al. [123] observe both mechanisms on ceramics subjected to high-speed particle impact. Based on surface observations they find that the importance of the micro-cracking process increases with an increase in the impact angle. This behavior leads to high erosion-loss on materials with low fracture resistance. The ceramic with the smallest grain size exhibits the highest erosion resistance. This material shows also the best plastic-deformation capability among the presented materials. Ramulu et al. [124] observe a micro-cutting of the matrix at low impact angles and subsurface cracking in ceramics at higher impact angles.

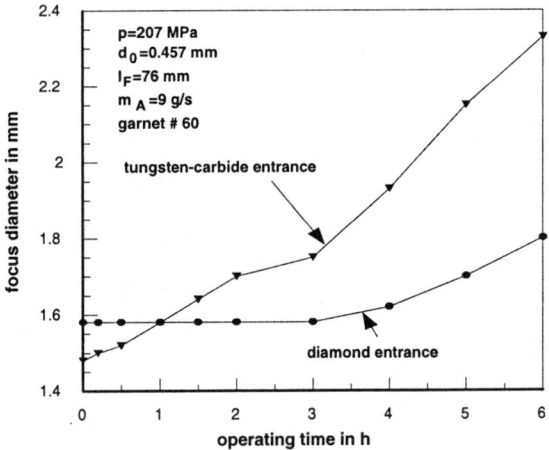

Figure 3.35 *Erosive-wear characteristics of a focus that consists of a material combination [110]*

3.6.4.2 'Two-Material Focus' Concept

The characteristic behavior discussed in section 3.6.4.1 supports the concept of a two-material-focus as developed by Hashish [110]. In the entry zone, use a material with some capability of plastic deformation because the probability of a large impact angle is high as shown in Figure 3.34. In contrast, at the exit stage of the tube use a very hard material to resist against the micro-ploughing. Figure 3.35 contains the wear characteristics of such a material combination. The wear of the focus considerably reduces.

3.6.4.3 Lifetime-Estimation Model

Typical lifetimes for focusing nozzles are about 100 h for high-resistant focus materials and 10 h for tungsten-carbide nozzles [26]. But, as already pointed out, the certain *lifetime* of focusing nozzles is defined for different conditions. Table 3.6 lists several definitions given in the reference literature.

Table 3.6 *Critical focus diameters for the lifetime definition*

Reference	Critical focus diameter
Hashish [110]	$1.25 \cdot d_F(t=0)$
Kovacevic and Beardsley [114]	$1.70 \cdot d_F(t=0)$
Mort [111]	20 mm
Nanduri et al. [116]	$(1.1\text{-}1.15) \cdot d_F(t=0)$

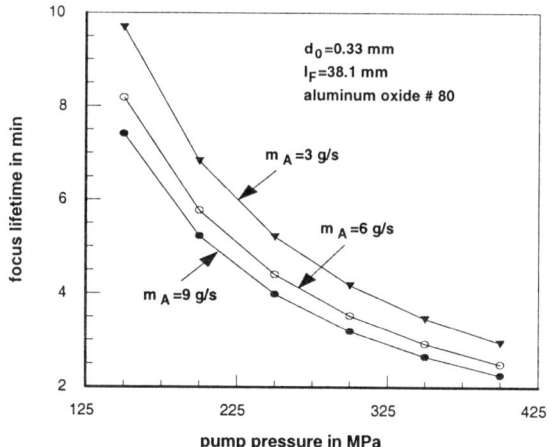

Figure 3.36 *Estimation of the focus-lifetime [114]*

Kovacevic and Beardsley [114] develop a lifetime model for focus-nozzle wear based on a multiple regression-analysis. The model delivers the relation

$$t_{TL} = \frac{5{,}587}{p^{1.215} \cdot \dot{m}_A^{0.245}}. \tag{3.47}$$

The lifetime definition is $t_{LT}=t(1.7 \cdot d_F)$. The pump pressure is in MPa and the abrasive-mass flow rate is in g/s which gives the lifetime in minutes. Figure 3.36 shows a graphical expression of Eq. (3.47).

3.7 Generation of Suspension-Abrasive Water Jets

3.7.1 General System Features

3.7.1.1 System Components

According to their generation, suspension-abrasive water jets divides into two systems as illustrated in Figure 3.37:

- bypass system,
- direct-pumping system.

Despite the different generation mechanisms, the most important process characterization of these systems are the pressures being generated. Bypass systems are limited to pressures up to p=200 MPa [125]; whereas, direct-pumping systems operate up to pressures of p=350 MPa [126].

Historically, the bypass system is first used for abrasive water-jet cutting operations in 1985 [127]. The first commercial system is introduced in 1986 [128]. Nevertheless, in oil well drilling, abrasive water jets based on the direct-pumping principle are experimentally used in the early 1970's [129].

Table 3.7 contains a comparison of both systems from the point of view of the process parameters being applied.

3.7.1.2 Bypass Systems

In the bypass system, part of the water-volume flow is used to bring the abrasive material out of the storage vessel and to mix it back into the main water-flow line.

Figure 3.38a shows the general structure of an early bypass-based system. The main components are a plunger pump, high-pressure abrasive storage tank, bypass line, and abrasive hopper. The pump delivers pressurized water that is split three ways. The majority of the flow bypasses the slurry-filled tank through an ejector before flowing to the nozzle.

3.7 Generation of Suspension-Abrasive Water Jets

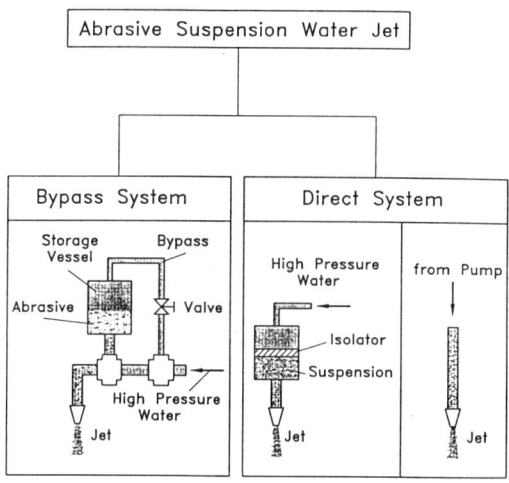

Figure 3.37 *Principles of generation of suspension-abrasive water jes [125]*

Table 3.7 *Technical characteristics of the generation of suspension-abrasive water jets*

Reference	Pump pressure [MPa]	Abrasive flow rate [g/s]	Slurry concent. [%]	Abrasive-grain size [μm]	Focus diameter [mm]
Bypass system					
Anderson [130]	28.5	108	-	-	2.8 - 4.8
Bloomfield and Yeomans [131]	35 - 69	2.8 - 173	12	-	0.3 - 2.8
Brandt et al. [125]	25 - 200	8.3 - 50		45 - 250	0.5 - 0.7
Guo et al. [132]	-	-	15	# 30 - # 100[1]	1.5 - 2.3
Laurinat et al. [133]	18	8.3 - 83		180 - 710	1.5 - 2.4
Liu and Ciu [127]	2 - 9	50	-	630	2.2
Liu et al. [134]	10 - 35	3.7 - 142	1.5 - 36	-	1.36
Shimizu and Wu [135]	20		18 - 24	75 - 212	1.0
Walters and Saunders [136]	15 - 69	28.3		150 - 500	1.0
Yazici and Summers [137]	21 - 35	20 - 150	-	300 - 1,250	2.0 - 2.8
You et al. [138]	35	33 - 117	13 - 17	# 28 - # 80[1]	-
				$0.3 \cdot d_F$	
Direct-pumping system					
Hashish [126][2]	104 - 345	1.0 - 11	6 - 48	# 80 - # 220[1]	0.23
Hollinger et al. [7] [2]	52 - 104	1.2 - 1.7	10	53 - 106	0.1 - 0.3

[1] Mesh designation [2] Polymer addition

68   3. Generation of Abrasive Water Jets

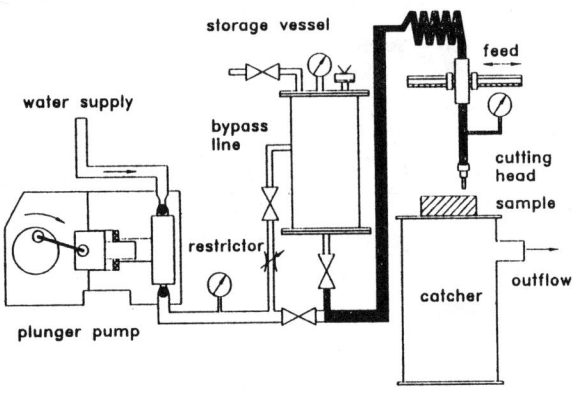

a - 'DIAJET'-system [128]

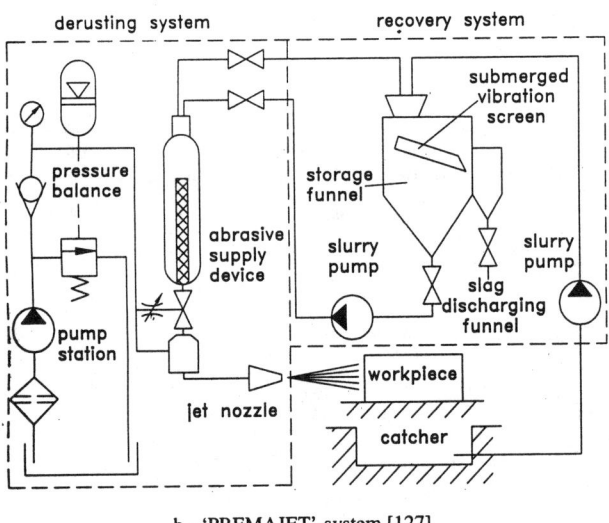

b - 'PREMAJET'-system [127]

Figure 3.38 *General structure of bypass suspension-abrasive water-jet systems*

Use the ejector to entrain slurry from the tank where the second part of the pressurized water is used to fluidize the abrasives. Apply the third part of the water at the top of the vessel to pressurize the largely undisturbed abrasive column. You et al. [82] make a modification of this basic design where the water does not flow into the tank but into a specially designed mixing chamber.

Liu and Cui [127] and Liu et al. [134] independently develop and use for cleaning and cutting-applications a very similar suspension-abrasive water jet bypass-system, but limited to pressures of about p=10 MPa (Figure 3.38b).

Liu and Cui [127], Liu et al. [134], and Shimizu and Wu [135] discuss several detailed problems, mainly related to the transport and calibration of the abrasive flow.

3.7.1.3 Direct-Pumping Systems

Figure 3.39 shows the structure of a suspension-abrasive water-jet equipment that bases on the direct-pumping system. In this system, the pre-mixed slurry is charged in a pressure vessel in which the high-pressure water is pumped to pressurize the slurry. A separator is used to prevent mixing of the slurry and the water. Hollinger et al. [7] and Hashish [126] use this system for cutting applications. The abrasive flow rate can not be varied independently of other process parameters including the pump pressure (Figure 3.40), nozzle geometry, and suspension characteristics.

Direct-pumping systems are usually mixed with high-viscous additives to suspend the abrasive particles in the storage vessel, and to reduce their settling velocity. Hashish [126], for example, uses an additive concentration of 3% by mass. Hollinger and Mannheimer [139] compare different additive types, such as a methyl-cellulose solution and polymeric 'superwater' in concentrations between 1.3% to 3.9% by mass.

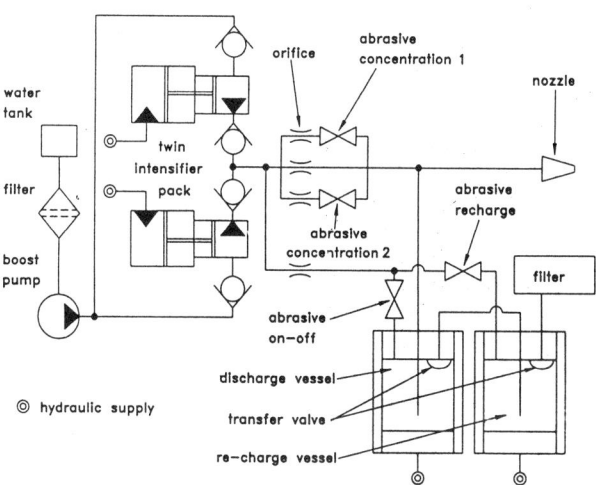

Figure 3.39 *Suspension-abrasive water-jet system based on direct-pumping [126]*

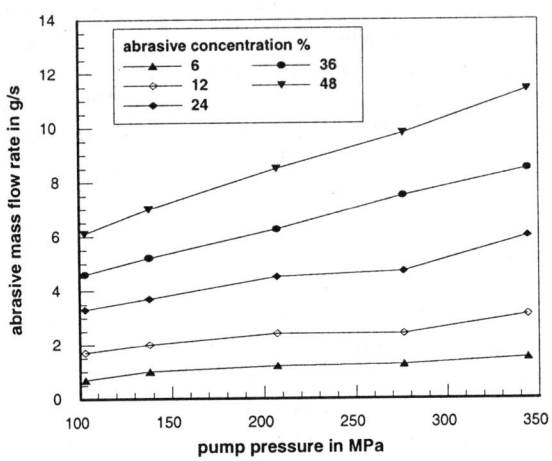

Figure 3.40 *Parameter relations for a direct-pumping system [126]*

3.7.2 Abrasive-Particle Acceleration

3.7.2.1 Acceleration-Nozzle Design

Figure 3.41 shows a typical suspension-abrasive water-jet nozzle system. The nozzle system basically consists of a pipe section, a conical designed acceleration section, and a focus section.

Laurinat et al. [133], Guo et al. [132], Shimizu and Wu [135, 140], Ye and Kovacevic [141], and Shimizu [142] conduct investigations into the nozzle-geometry influence on the abrasive acceleration and on the cutting process.

3.7.2.2 Simple Momentum-Transfer Model

Neglecting the slip between the water and abrasive particle in a suspension-abrasive water jet nozzle, use Bernoulli's law (Section 3.1), the average velocity of an escaping abrasive particle approximates

$$v_P = \mu_{Sus} \cdot \sqrt{\frac{2 \cdot p}{\rho_{mix}}}. \tag{3.48}$$

Shimizu and Wu [140] verify the square-root relation between abrasive-particle velocity and pump pressure via numerical simulations. Via high-speed photography, Shimizu [142] experimentally estimates the momentum-transfer parameter. For a pump pressure of 12 MPa, he finds values between $\mu_{Sus}=0.9$ and $\mu_{Sus}=0.95$.

3.7 Generation of Suspension-Abrasive Water Jets

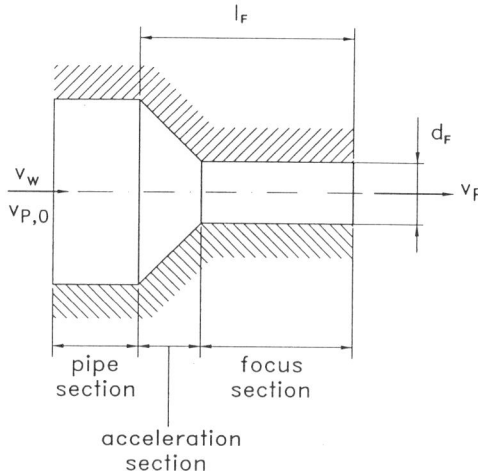

Figure 3.41 *Structure of a nozzle system for a suspension-abrasive water jet*

The values depend on the focus design and abrasive-particle size. For larger abrasive diameters, the momentum transfer improves.

Assuming a pipe flow in the nozzle system, estimate the momentum-transfer parameter,

$$\mu_{Sus} = 1 - \frac{p_V}{p}, \qquad (3.49)$$

with,

$$p_V = \xi_{pipe} \cdot \frac{\rho_{mix}}{2} \cdot v_{pipe}^2 \cdot \frac{l_F}{d_F}. \qquad (3.50)$$

The flow velocity of the suspension inside a pipe is

$$v_{pipe} = \frac{4 \cdot \left[\frac{\dot{m}_W}{\rho_W} + \frac{\dot{m}_A}{\rho_P}\right]}{\pi \cdot d_F^2}. \qquad (3.51)$$

Brauer [75], among others, tabulates the pipe-flow resistance parameter, ξ_{pipe}. The mixture density, ρ_{mix}, is

$$\rho_{mix} = \rho_P \cdot \frac{1+R}{\Omega + R}, \qquad (3.52)$$

with $R = \dot{m}_A / \dot{m}_W$, and $\Omega = \rho_P/\rho_W$.

3.7.2.3 Numerical Simulations

Several authors carry out numerical simulations to calculate the abrasive-particle velocity in a more detailed manner, Guo et al. [132] as well as Shimuzu and Wu [135, 140] use the following simplifications to write the equations of the motion of a single particle in a fluid separately in the acceleration section and in the focus section: there is no slip between the water velocity and the particle velocity, the flow in the nozzle is one-dimensional and incompressible, the flow is not effected by the existence of the particles, and the particles are spherical. Shimizu and Wu [135, 140] solve the equation of motion for a single grain in a fluid, as presented in Eq. (3.28), for the suspension-abrasive water jets in a pressure range up to 200 MPa. Figure 3.42 gives some results.

Figure 3.42a shows the particle acceleration in the acceleration sections of two different designed abrasive-suspension water jet nozzles. In this section, the water as well as the abrasive particles are heavily accelerated. Typical acceleration ratios are n_V=10-15.

In contrast, the acceleration process is not very pronounced in the focus section that is seen in Figure 3.42b. At the beginning of the focus section, the velocity of the particles accelerated in nozzle # 1 is slightly smaller than that of the particles entrained into nozzle # 2. Nevertheless, the nozzle # 1 accelerates the abrasives more rapidly in the focus section maybe due to larger drag forces acting on the particles because of the larger relative velocity between the abrasive and water. In Figure 3.42b, a decrease in the abrasive velocity at a certain focus length is noticed. Earlier, Laurinat et al. [133] observe and qualitatively discuss this effect. Friction between the suspension flow and the focus wall causes the velocity drop. As concluded from Figure 3.42b, the critical focus length when the particle velocity has a maximum value is about $l_{Fcr}=10 \cdot d_F$. The critical value tends to increase with an increase in the abrasive-particle size, but the absolute values of the particle velocities decrease with an increase in the particle diameter (Figure 3.42c). Interestingly, the optimum focus length is almost independent on the pump pressure [140]. Figure 3.42d is a plot between the abrasive-material density and velocity ratio. As the abrasive-material density increases, the velocity ratio notably decreases.

You et al. [143] also perform a numerical simulation of the abrasive-acceleration process at low pump pressures. They evaluate the efficiency of abrasive-particle acceleration by a ratio $R_A = v_P^2/p$, and find that a maximum of R_A=0.55 occurs at l_{Fcon}=5·d_F. The outflow coefficient is between α=0.9 and α=0.95, which is in agreement with values that Shimizu [140] reports.

3.7 Generation of Suspension-Abrasive Water Jets   73

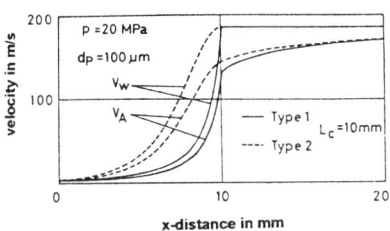

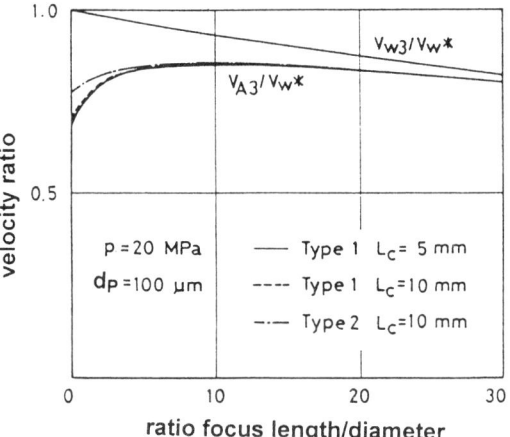

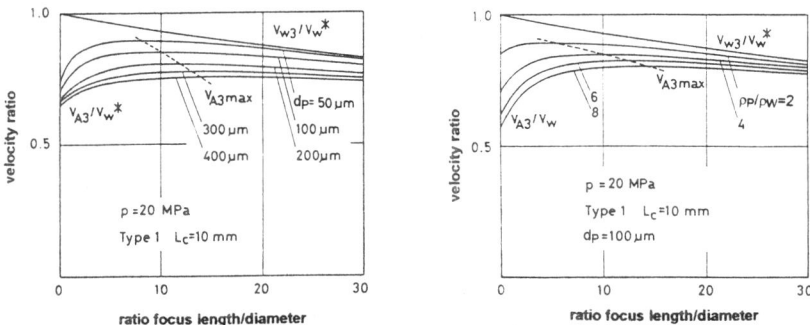

Figure 3.42 *Water and abrasive-particle velocities in a nozzle-system for suspension-abrasive water jes [135, 140]*
a - Acceleration section, b - Focus section, c - Abrasive diameter, d – Abrasive-material density

3.7.2.4 Finite-Element Modeling

Ye and Kovacevic [141] alternatively use a finite-element-based software for the numerical modeling of the flow inside a suspension-abrasive water-jet nozzle system. They apply a two-dimensional model to consider the radial particle movement in the nozzle.

Their results confirm that the acceleration mainly occurs in the tapered acceleration section, and that smaller abrasive particles more effectively accelerate (Figure 3.43). Figure 3.44 illustrates the complete flow characteristics of an abrasive particle that is accelerated in a suspension-abrasive water-jet nozzle system, including axial velocity, radial velocity, and particle trajectory.

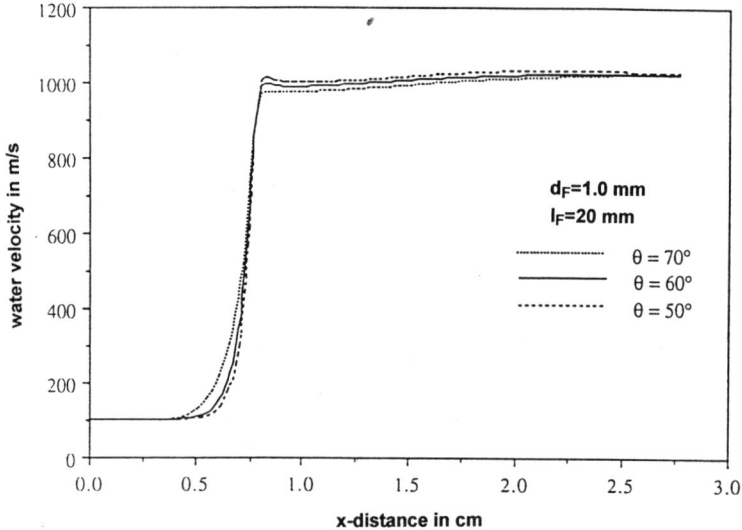

Figure 3.43 *Numerically simulated abrasive-particle velocities on the centerline of a nozzle-system for suspension-abrasive water jes [141]*

Ye and Kovacevic [141] notice that the abrasive-particle size as well as the initial particle location have a major influence on the particle trajectories. Small particles follow the motion of the water flow well; whereas, the larger abrasive grains with initial locations of larger radial distances obviously have a higher probability to impact upon the focus wall. A comparison of different taper angles in the nozzle acceleration-zone show that with larger angles the acceleration of water and abrasives is smoother, and the abrasive particles have a relatively low chance to collide with the focus wall.

3.7 Generation of Suspension-Abrasive Water Jets 75

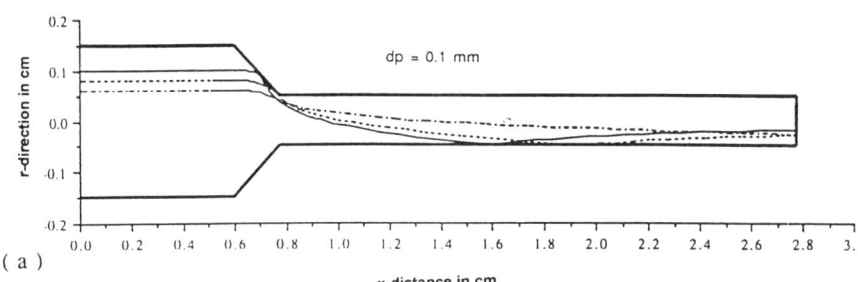

(a)

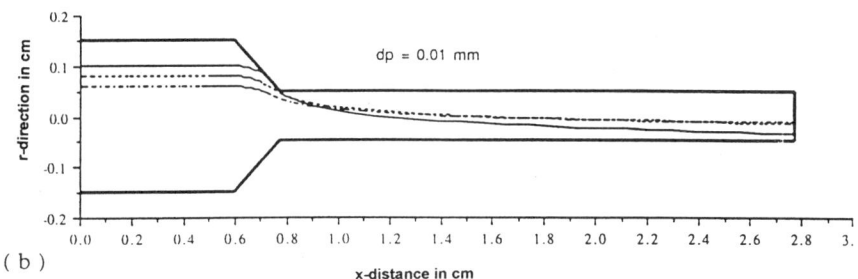

(b)

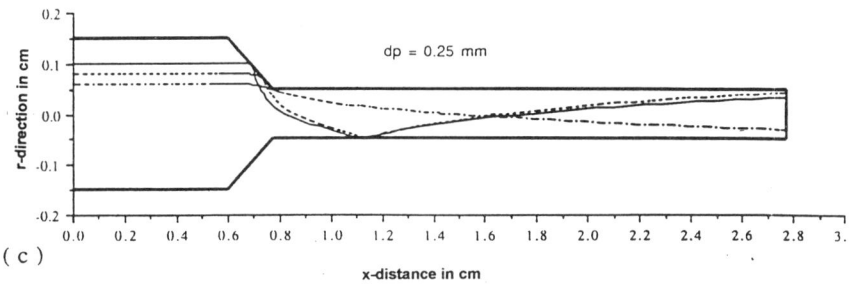

(c)

Figure 3.44 *Numerically simulated abrasive-particle trajectories in a nozzle-system for suspension-abrasive water jets [141]*

3.7.2.5 Acceleration-Nozzle Wear

Figure 3.45 illustrates some aspects of the suspension-abrasive water jet nozzle wear. Figure 3.45a shows the influence of the focus diameter on the wear rate. As the focus diameter increases, the wear rate decreases. Yazici and Summers [137] point out that after 45 minutes the focus diameter increases enough to reduce the available pump pressure up to a level of 80 %. Figure 3.45b illustrates the wear capability of several abrasive types. As expected, the very hard aluminum-oxide generates the highest wear rate. Figure 3.45c plots the pump pressure influence. Figure 3.45d shows the influence of the focus-section geometry. The larger the ratio between focus-length section to focus diameter, the more the wear reduces.

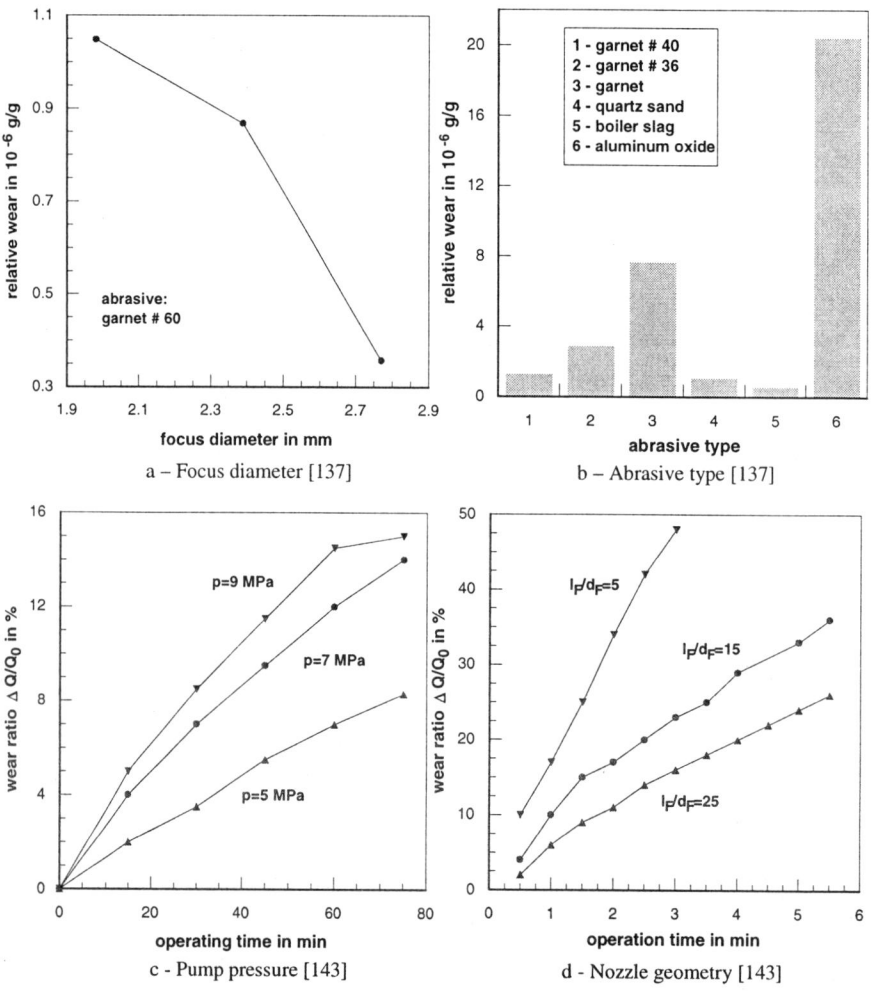

Figure 3.45 *Wear performance of a nozzle-system for suspension-abrasive water jets [137, 143]*

4 Structure and Hydrodynamics of Abrasive Water Jets

4.1 General Structure of Injection-Abrasive Water Jets

4.1.1 General Structural Features

The structure of abrasive water jets is evaluated in terms of several features:

- water volume or mass distributions,
- solid particle volume or mass distributions,
- air volume or mass distributions,
- velocity distributions,
- degree of turbulence.

4.1.2 Optical Examinations

Geskin et al. [144] perform high-speed photographs to investigate the conditions of the formation of abrasive water jets (section 10.4). The photographs show a jet surrounded by an array of droplets and solid particles. The jet is subjected to oscillations in axial as well as in radial directions. These oscillations grow into large disturbances along the jet axis, and eventually destroy the jet continuity. The authors do not find regions of preferable abrasive-particle concentrations.

Himmelreich [47] performs light-section-procedure experiments as reported in section 10.4 to observe the structure of injection-abrasive water jets. Figure 4.1 shows photographs of abrasive water jets with different abrasive-mass flow rates. The jet diameter significantly increases as the abrasive-mass flow rate increases. For the lowest abrasive-mass flow rate, the core structure of the water jet is still visible. Several single particle trajectories are distinguished at the edge and at the focus exit. For the 6 %-abrasive mixture, the jet diameter significantly increases. This increase is a discontinuous process. At a distance of $x/d_0=150$, a jump in the jet diameter is notable. The particle trajectories directly under the focus exit disappear. A further increase in the abrasive-mass flow rate transfers the beginning of the jet disturbances in the upper region of the picture; whereas, the upstream portion of the jet consists mainly of discrete particle trajectories. Again, no particle trajectories are visible at the focus exit. First, the coherence of the injection-abrasive water jet in the distance of several millimeters from the focus exit is satisfactory. Second, an increase in the abrasive-mass flow rate improves the jet structure immediately after the focus exit.

78   4. Structure and Hydrodynamics of Abrasive Water Jets

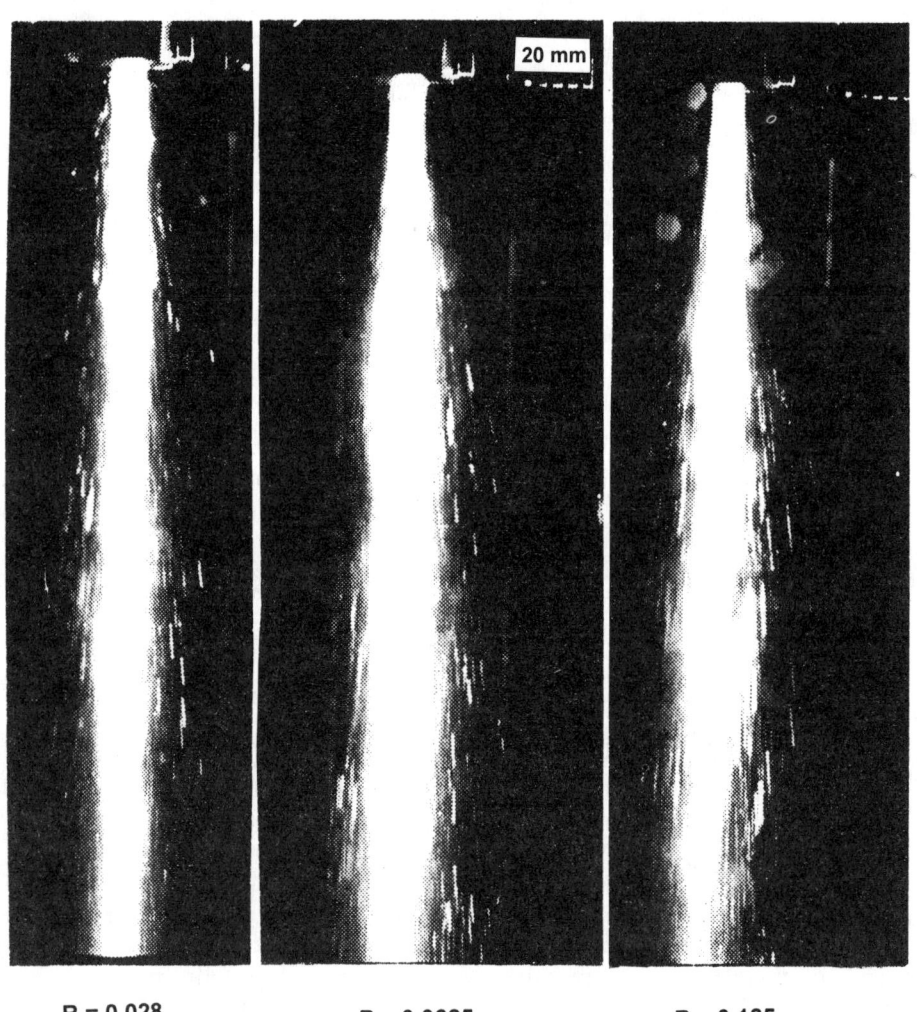

R = 0.028   R = 0.0625   R = 0.125

Figure 4.1 *Photographs from light-section-procedure experiments [47]*

4.2 Phase Distributions in Injection-Abrasive Water Jets

4.2.1 Average Abrasive-Density Distributions

Geskin et al. [144] and Simpson [90] use the flow-separation-technique as illustrated in section 10.6. Geskin et al. [144] find that the time average of the abrasive density over the jet cross-section has a constant value. The relation between the abrasive weight-percentage estimated for a certain cross-area and the cross area fits into a straight line as shown in Figure 4.2. Reasonable variations in the abrasive-mass flow rate, abrasive type, orifice diameter, and focus diameter do not alter this result.

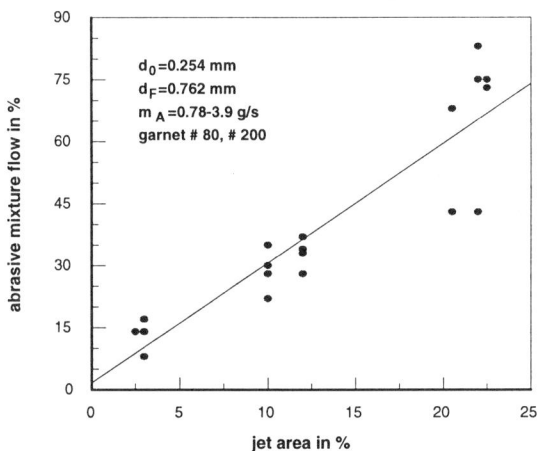

Figure 4.2 *Relation between the percentile jet area-range and percentile abrasive-mixture flow [144]*

4.2.2 Radial-Zone Model

Simpson [90] use the flow-separation technique to establish the abrasive-mass flow for each annular cross-section and the abrasive-mass flow per unit area of each annular cross-section of the abrasive water jet. Figure 4.3 shows some typical results. These graphs illustrate the abrasive distribution outward from the center of the abrasive water jet.

In his investigation, Simpson [90] considers three main zones within an abrasive water jet: a core zone that is the circular area equivalent to the water jet diameter, an inner zone as the annular area between the water-jet diameter and the focus diameter, and an outer zone that is the annular area between the focus diameter and the outer limit of the abrasive particles.

80   4. Structure and Hydrodynamics of Abrasive Water Jets

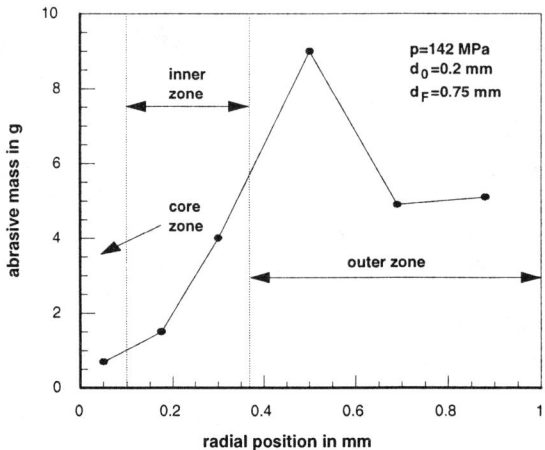

Figure 4.3 *Relation between the jet radius and abrasive-particle distribution in an injection-abrasive water jet [90]*

For relatively fine abrasive particles (below $d_P=150$ μm), the abrasive mass entrained in the core zone is small, while the abrasive mass in the inner zone increases as the distance from the center of the jet stream also increases. This trend continues until a maximum is reached toward the outer limits of the inner zone, with no abrasives detected in the outer zone.

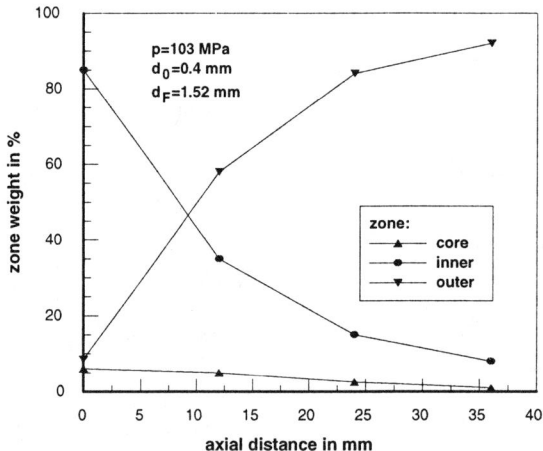

Figure 4.4 *Relation between the standoff distance (jet length) and abrasive-particle distribution in an injection-abrasive water jet [90]*

4.2 Phase Distributions in Injection-Abrasive Water Jets

A similar pattern emerges with larger abrasive grains in the range $d_P=150$ μm to $d_P=300$ μm, but instead of reaching a maximum near the outer limits of the inner zone, the abrasive-mass distribution tapers off due to abrasive entrainment in the outer zone. This result indicates that the larger abrasive particles cannot penetrate the core and inner zone as readily as the smaller particles.

Simpson [90] investigates also the influence of the standoff distance (jet length) on the abrasive-particle distribution. As shown in Figure 4.4, the proportion of abrasives in the core zone does not vary much over the first 12 mm, but drops beyond this point. In contrast, the abrasive number in the inner zone is sharply reduced over the first 12 mm, and continues to drop as the jet lenght increases and more abrasive particles spread into the outer zone.

4.2.3 Phase Estimations by X-Ray Densitometer

4.2.3.1 Water-Phase Distribution

Neusen et al. [145] employ a scanning-x-ray densitometer as discussed in section 10.6 to investigate the structure of water jets and injection-abrasive water jets. They detect a saddle-shape distribution of the water phase over the jet area as illustrated in Figure 4.5. The water concentration generally reaches a maximum between $1/2$ and $3/4$ of the jet radius. Also, the magnitude of the maximum water concentration is in the range of 4 percent to 8 percent. Neusen et al. [145] conclude from these results that no water 'core' remains in the water jet as it exits the focus. Neither is the water uniformly or normally distributed across the jet diameter.

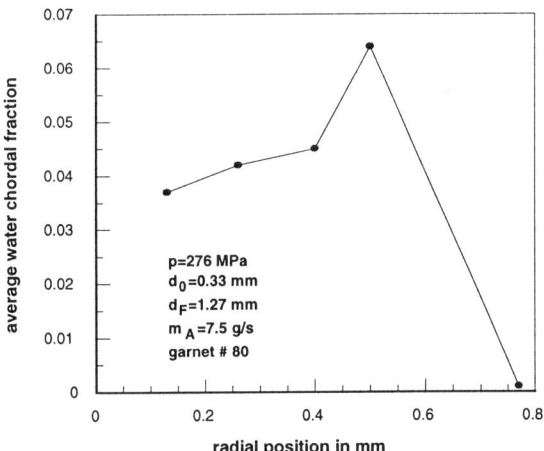

Figure 4.5 *Relation between the water-jet radius and average water fraction in an injection-abrasive water jet [145]*

4.2.3.2 Abrasive-Phase Distribution

The x-ray-measurements confirm the results of Simpson [90] who find a saddle-shaped concentration for the abrasive-particle distribution over the jet-cross area. Figure 4.6a shows that the abrasives are more concentrated in the outer radial locations. However, with an increase in the pump pressure or an increase in the abrasive mass flow rate, more abrasives are present in the central region of the abrasive water jet (Figure 4.6b). Due to experiments with the discrete abrasive water-jet milling, Öjmertz and Amini [146] independently estimate the saddle-shaped radial distribution of the abrasive particles in an abrasive water jet.

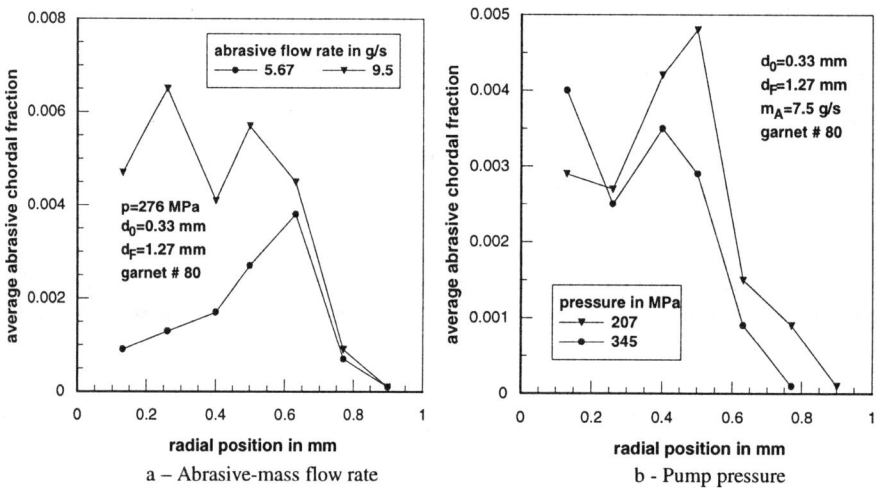

a – Abrasive-mass flow rate   b - Pump pressure

Figure 4.6 *Parameter influence on the abrasive-particle fraction in an injection-abrasive water jet [145]*

4.2.3.3 Air Content

Tazibt et al. [50] measure the air content in an injection-abrasive water jet. Figure 4.7a illustrates that the air sucked with the abrasive particles occupies more than 90 percent of the volume of an abrasive water jet. Figure 4.7b gives a typical phase distribution of an injection-abrasive water jet.

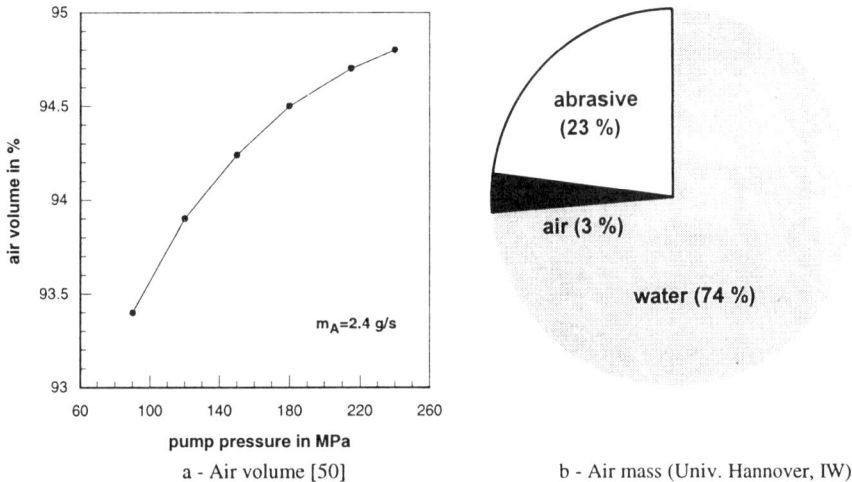

a - Air volume [50]   b - Air mass (Univ. Hannover, IW)

Figure 4.7 *Air content in an injection-abrasive water jet*

4.3. Abrasive-Particle Velocity Distribution in Injection-Abrasive Water Jets

4.3.1 Radial Velocity-Profile

Himmelreich [47] and Himmelreich and Riess [147] use the laser-anemometer-technique (section 10.5.3) to systematically investigate the radial velocity-distribution of the abrasive particles in an injection-abrasive water jet. Figure 4.8 illustrates a typical result ot the influence of the abrasive addition on the jet structure. Notice three effects that the abrasive grains cause: an increase in the jet diameter, a decrease in the jet velocity at the centerline, and an increase in the turbulence. Nevertheless, the abrasive water jet shows a smooth velocity profile indicating comparatively good mixing-and-acceleration conditions.

Figure 4.9 illustrates the influence of the focus geometry on the radial abrasive-velocity profile. The particle velocity is very sensitive to changes in the focus diameter. A very significant increase in the velocity to the outer locations is observed for the large focus diameter. From this profile, a severe disintegration of the jet beyond the focus is concluded. The increase in the border-zone velocity for the medium focus diameter is attributed to single abrasive grains and droplets that confirms the high-speed film observations by Geskin et al. [144]. Notice a pipe-flow profile for the small focus diameter. This result indicates that the effects of wall friction and particle- focus-wall interaction control the mixing process.

84   4. Structure and Hydrodynamics of Abrasive Water Jets

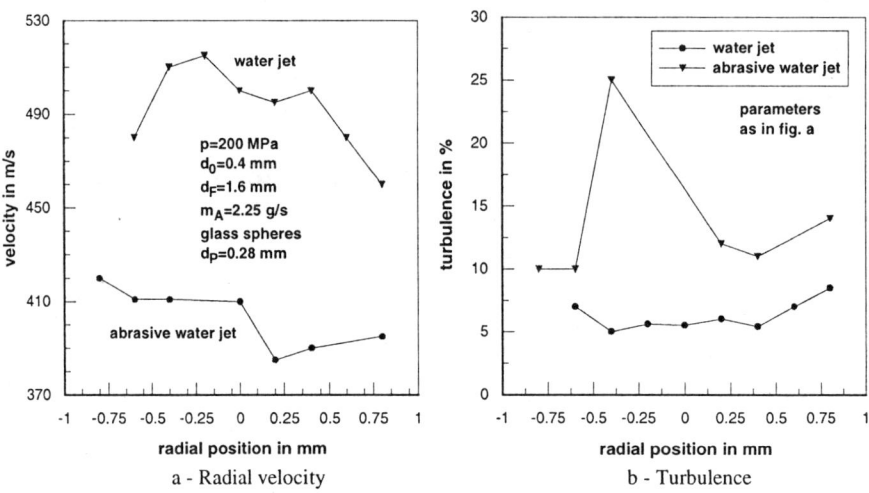

Figure 4.8 *Radial velocity and turbulence profiles of a water jet and an injection-abrasive water jet [147]*

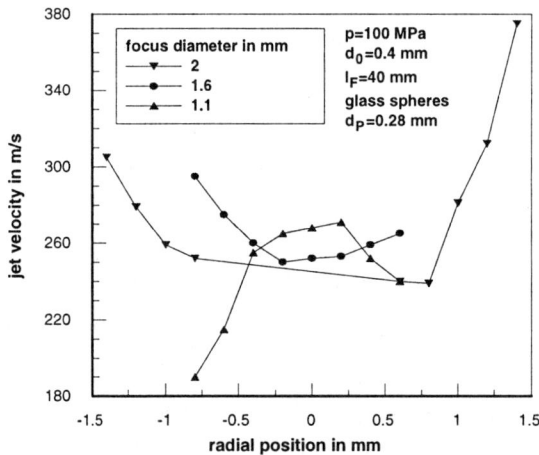

Figure 4.9 *Radial profiles of the abrasive-particle velocity in injection-abrasive water jets for different focus diameters [147]*

Himmelreich [47] observes that the radial velocity-profile is more regular for higher pump pressures. Also, the influence of the focus diameter on the average abrasive-particle velocity is reduced at a higher pump pressure.

4.3.2 Turbulence Profile

Another parameter to characterize the abrasive water jet-structure is the turbulence,

$$T_U = \frac{\sigma_{v_P}}{\overline{v}_P}. \tag{4.1}$$

The degree of turbulence in an injection-abrasive water jet depends on several process and geometry parameters.

Figure 4.10a illustrates the influence of the abrasive-mass flow rate and focus diameter on the turbulence. For the large-diameter focus (d_F=2.0 mm), the deviations in the abrasive-particle velocity are between 25 % and 30 % of the average velocity, indicating an inefficient mixing performance. Also, for this focus diameter, the turbulence degree is very sensitive to the abrasive-mass flow rate in the range of small abrasive-mass flow rates. For the medium-diameter focus (d_F=1.6 mm), the turbulence reduces to about 10 % for all abrasive-mass flow rates. The small-diameter focus (d_F=1.1 mm) shows turbulence degrees between 11 % and 14 % and just a weak dependence on the abrasive-mass flow rate. The high degree of turbulence compared to the 1.6 mm-diameter focus is due to shear stress on the focus wall which is even observed in the case of the plain water jet. For the small-diameter focus, this influence of the shear stress is more predominant; whereas, for the large-diameter focus, the mixing process between the water jet and abrasive particles is responsible for the turbulence generation.

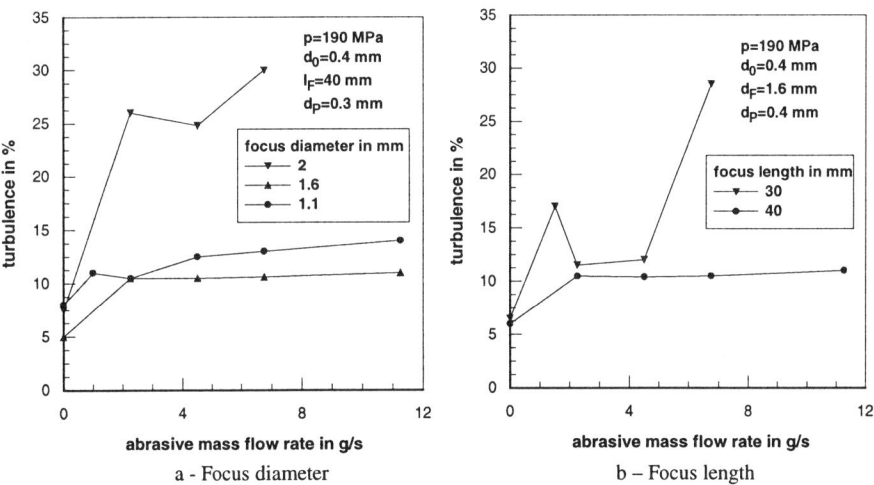

Figure 4.10 *Parameter influence on the turbulence degree in an abrasive water jet [47]*

Figure 4.10b illuminates the influence of the focus length on the turbulence degree. The use of a short focus (l_F=30 mm) leads to a comparatively high turbulence (ca. 30 %) that indicates an inefficient mixing process. Figure 4.10 shows the radial turbulence-profile in an abrasive water jet. Figure 4.11 shows that the turbulence in an abrasive water jet is not significantly influenced by the standoff distance by using a small-diameter focus. For the focus with the larger diameter, the jet starts to disintegrate at a relative standoff distance of $x/d_0=170$. An increase in the pump pressure reductes this disintegration distance.

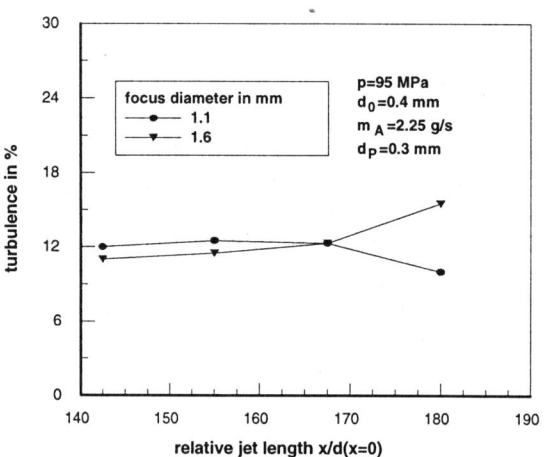

Figure 4.11 *Relation between the turbulence degree and non-dimensional jet length [47]*

4.3.3 Statistical Abrasive-Particle Velocity Distribution

Chen and Geskin [46] and Neusen et al. [48] experimentally investigate the statistical distribution of the abrasive-particle velocity. Both authors find a velocity distributions as given in Figure 4.12. Momber [148] suggests a Gaussian-Normal-Distribution (GND) to mathematically characterize the distribution of the abrasive-particle velocity,

$$f(v_P) = \frac{1}{\sqrt{2\cdot\pi\cdot\sigma_{VP}}} \cdot \exp\left[\frac{-(v_P - \bar{v}_P)^2}{2\cdot\sigma_{VP}^2}\right]. \qquad (4.2)$$

Typical values are $v_P=250$ m/s for the average particle velocity, and $\sigma_{VP}=41.4$ m/s for the standard deviation of the abrasive-particle velocity, repectively [48].

Chen and Geskin [46] show that the patterns of the probability distribution of the abrasive-particle velocity are similar for different abrasive-particle input diameters, which may be attributed to the abrasive-particle comminution during the mixing and acceleration process (section 3.5).

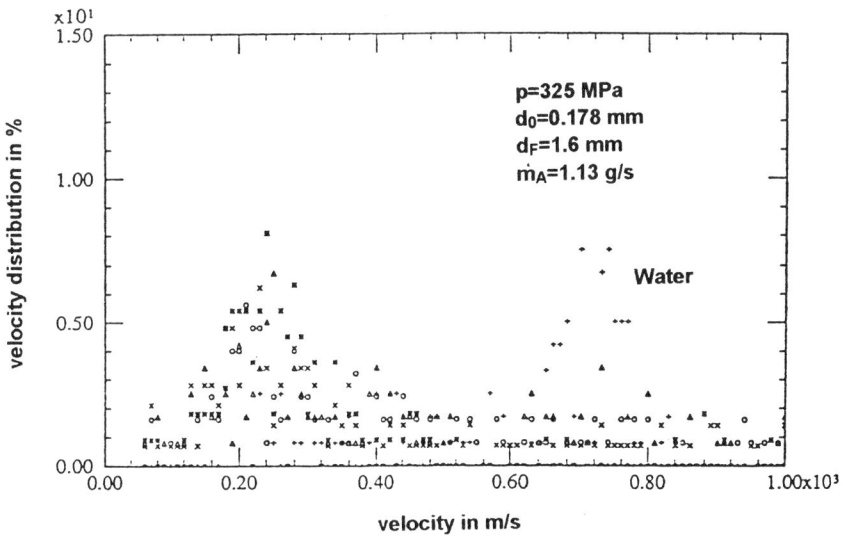

Figure 4.12 *Statistical distribution of the abrasive-particle velocity [46]*

4.4 Structure of Suspension-Abrasive Water Jets

Shimizu and Wu [135] investigate the structure of abrasive water jets that are generated in a suspension-nozzle system using short-time photography. Figure 4.13 shows two exemplary photographs for two different abrasive sizes. The exposure time is about 1.5 µs.

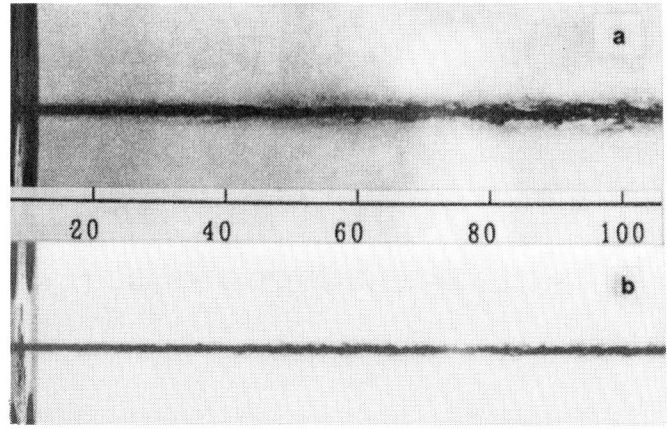

Figure 4.13 *Structures of suspension-abrasive water jets moving in air [135]*
a - Abrasive-particle size mesh # 100   b - Abrasive-particle size mesh # 220

In the case of mesh # 100 particles, the jet diameter just downstream the focus outlet increases with an increase in the standoff distance. At a distance of about $x=25 \cdot d_0$, the jet becomes unstable and the jet breakup finally occurs at a distance of about $x=80 \cdot d_0$. In contrast, the mesh # 220 particles form a jet that is rather compact, and the structure remains stable up to a standoff distance of about $x=300 \cdot d_0$. Larger abrasive particles contribute to the enlargement of turbulent motion of the water flow and promote the jet breakup at comparatively short standoff distances.

5 Material-Removal Mechanisms in Abrasive Water-Jet Machining

5.1 Erosion by Single Solid-Particle Impact

5.1.1 General Aspects of Solid-Particle Impact

The impact of single solid-particles is the basic event in the material removal by abrasive water jets. Therefore, a compressed review on the material erosion by solid-particle is given in this paragraph.

The literature about solid-particle erosion is extensive. Engel [149], Adler [150], and Preece [151] present general reviews about earlier investigations. More recently, Elleerma [152] and Meng and Ludema [153] analyze the state-of-the-art modeling of the solid-particle erosion.

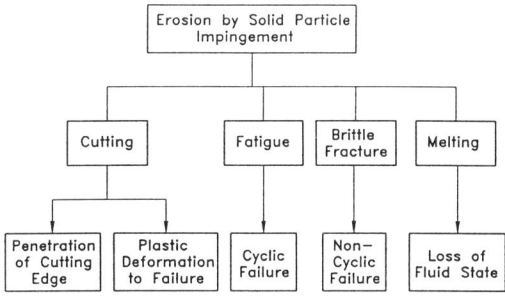

Figure 5.1 *Mechanisms of material-removal by solid-particle erosion [153]*

Meng and Ludema [153] define four sub-mechanisms by which solid-particles separate material from a target surface (Figure 5.1). These mechanisms are cutting, fatigue, melting, and brittle fracture. Clearly, these mechanisms generally do not act separately, but in combination. Their importance for the particular erosion process depends on several factors, such as the impact angle, the particle kinetic energy, the particle shape, target-material properties, and environmental conditions. Nevertheless, all four mechanisms are observed during the abrasive water-jet cutting and at least three of them are applied for the material removal modeling of the abrasive water-jet machining process (chapters 6 and 9).

The solid-particle erosion process is generally characterized by a non-dimensional erosion-number, E_N,

$$E_N = \frac{m_M}{m_P}.\tag{5.1}$$

Thus, the removed volume per solid particle is

$$V_M = \frac{E_N \cdot m_P}{\rho_M}.\tag{5.2}$$

The following sections briefly review selected solutions for Eq. (5.2) for ductile as well as brittle-behaving materials are.

5.1.2 Erosion of Ductile-Behaving Materials

5.1.2.1 Generalized Erosion Equation

Magnee [154] suggests a generalization of solid-particle erosion models for ductile-behaving materials as

$$V_M = \frac{C_1 \cdot v_P^2 \cdot m_P}{2 \cdot \varepsilon_M} \cdot \lambda \cdot f\left(\frac{H_P}{H_M}\right) \cdot f(\varphi).\tag{5.3}$$

This equation is illustrated on two examples that play a role in chapter 6 where the abrasive water-jet cutting models are discussed.

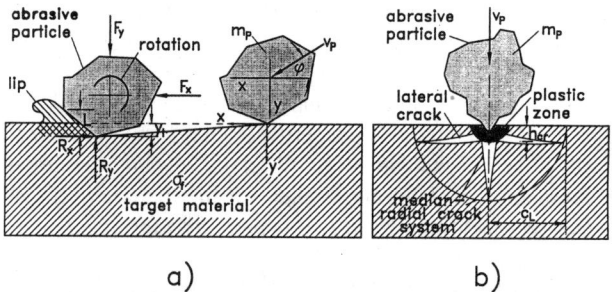

Figure 5.2 *Solid-particle material removal characteristics*
a - 'Micro-cutting' [155] b - Lateral cracking [156]

5.1.2.2 'Micro-Cutting' Model

An early and often used consideration of the material erosion by micro-cutting is due to Finnie [155]. He discusses the process by assuming a plastic response character of the material determined by its flow stress. Figure 5.2a gives the basic geometrical and kinematic parameters of this model.

After calculating the trajectory of a single particle during the removal process, Finnie [155] derives

$$V_M = \frac{m_P \cdot v_P^2}{\sigma_f \cdot K \cdot \psi} \cdot f(\varphi) \tag{5.4}$$

$$f(\varphi) = \sin(2 \cdot \varphi) - \frac{6}{K} \cdot \sin^2 \varphi \quad \text{for } \tan\varphi < \frac{K}{6},$$

$$f(\varphi) = \frac{K \cdot \cos^2 \varphi}{6} \quad \text{for } \tan\varphi > \frac{K}{6},$$

$$K = \frac{F_Y}{F_X} \cong 2, \quad \psi \cong 2.$$

Eq. (5.4) has the structure of Eq. (5.1) when the material's flow stress is replaced by the hardness of the eroded material. The parameter K is the ratio of vertical to horizontal force, and $\psi = L/y_t$ (Figure 5.2a). Hashish [157], Zeng and Kim [158] and Mazurkiewicz [159] use Eq. (5.4) for the modeling of abrasive water-jet cutting processes. Bitter [160, 161] critically discusses, that Eq. (5.4) causes several problems concerning the effect of flow stress, the particle velocity exponent, and the applicability of the model for attack angles near $\varphi=90°$. Later, Finnie and McFadden [162] improve the model leading to a particle velocity exponent of about 2.5.

5.1.2.3 Extended 'Cutting-Deformation' Model

Bitter [160, 161] who divides the entire material-removal process into two modes called by him 'cutting wear' that happen at low-impact angles, and 'deformation wear' that occurr at high-impact angles, develops a more general model. By considering the energies involved in the erosion process, Bitter derives two formulas for both material-removal modes:

$$V_{Mcut} = \frac{m_P \cdot [v_P^2 \cdot \cos^2 \varphi - C_2 \cdot (v_P \cdot \sin\varphi - v_{el})^{\frac{3}{2}}]}{2 \cdot \varepsilon_{Mcut}}, \tag{5.5a}$$

$$C_2 = f(E_M, E_p, \rho_M, v_M, v_P)$$

for the 'cutting wear' mode, and

$$V_{Mdef} = \frac{m_P \cdot (v_P \cdot \sin \varphi - v_{el})^2}{2 \cdot \varepsilon_{Mdef}} \tag{5.5b}$$

for the 'deformation wear' mode. For certain conditions, Eq. (5.5a) is identical to Eq. (5.4). Neilson and Gilchrist [163] simplify this model. The disadvantage of the model is that it requires experimentally determined parameters for a complete application.

5.1.2.4 'Ploughing-Deformation'-Model

Hutchings [164] introduces an alternative discussion of the micro-cutting processes during the solid-particle erosion. Based on high-speed photographs and SEM-observations, he defines two modes of material-removal due to micro-cutting, such as cutting-deformation and ploughing-deformation (Figure 5.3). The ploughing-deformation mode dominates the material removal by spherical particles; whereas, cutting-deformation is significant for sharp-edged, angular particles. Hutchings [164] makes a further subdivision of the cutting deformation into Type I cutting-deformation and Type II cutting-deformation, depending on the direction of the particle rotation (see Figure 5.3). For forward rotating particles Type I is dominant, and for backward rotation Type II is valid.

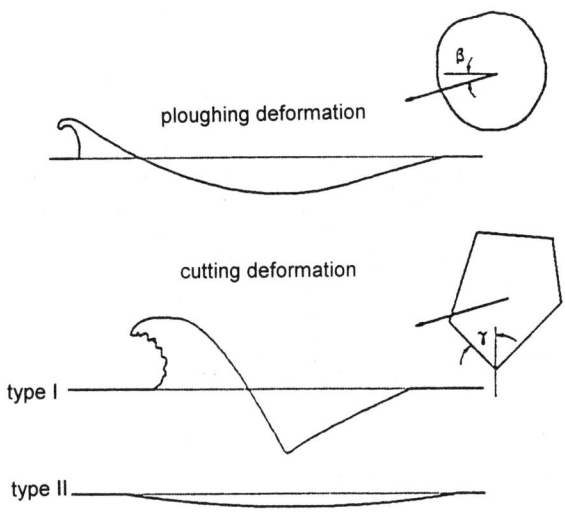

Figure: 5.3 *'Cutting' and 'ploughing' by solid particles at oblique impact angles [164, 165]*

5.1.2.5 Low-Cycle Fatigue and Thermal Effects

Hutchings [166] developed a solid-particle erosion model for normal impact that is based on low-cycle fatigue; although earlier attempts are made by Edington and Wright [167].

The plastic work associated with the deformation of a material around the impact site largely degrades into heat. Hutchings and Levy [168] give a recent review about the thermal effects in solid-particle erosion of ductile materials. It appears from their discussion that thermally determined features of the solid-particle erosion process are not a necessary assumption under all impact conditions. With smaller particles at low impact velocities, thermal effects are not important.

5.1.2.6 Comparison of Models for Ductile-Behaving Materials

Elleermaa [152] performs a critical review of solid-particle erosion models developed for the estimation of the materials removal in ductile-behaving materials. As she shows (Table 5.1), Finnie's model especially exhibits a comparatively high divergence from experimentally estimated values, illustrating that a simple micro-cutting process does not cover the complexity of the material-removal process. Despite these limitations, Finnie's as well as Bitter's model are extensively used for the modeling of material-removal processes in abrasive water jetting (chapter 6).

Table 5.1 *Comparison of solid-particle erosion models for several ductile materials [152]*

Theory	Sums of least squares[1]				
↓ Material →	Al	Ti	Fe	20/1[2]	20/2[3]
Lebedev	3.72	4.99	4.07	2.81	2.59
Finnie	3.30	5.15	4.56	4.92	4.33
Bitter	1.49	2.56	2.82	0.68	0.43
Nepomnyashchy	$1.2 \cdot 10^3$	39.25	98.55	3.12	28.74
Beckmann	0.92	2.31	1.97	0.25	0.39
Abramov	5.86	3.35	5.99	3.84	4.69
Peter	1.29	7.10	4.96	3.56	0.92

[1] n=24 [2] AISI 1020, HV 434 [3] AISI 1020, HV 193

5.1.3 Erosion of Brittle-Behaving Materials

5.1.3.1 Generalized Erosion Equation

Erosion models for brittle-behaving materials are relatively well established. The most important material parameters that control the erosion process are identified based mainly on developments in indentation fracture mechanics. The general equation for the material removal in brittle-behaving materials, especially in ceramics, by an impacting solid particle is

$$V_M = \frac{\pi}{4} \cdot (2 \cdot c_L)^2 \cdot h_{Cr} = \pi \cdot c_L^2 \cdot h_{Cr}. \tag{5.6}$$

The schematics in Figure 5.2b illustrates the essential features of Eq. (5.6). Eq. (5.6) can be solved by finding relations between the crack-formation geometry and the main process parameters of the erosion process.

5.1.3.2 Elastic Model

Sheldon and Finnie [169] propose a purely elastic model and assume that erosion occurs entirely by crack propagation and chipping as a result of contact stresses during impact. These stresses cause cracks to growth from pre-existing cracks in the material surface. Nevertheless, the model neglects the lateral crack formation which is the main cause of material removal during solid-particle erosion.

5.1.3.3 Elastic-Plastic Models

The erosion model developed by Evans et al. [170], and the elastic-plastic theory of Wiederhorn and Lawn [171] calculate the volume loss from the depth of particle penetration and the maximum size of the lateral cracks that forms during impact (Figure 5.2b). They assume that the depth of the lateral cracks is proportional to the radial-crack size.

5.1.3.4 Grain-Ejection Model

Ritter [172] notices on polycrystalline alumina ceramics a material removal by individual grain ejection from the surface. This process is modeled by arguing that a fraction of the solid-particle kinetic energy is used in grain-boundary cracking.

5.1.3.5 Comparison of Models for Brittle-Behaving Materials

Most of the models developed for the erosion by brittle fracture follow the general relation,

$$V_M = C_1 \cdot \left(\frac{d_P}{2}\right)^{C_2} \cdot v_P^{C_3} \cdot H_M^{C_4} \cdot K_{Ic}^{C_5} \cdot E_M^{C_6}. \tag{5.7}$$

Table 5.2 lists the power exponents of the different models related to Eq. (5.7)

Table 5.2 *Dry solid particle erosion models for brittle-behaving materials, Eq. (5.7)*

Basic	Reference	c_2	c_3	c_4	c_5	c_6
Elastic model	Sheldon and Finnie [169]	f(M)	f(M)	-	-	-
Quasi-static lateral crack model	Wiederhorn and Lawn [171]	3.67	2.45	0.11	-1.33	-
Dynamic lateral crack model	Evans et al. [170]	3.67	3.17	-0.25	-1.33	-
Modified lateral crack model	Marshall et al. [173]	3.50	2.33	-1.42	-1.00	1.25
Grain ejection model	Ritter [172]	3.00	2.00	-	-2.00	1.00
Indentation model	Buijs [156]	-	-	-1.42	-1.00	1.25

Whereas, the elastic theory of Sheldon and Finnie assumes that the power exponents are functions of the target-material properties (especially of the Weibull-distribution parameters), the elastic-plastic models treat the power exponents independently of the target material. The models listed in Table 5.2 are directly valid for the case of normal impact of the solid particles.

5.2 Micro-Mechanisms of Abrasive-Particle Material Removal in Abrasive Water Jet Machining

5.2.1 Observations on Ductile-Behaving Materials

5.2.1.1 SEM-Observations

Kovacevic et al. [174], Webb and Rajukar [174], Kovacevic [176], and Arola and Ramulu [177] perform systematic SEM-observations for ductile-behaving materials. In stainless steel, Kovacevic et al. [174] and Kovacevic [176] detect separated wear tracks plowed by single abrasive grains due to scratching and scooping. The widths of these tracks are related to the size of the abrasive particles but non-uniform. This non-uniformity attributes to the particle-size distribution of the abrasives as they leave the focus and hit the material surface. The tracks are vertical on the top region of the cut but start to incline at deeper locations. Webb and Rajukar [174] make the same observations in an inconel cut by an abrasive water jet. Arola and Ramulu [177] detect chip formation and plowing as important material-removal mechanisms in a titanium cut by an abrasive water jet. Also, they find that the length of the path of the single abrasive particles in the workpiece decreases, and the randomness of the path orientation increases as the depth of cut increases. Ramulu et al. [126] observe in an aluminum cut by an abrasive water jet under shallow impact angles ($\varphi=5°-20°$) that abrasive-particle impacts are randomly oriented and appear to have different impact angles.

96   5. Material-Removal Mechanisms in Abrasive Water-Jet Machining

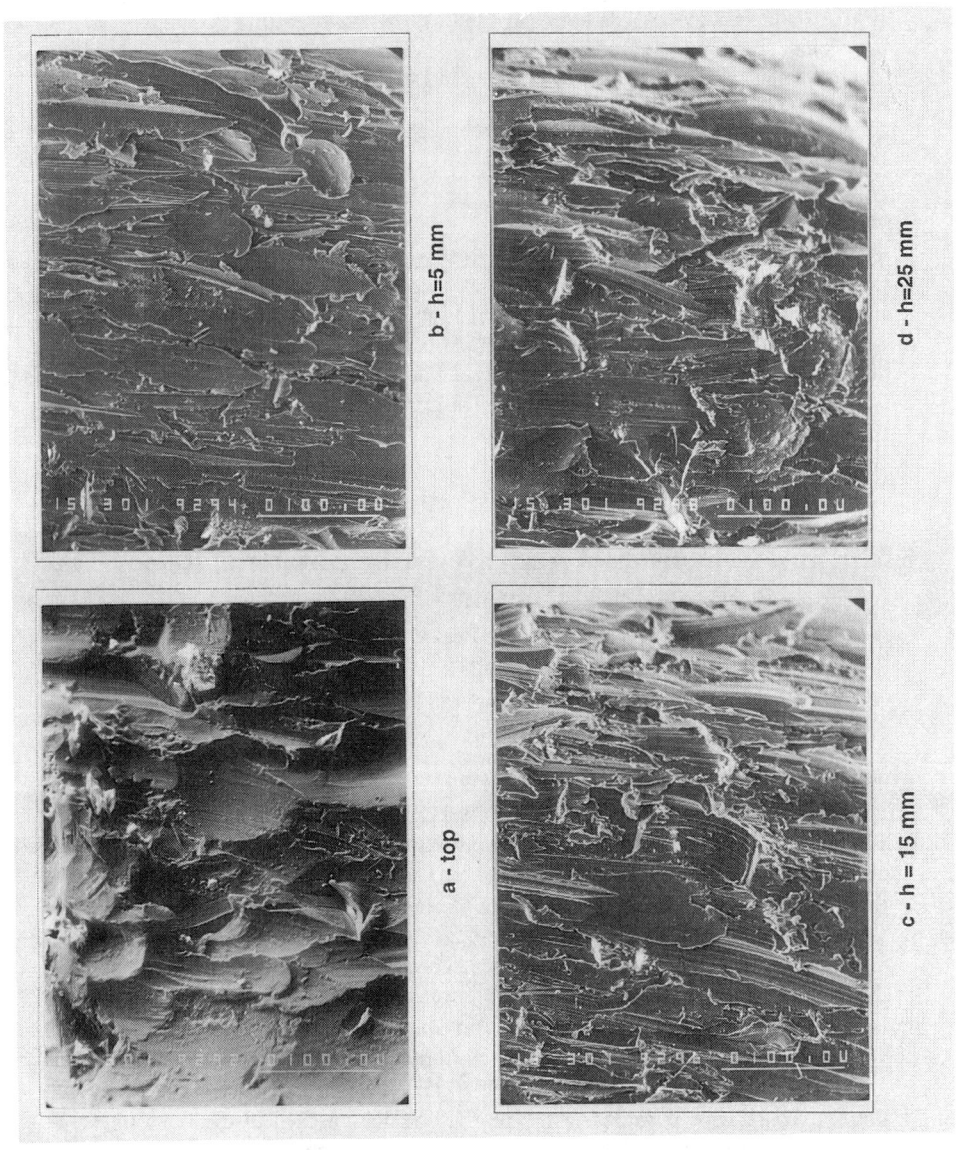

Figure 5.4 *SEM-images from different locations of the cutting front [180]*

5.2 Micro-Mechanisms of Abrasive-Particle Material-Removal 97

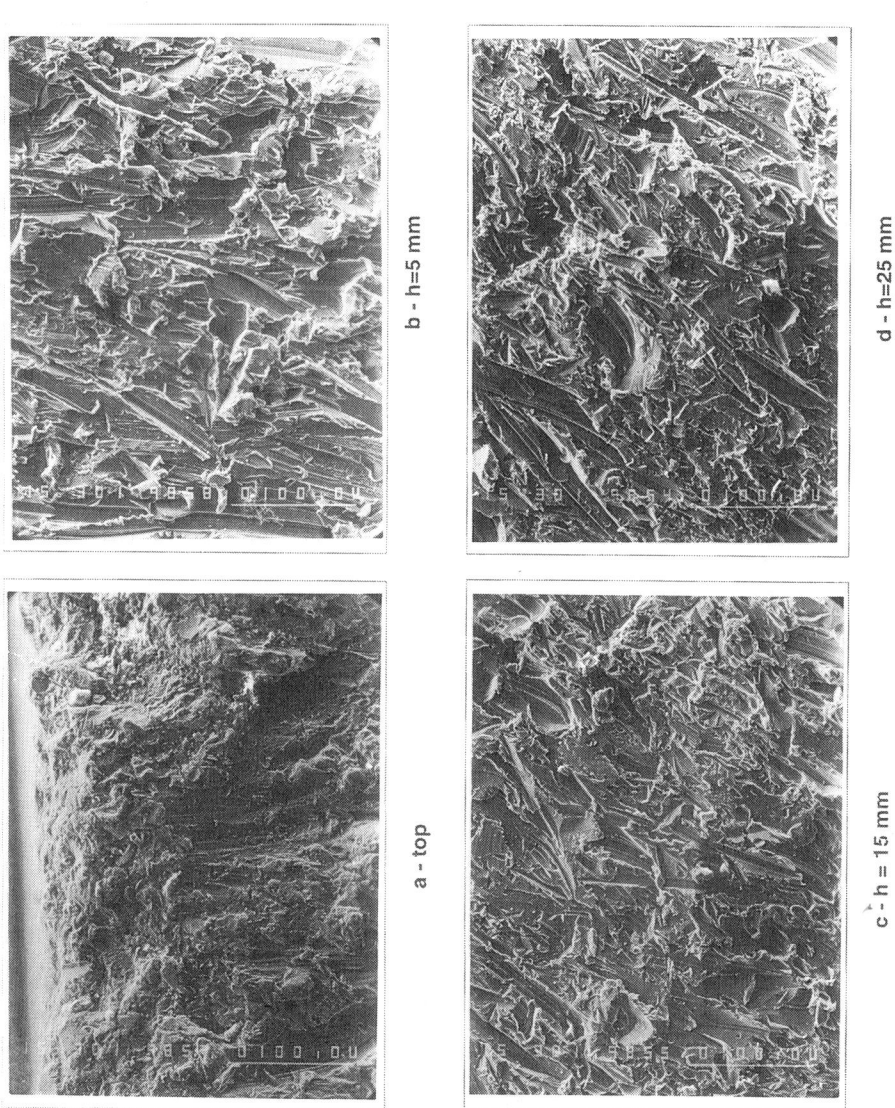

Figure 5.5 *SEM-images from different locations of the cutting surface [180]*

98   5. Material-Removal Mechanisms in Abrasive Water-Jet Machining

Zeng and Kim [178] observe material removal by a fully plastic flow in stainless steel subjected to an abrasive water jet at low ($\varphi=20°$) impact angles. Also, they notice smaller craters distributed around the major crater that are produced by fractured abrasive particles. Summers et al. [179] indicate clear evidence of micro-cutting and shearing in ductile metals eroded by a suspension-abrasive water jet.

Guo [180] uses SEM-microscopy to intensively examine the erosion sites in several metals, such as aluminum-alloy, titanium alloy, and ferrite steel, on the cutting front as well as on the generated kerf wall. Figure 5.4 shows the results from the cutting-front observations. In the very immediate top zone (Figure 5.4a), up to a depth of 300 μm to 400 μm, the wear tracks are comparatively bright, short and deep. Very clearly, the formation of material lips is noticed at the end of the tracks. In the second stage (Figure 5.4b), the tracks are longer and shallower and do show a uniform orientation, preferably in the jet direction. As the jet moves forward into the material (Figure 5.4c,d), the general micro-mechanism does not change, but the orientation of the wear tracks becomes non-regular. Even if the tracks are deeper and lip formation can be observed again, micro-cutting is still the dominating material-removal process. The photographs taken from the kerf wall of the same specimens (Figure 5.5) exhibit identical characteristics.

Figure 5.6 that shows a polished cross section of a steel specimen subjected to an abrasive water jet, supports the assumption of a micro-machining process in the abrasive water jet cutting of ductile-behaving materials as described previously for dry solid-particle erosion. The damage profile is very similar to that for micro-ploughing shown in Figure 5.3. The marked lip on the left side of the crater is later removed by other impacting particles. In the result, a chip is generated. The summary of all generated microchips determines the removed material mass.

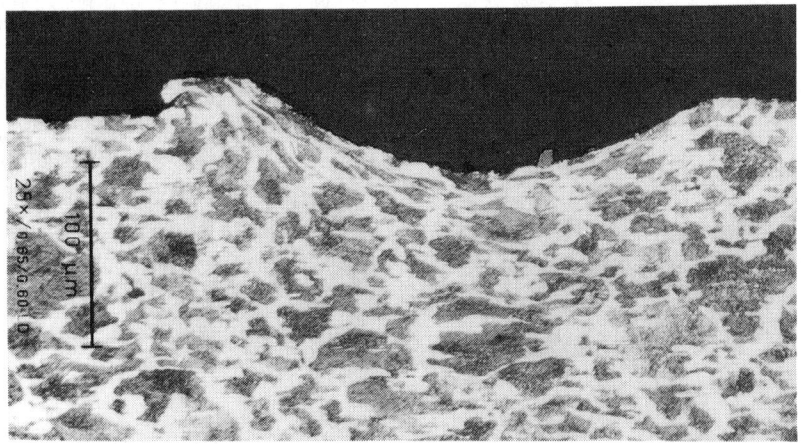

Figure 5.6 *Polished cross section of a steel specimen indicating micro-ploughing (Univ. Hannover, IW)*

5.2.1.2 Stress Measurements

In order investigate the formation of internal stresses due to plastic deformation, Guo [180] performs stress measurements on the surfaces in horizontal as well as in vertical directions. Figure 5.7 shows the results. Internal compressive stresses are notable in both figures. Interestingly, the stresses in vertical direction show some orientation-dependence in the upper region: the stresses perpendicular to the velocity vector of the abrasive particles (x-direction) are larger than those parallel to the vector (z-direction). Probably, shearing compensates the stresses generated in the abrasive particle impact direction. Thus, one reason for the material separation of material particles in ductile-behaving materials is ductile shearing. Arola and Ramulu [181] who examine aluminum specimens also suggest ductile shearing. The situation changes as the a depth level of 50 % of the final depth of cut is reached. In this case, there is no preferred direction in stress orientation. Internal stresses measured perpendicularly to the kerf wall in the very upper region of the cut also show a typical orientation in the x-direction over the entire range. Generally, from both figures, no significant relation is noticed between the depth of cut and generated compressive stresses.

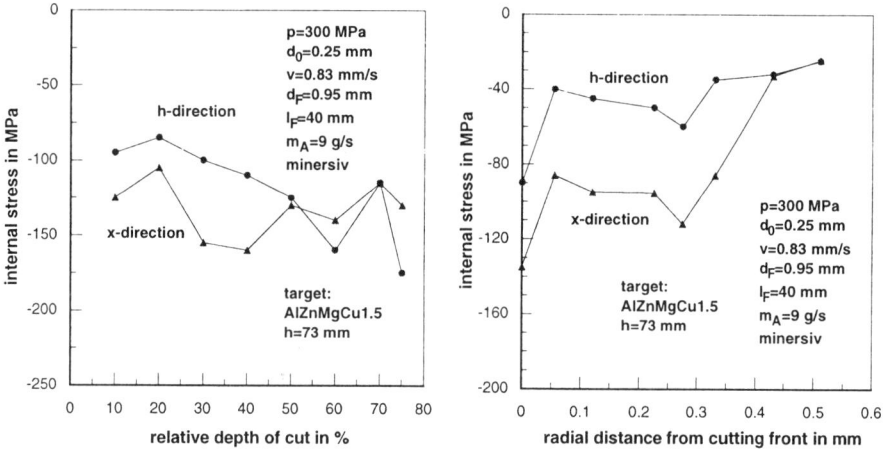

Figure 5.7 *Stresses generated during the abrasive water-jet cutting [180]*

5.2.2 Observations on Composite Materials

5.2.2.1 SEM-Observations on Metal-Matrix Composites

Neusen et al. [182] and Savrun and Taya ([183] perform early systematic SEM-studies about the micro-mechanisms of material-removal in composite materials. Neusen et al. [182] subject metal-matrix composites to an abrasive water jet and observe abrasive wear tracks very similar to those described above for ductile materials. The reinforcing ceramic particles are also cut because of their good bond to the matrix. Savrun and Taya [183] and Ramulu et al. [126] detect an erosion of the matrix material by micro-cutting, but the ceramic grains are pulled out rather than cut. Savrun and Taya additionally observe micro-melting in some areas of the metal-matrix material. Hamatani and Ramulu [184] detect scooping and ploughing even at the bottom of metal-matrix composites. This result indicates that these mechanisms dictates the material-removal process over the entire cutting front.

Investigations by Schwetz et al. [113] on boron-carbide ceramics and metal-matrix composites show that the micro-damage in a particular material depends mainly on its microstructure. In a tungsten-carbide composite, for example, the cobalt binder is removed; then, the tungsten-carbide grains are plucked out in a grain-by-grain mode. Neither plastic deformation nor cracking is observed. In contrast, in fine grained boron-carbide, a brittle fracture is noticed. The predominant features of this failure mode are local pits formed from lateral cracking. In a typical matrix-grain compound, such as a boron-carbide reinforced by titanium-boride, the material is eroded by a mixed inter-/trans-granular mechanism. Because of the weak interfaces, the grains are separated from the matrix. Then, the boron-carbide matrix trans-granularly cracks.

5.2.2.2 SEM-Observations on Fiber-Reinforced Composites

Arola and Ramulu [181, 185] and Ramulu and Arola [186, 187] examine the behavior of graphite-fiber reinforced epoxies cut by an abrasive water jet. The material-removal process is generally dictated by broken fibers or fiber pullout. These features of the micro-mechanism are found over the entire cutting front together with large removal pockets in the top area of the cut. Certainly, the size of the pockets generated in the matrix material decreases with an increase in the depth of cut. The material response is determined by the brittle properties of the fibers. A combination of micro-machining and the brittle fracture of the fibers is observed. The fracture surfaces of the fibers and the surrounding matrix appear to be machined, contrary to regions of macro-fracture induced by cantilever bending. Features of post-machined fibers and an interstitial matrix indicate that interdependent fracture of the constituent occurs during material removal. Figure 5.8 shows a micrograph of a fiber-reinforced epoxy cut by an abrasive water jet. In the matrix, abrasive wear tracks are observed similar to ductile-behaving materials. The wear-track angle increases with an increase in the depth of cut; whereas, severity and track depth decreases. The width of the wear tracks is about 20 µm to 50 µm (1-2 fiber diameters), and their length is about 100 µm to 200 µm. Both the width and the length augment with an increase in the pump pressure (abrasive particle velocity) and the particle diameter. The wear-track density is very high on top of the cut.

5.2 Micro-Mechanisms of Abrasive-Particle Material-Removal   101

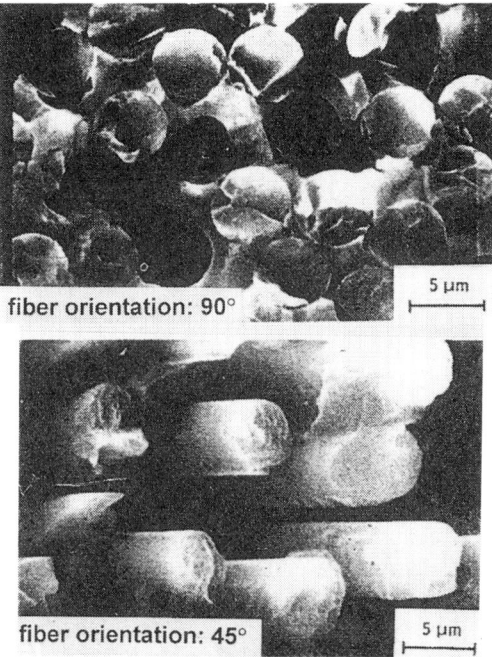

Figure 5.8 *Material-removal characteristics in a fiber-reinforced epoxy*
(Univ. of Washington, Seattle)

5.2.3 Observations on Brittle-Behaving Materials

5.2.3.1 SEM-Observations on Polycrystalline Ceramics

Zeng and Kim [158, 178] conduct an extensive SEM-study into the behavior of polycrystalline (aluminum oxide) ceramic. Through grooving experiments, sweeping experiments, and cutting experiments by an abrasive water jet under low ($\varphi=20°$) and perpendicular impact angles, they identify a mixed material-removal mode that consists of brittle-fracture phenomena and plastic deformation. For the low impact angle, they mainly indicate scratching marks by single abrasive grains but also some inter-granular cracking. In contrast, inter-granular fracture dominates the material removal at perpendicular impact angles. Under this condition, it is observed that some material grains undergo densification that indicate the effects of plastic deformation. Traces of plastic flow are present, but considerably smaller and shorter than those for low impact angles.

5.2.3.2 SEM-Observations on Refractory Ceramics

Momber et al. [188, 189] find in an abrasive water jet cutting study on refractory ceramics, that the micro-failure mechanism changes from a trans-granular fracture on the top of the cut to an inter-granular removal mode on the bottom. Figure 5.9a shows a SEM-photograph of the upper zone of a magnesia sample cut by an abrasive water jet. In the right top region of the figure, a very smooth cut surface is notable that includes open pores with sharp edges. The left side of the figure shows periclase fragments. Obviously, the periclase grains are fractured due to the abrasive-particle action. In this range, open pores with sharp edges are present. The micro-mechanism of material removal is characterized by trans-granular cracking through a matrix and through inclusions. The situation is very different in Figure 5.9b that shows the bottom zone of the same cut. In the picture, the periclase grains are completely intact, but the matrix between them is removed. A further magnification shows that penetrating cracks are stopped by the hard inclusion grains. No open pores are found. This result indicates an inter-granular removal mode. Nevertheless, a systematic inspection of the entire cutting front does not give any indication of an abrupt change in the microscopic material-removal mode.

For materials with directly connected inclusion grains, such as magnesia-chromite ceramics, a mixed material-removal mode is observed. Figure 5.10a shows that the inclusion grain (periclase) is either exposed due to the matrix removal or is simultaneously cut with the surrounding matrix. Obviously, the mode that specifically acts depends on the particular local situation on the impact site. In contrast, in a material with very pronounced interfaces between inclusions and a matrix material, such as bauxite-ceramics, the hard inclusions (corundum) resist the abrasive-particle attack and are pulled out almost intact (Figure 5.10b).

5.2.3.3 Acoustic-Emission Measurements on Brittle-Behaving Materials

Momber et al. [190, 191] apply the acoustic-emission technique for an on-line observation of the micro-failure of refractory ceramics and pre-cracked multiphase materials during the abrasive water-jet cutting. In their investigation, they use fine-grained concrete with a soft cement matrix as well as coarse-grained concrete with a high-strength matrix. Conventional optical observations show that the first type of material fails by eroding the matrix and separately pulling out the inclusion grains. This case is considered to be a more or less continuous erosion process. In contrast, the second material group is removed by an intense spalling fracture with cracks running through the matrix as well as through the inclusions. Figure 5.11 shows the corresponding time-domain acoustic-emission signals. Whereas, the signal for the erosion mode shows a continuous behavior, the signal acquired from the spalling-fracture mode is characterized by burst emissions that indicate a sudden energy release caused by inclusion fractures.

5.2 Micro-Mechanisms of Abrasive-Particle Material-Removal 103

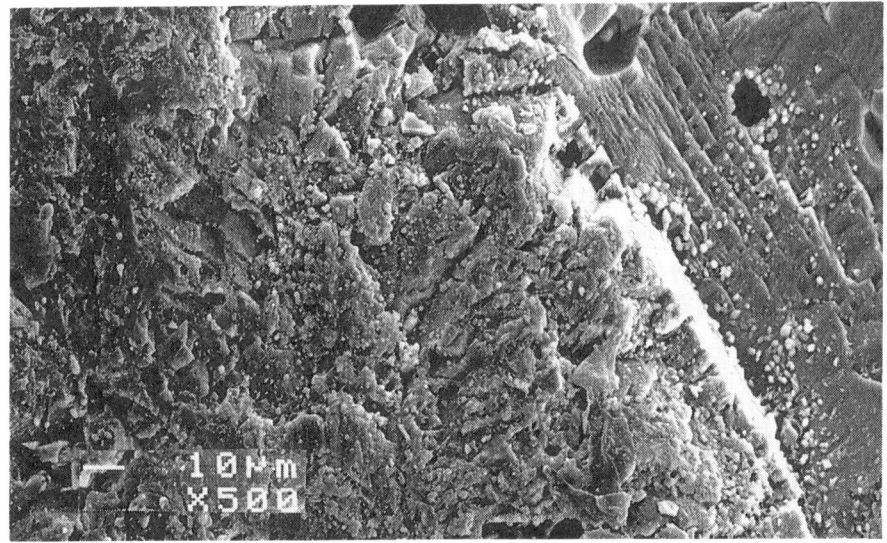

a - Top region

b - Bottom region

Figure 5.9 *SEM-images of a sintered-magnesia specimen cut by an abrasive water jet*

104   5. Material-Removal Mechanisms in Abrasive Water-Jet Machining

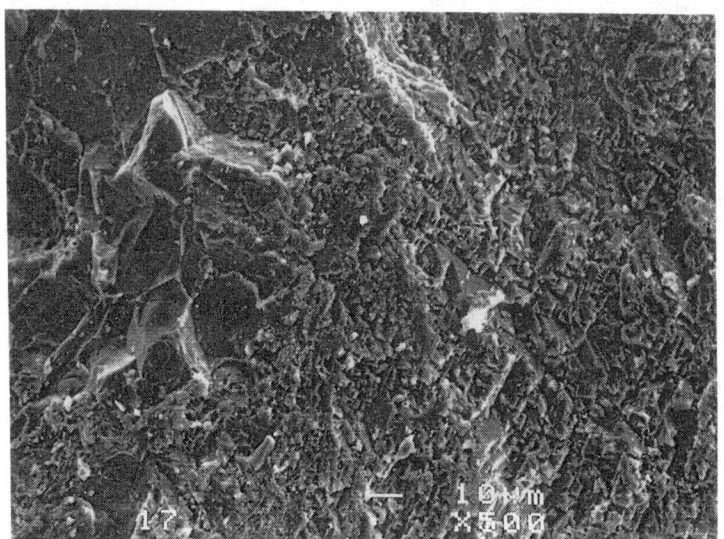

a - Mixed material-removal mode

b - Grain-ejection mode

Figure 5.10 *SEM-image of magnesium chromite eroded by an abrasive water jet*

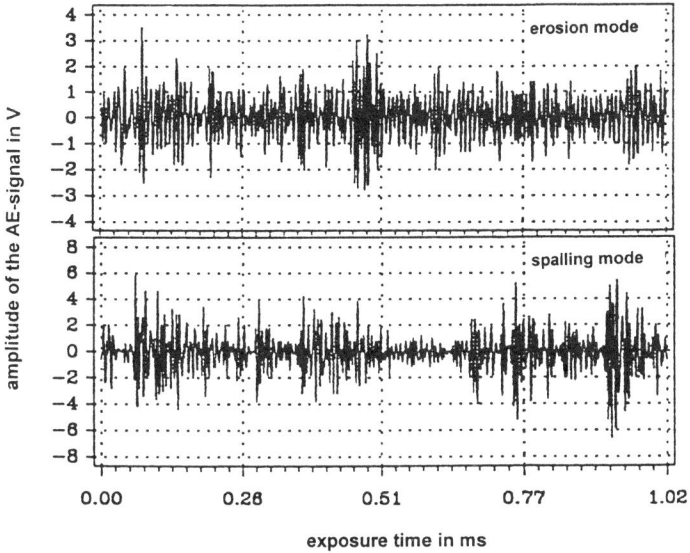

Figure 5.11 *Acoustic-emission signals, acquired in pre-cracked multiphase materials during the cuttting by an abrasive water jet [190]*

5.2.3.4 Photoelasticity Investigations on Brittle-Behaving Materials

Ramulu [192] who carries out photoelastic studies on materials subjected to an abrasive water-jet impact, supposes a different concept for the material removal in brittle materials. During piercing experiments in polycarbonate specimens, he detects stress-wave propagation immediately after the jet impact (Figure 5.12). These stress waves are probably due to the impact of the water phase. Cracks that are generated during this event are eroded by the abrasive particles that follow the water impact with a short delay.

5.2.3.5 Microboiling in Ceramics and Metal-Matrix Composites

Kahlmann et al. [193] who find molten garnet smeared over the surfaces of materials cut by an abrasive water jet, discusses an alternative material-removal mechanism for metal-matrix composites and ceramics. The smeared garnet that follows the direction of the jet flow, contains small holes with diameters <1.0 mm as a result of micro-boiling. The iron distribution, measured by dispersive spectrometry-mapping, increases with an increase of the slope of the cutting front. Because garnet melts at very high temperatures ($\cong 1{,}280$ °C), the authors assume the existence of a high, localized temperature that is generated by friction between the abrasive particles and the cutting front. This heated zone suddenly cools down by the following excess water-flow that creates fracture due to thermal spalling. It is observed that whisker-reinforced ceramics are more resistant against this damage mode. Because of their high thermal conductivity, the whiskers act as microscale heat sinks that draw heat away from the surface into the specimen.

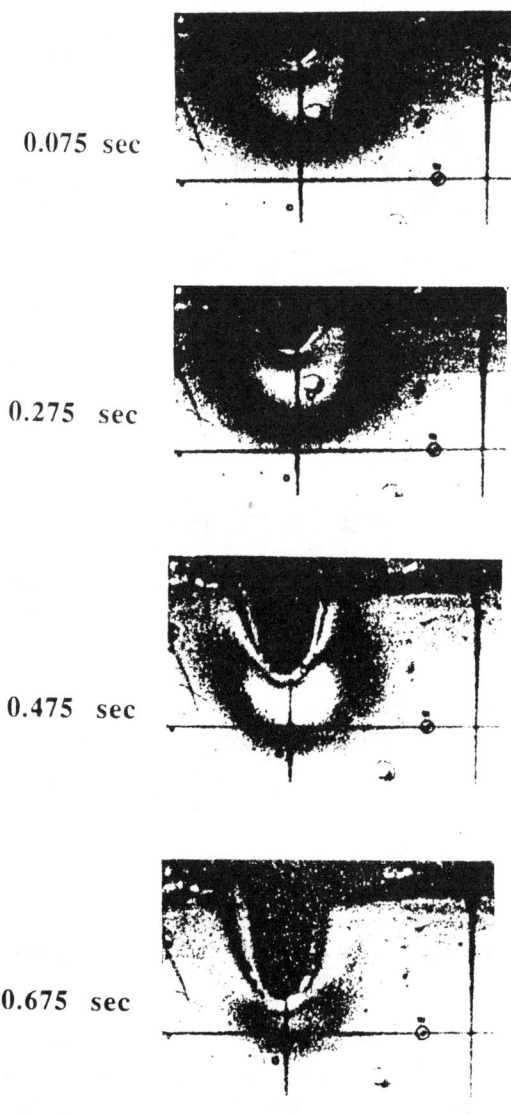

Figure 5.12 *Photoelastic records of a polyethylene specimen impacted by an abrasive water jet (Univ. of Washington, Seattle)*

5.2.3.6 Observations on Glass

For homogeneous brittle materials, such as glass, no systematic inspection of the micro-mechanisms involved in the material-removal is yet performed. Figure 5.13a shows a micrograph of an abrasive-particle impact on a glass surface. This figure illustrates that lateral cracking is one possible material-removal mechanism. Acoustic-emission signals from glass show comparatively continuous structures (Figure 5.13b) illustrating the generation of a continuously growing microcrack network.

a - Optical micrograph of a crater created in glass

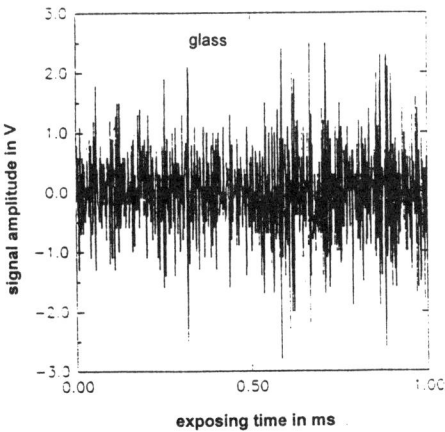

b – Time-domain acoustic-emission signal

Figure 5.13 *Micromechanisms of material removal in glass (UK, CRMS, Lexington)*

5.3 Material Removal by the High-Speed Water Flow

5.3.1 General Observations

The role of the high-speed water flow during the material removal by an abrasive water jet is a phenomenon that is not yet completely understood. Some authors consider the water to be a carrier-and-acceleration medium for the entrained abrasive particles [157]. A wide range of metallic materials that can not be cut by plain water jets proves this assumption.

Nevertheless, several authors present experimental evidence for a direct contribution of the water flow on the material-removal process in ductile-behaving as well as in brittle-behaving materials. Summers et al. [179] impact steel samples with jets of plain water at high pump pressure (p=210 MPa). Through SEM-observation, the authors find a clear track across the sample with pittings at the surface and water-jet penetration along the crystal boundaries. Ramulu [192] suggest a combined action of the abrasive particles and a 'water wedge' during the piercing of brittle compound materials by an abrasive water jet. Whereas the impacting abrasive particles create micro-cracking, the high-speed water enters these cracks and wides them due to hydrostatic forces.

5.3.2 Observations on Pre-Cracked Materials

5.3.2.1 Effect of 'Water Wedging'

Pre-cracked materials and materials containing a certain degree of instabilities, such as micro-cracks and pores, can be cut effectively with plain water jets at commercial pressure ranges [194, 195]. In a classical experiment, it is shown that a pre-cracked material can not be damaged by a plain water jet if the surface is covered by a very thin metal layer that prevents the high-velocity water flow from entering the cracks and pores. If the layer is removed, the material fails by internal stresses that are created in the material by the penetrating water stream [196].

Momber and Kovacevic [195, 198] and Momber et al. [197] suggest that the water enters a crack with high velocities and generates stresses on the crack walls. As far as the intensity of these stresses exceeds the critical local material-resistance parameters, such as the fracture toughness, the crack grows. An intersection of several cracks forms a micro-crack network and wear particles. Figure 5.14 illustrates this mechanism. Wiedemeier [3] and Momber and Kovacevic [195] show for the water-jet cutting of brittle materials that

$$v_{0Cr} \propto K_{Ic} = \frac{K_{Ic}}{\sqrt{2 \cdot \pi \cdot l_{Cr}}}. \tag{5.8}$$

In the equation, v_{0Cr} is the critical threshold velocity of the water flow. For an undamaged material with $l_{Cr}=0$, the threshold water-jet velocity, v_{0Cr}, is infinite and the material is not cut. But the water threshold-velocity significantly reduces for cutting a material that contains flaws and cracks.

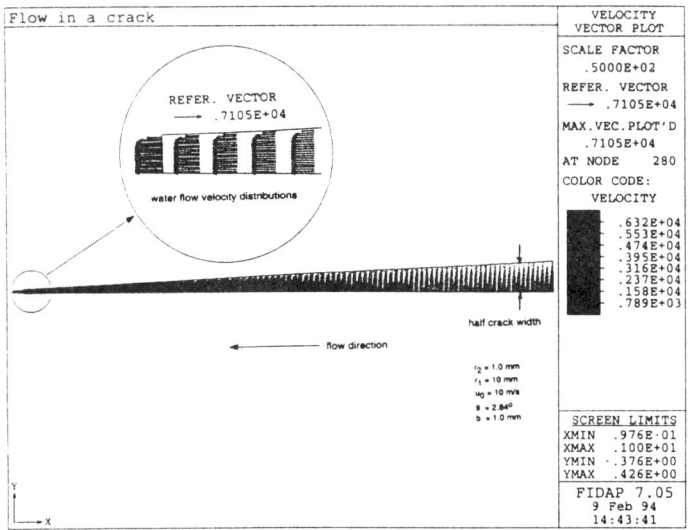

Figure 5.14 *Water wedging in a pre-cracked material by a plain water jet [197]*

5.3.2.2 'Transition-Velocity' Concept

Momber et al. [199] perform a systematic investigation on gray cast iron into the influence of the water flow on abrasive water jet material-removal processes. In this experiment, the authors collect the removed wear particles and calculate a contact number according to Eq. (5.28).

Gray cast iron contains some instabilities because of the graphite flakes in the structure. Nevertheless, cast iron usually can not be cut by plain water jets. This situation changes if a certain number of surface cracks is present that can be penetrated by the high-speed water flow. Balan et al. [200] observe the formation of surface cracks during the erosion of gray cast iron by solid particles. These cracks are still present during the post-erosion SEM-inspection. This observation indicates that no additional force is present to widen and intersect the cracks. Therefore, several particle impacts are necessary to remove material debris under the condition of dry solid-particle erosion.

This condition means $C_N<1$, which is the situation in the abrasive water jet cutting of gray cast iron with jet velocities below a 'transition' velocity. For cast iron, Momber et al. [199] estimate a transition velocity of $v_T \approx 200$ m/s. Below this value, the water velocity is not high enough to widen the cracks that are generated by the abrasive-particle impact. Thus, Eq. (5.8) is not satisfied. As the jet velocity increases beyond the transition velocity, $v_P > v_T$, it can first be assumed that the length or the number of the generated cracks increase due to the higher kinetic

5. Material-Removal Mechanisms in Abrasive Water-Jet Machining

energy of the impacting abrasive particles. Additionally, the water-flow velocity increases. A certain combination of both effects fulfills Eq. (5.8) that allows the water flow to directly contribute to the material-removal process. The water opens isolated cracks and creates a crack network. Okada et al. [201] shows for the cavitation erosion of cast iron that the embedded graphite flakes support this process. Assuming dynamic fragmentation, the average fragment size reduces with an increase in the loading intensity [101] (Table 3.3). With an increase in the material-removal rate and a decrease in the average wear-particle size, the number of removed target particles increase. Figure 5.15 shows that this increase yields larger contact numbers for higher pump pressures.

5.3.2.3 Pocket Formation in Soft Materials

Momber et al. [190, 191] observe a reduction of damping effects as well as the formation of extremely large pockets in the very lower region of the cuts in soft concrete materials subjected to an abrasive water jet (Figure 5.16). The authors suppose that the matrix material is washed away by the high-speed water flow.

Nevertheless, the influence of the water flow in the course of abrasive water jet cutting is still an unsolved problem and further work is required to clarify this problem.

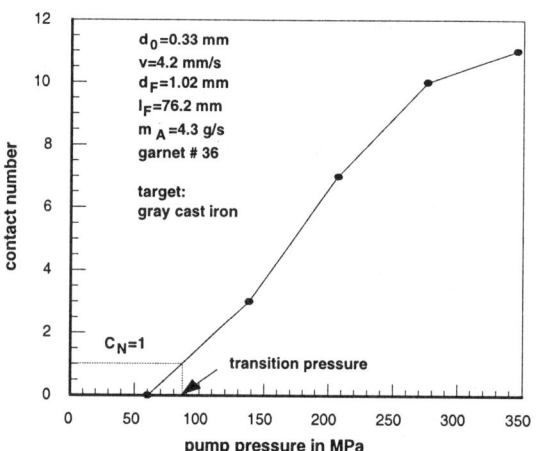

Figure 5.15 *Relation between the pump pressure and contact number for cast iron [199]*

Fig. 5.16 *Pocket formation in soft concrete [190]*

5.4 Macro-Mechanisms of Abrasive Water-Jet Material Removal

5.4.1 Some Observations of the Surface Topography

5.4.1.1 General Statement

The geometry of the generated cutting front that is generally curved as shown in Figure 5.17, and the topography of the generated surface are two major features of the macroscopic, abrasive water jet material-removal regime. Thus, insight into the character of at least the macroscopic, material-removal process is obtained from analysis of the surface structures of the abrasive water jet cut specimens. The next section discusses aspects of cutting front inspections.

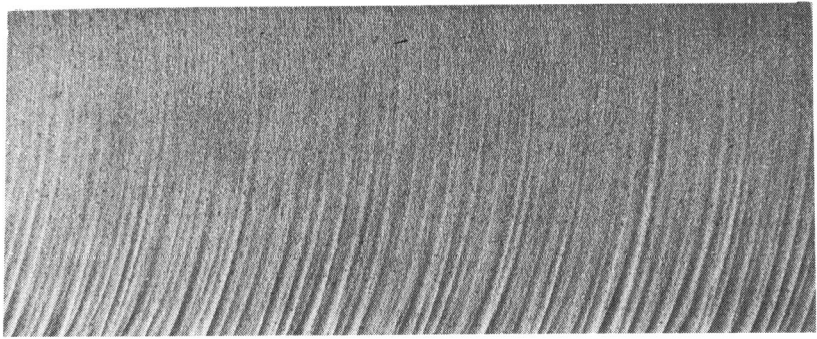

Figure 5.17 *Geometry of the cutting front generated during abrasive water-jet cutting (Univ. Hannover, IW)*

112   5. Material-Removal Mechanisms in Abrasive Water-Jet Machining

5.4.1.2 Surface-Profile Inspections

Despite the surface topography, Figure 5.17 shows surface profiles of an aluminum sample. From the first view, the most striking feature is the existence of two different regions. One region, covering the upper part of the specimen, shows a smooth surface; whereas, the second region, located in the lower part, exhibits some regularly-appearing surface marks. Figures 5.18 and 5.19 show surface profiles measured at different depths, and the corresponding power spectra. Especially the power spectra plots clearly represents the surface characteristics. In the upper part, the spectrum distributes in a wide range of frequencies and does not show preferred frequency peaks. Therefore, the surface profile is almost random and does not possess a characteristic wavelength. With increasing depth of cut, the spectral distribution concentrates in a narrow range of frequency that demonstrates a dominant harmonic component in the surface profile.

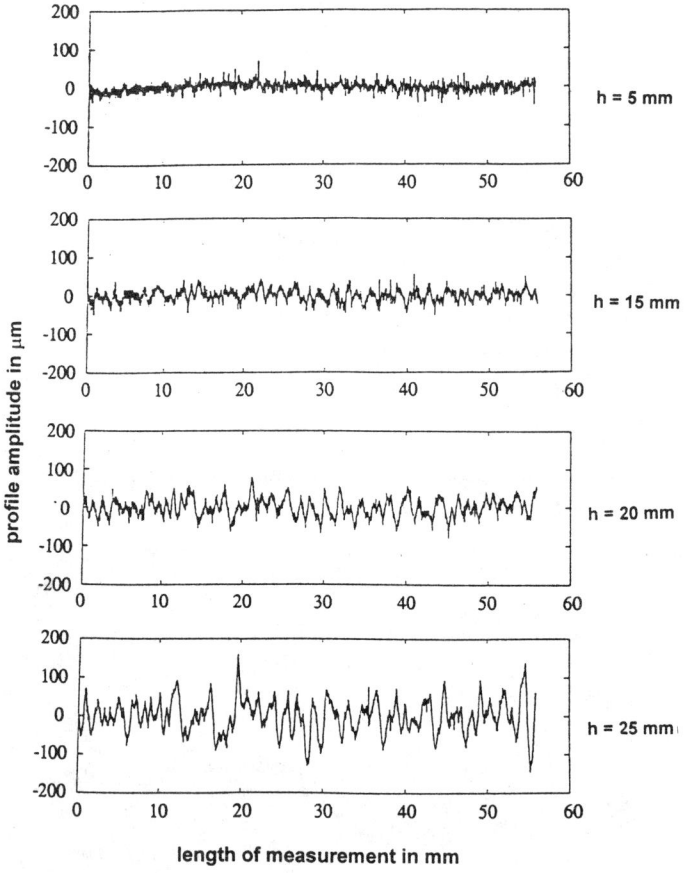

Figure 5.18 *Time-domain signals of the surface profile for aluminum [180]*
p=240 MPa, d_0=0.25 mm, v=1.67 mm/s, d_F=0.9 mm, l_F=40 mm, m_A=8 g/s, minersiv

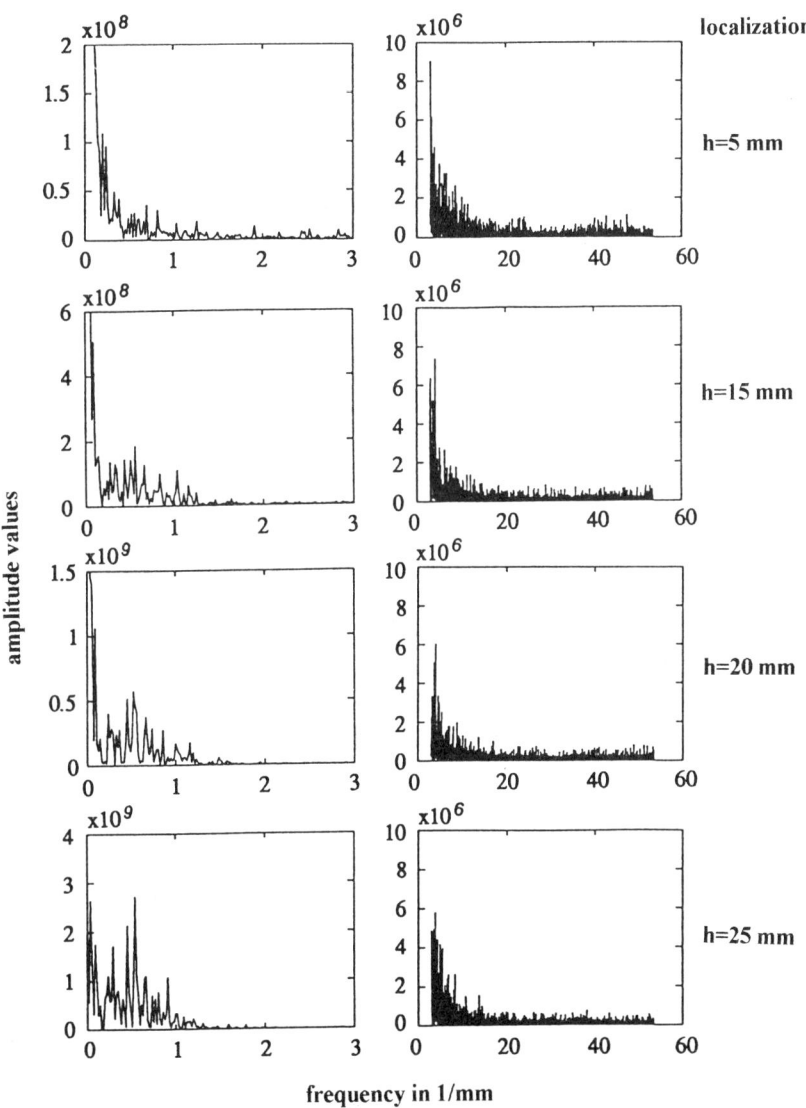

Figure 5.19 *Amplitude spectra for aluminum cut by an abrasive water jet [180]*
(parameters as in Fig. 5.18)

Guo et al. [203] shows that the highest peaks of the power spectrum are at about the same frequency for different depths of cut in the lower region. At a certain depth, the characteristics of the surface changes from a random profile to surface profiles with dominant components.

5.4.1.3 Wavelength-Decomposition

More information is obtained by the wavelength decomposition of the surface profile models, as carried out by Webb and Rajukar [175], Mohan et al. [204], and Kovacevic et al. [205, 206] for auto-regressive models, and by Guo et al. [203] for Fourier-synthesis. The surface profile is generally characterized by two typical wavelengths, one wavelength with high values and one wavelength on a lower-length level. Because the higher wavelength always absorbs most of the signal energy (about 80% to 90%), it is called the 'primary' wavelength. Therefore, the smaller wavelength is the 'secondary' wavelength. Tables 5.3 and 8.7 list some examples of estimated wavelengths of surface profiles generated by an abrasive water jet.

Table 5.3 *Results of profile wavelength-decomposition of surfaces generated by an abrasive water jet [175]*

d_P in mm	Traverse rate in mm/s					
	1.27	2.96	4.66	1.27	2.96	4.66
	Wavelength in mm					
	Primary wavelength*			Secondary wavelength		
0.125	4.48	2.50	1.66	0.15	0.11	0.12
0.180	4.49	3.64	2.65	0.19	0.07	0.20
0.300	4.87	3.54	-	0.29	0.21	-

* Focus diameter d_F=1.1 mm

The results in Table 5.3 show that the secondary wavelengths remain fairly consistent for the different cutting conditions. This consistency indicates that these wavelengths are associated with surface variations left by the single abrasive particles, not by the global action of the jet stream. Table 5.3 evidences that the secondary wavelengths are almost identical to the average particle diameter. In contrast, the primary wavelengths are more related to the focus diameter or jet diameter, respectively. In Table 5.3, the primary wavelength is between two to four times the focus diameter. This result agrees with observations from Guo et al. [203] who find a strong correlation between the focus diameter and the length of the significant wave components. Kovacevic et al. [206] obtain comparative results. Section 8.2.3 discusses this result in more detail. Noticeable from Table 8.6, that the contribution (power amount) of the primary wavelength increases as the depth of cut increases. In contrast, the secondary wavelength becomes less important with an increase in the depth of cut. Therefore, the upper stage of the specimen surface is the result of the action of the randomly-localized, impacting abrasive particles. The lower region is generated by the global action of the jet flow in the kerf that contains several physical phenomena as will be discussed later.

Several references make approaches in explaining these phenomena as discussed in the following sections.

5.4 Macro-Mechanisms of Abrasive Water-Jet Material Removal 115

5.4.2 Two-Dimensional Model of the Integral Material Removal

5.4.2.1 Traverse-Direction Stages

Hashish [157, 207] carries out the first systematic observations of macroscopic features of the abrasive water jet cutting-process. Based on high-speed photographs in transparent materials (plexiglass, lexan), he suggests a two-dimensional structure of the cutting process as shown in Figure 5.20.

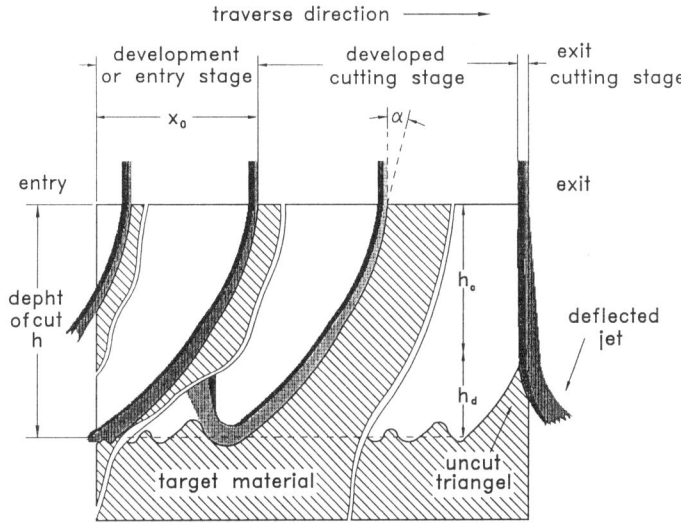

Figure 5.20 *General process-structure of the abrasive water-jet cutting, adapted from [207]*

Along the traverse (x-) direction, Hashish [207] divides between an entry stage, a cutting stage, and an exit stage. In the entry stage, the depth of cut is lower than the maximum possible depth of cut and lower than the workpiece thickness, respectively. During this stage, the jet enters the workpiece in the y-direction, and the material-removal process introduces. If the final depth of cut is reached, the cutting stage starts. This stage is characterized by a cyclical removal process as discussed later. At the end of the workpiece, the jet reflects against the traverse direction that leads to the generation of the typical uncut triangle as marked in Figure 5.20. This stage is the exit stage. Kovacevic et al. [208] show by using infrared-thermography that the range around the uncut triangle is in a stage of 'overheating' that indicates a cumulative heat generation process (Figure 10.27). Mohan et al. [209] find an accumulation of acoustic-emission signals in the entry stage as well in the exit stage (Figure 10.22) that indicates hydrodynamic activities in these ranges, such as turbulence, cavitation, and pressure pulses. The surface structure of the workpiece being cut documents all three stages. It is also concluded from these studies that the abrasive water jet cutting-process is cyclic, consisting of a set of almost identical cutting cycles as illustrated in Figure 5.20.

5.4.2.2 Penetration-Direction Stages

In the penetration (y-) direction, a single, complete cutting cycle (Figure 5.20) also divides into several different stages. Hashish [207] adapts the solid-particle erosion concept developed by Bitter [160, 161] and divides one cutting cycle into two distinctive stages. Hashish calls this first stage the 'cutting-wear zone'. This stage is marked as h_c in Figure 5.20. According to Hashish [207] and Blickwedel [69] the stage is characterized by material removal due to individual abrasive particles hitting the surface at small impact angles (section 5.1). The material is removed by micro-cutting. This performance is assumed to be a steady-stage process where the material-removal rate is equal to the material-displacement rate. A geometrical expression of this zone is the roughness of the cut-wall surface (section 8.2).

If a certain impact angle is reached, the material-removal regime changes to a 'deformation wear', (called by Hashish) that is characterized by particles impacting at large angles. This stage is marked as h_d in Figure 5.20. In this stage, the abrasive water jet penetrates the material at a decreasing rate as the depth of cut increases. The material is removed by a different erosion mode that is associated with multi-pass particle bombardment, surface hardening by plastic deformation, and crack formation. The so-called 'striations marks', or waviness zone, geometrically characterizes this zone. Section 8.2 discusses this aspect of the cutting front. In a final stage, near the bottom of the cut, the cutting process turns into a 'drilling' process with an impact angle of $\varphi=90°$.

Nevertheless, Hashish [207] does not give an explanation about what physical phenomena causes the existence of two different material-removal regimes. Also, as discussed in section 5.2, there is no experimental evidence for a change in the general microscopic material-removal mode with the depth of the cut neither in ductile-behaving nor in brittle-behaving materials.

5.4.2.3 Further Development of the Model

Blickwedel [69] who uses sloped specimens to simulate the different process conditions on the cutting front, further developes Hashish's two-stage model. Using a high-speed camera, Blickwedel observes the existence of more than one step on the cutting front as shown in Figure 5.21. He localizes the beginning of the step formation (the beginning of the 'deformation-wear' zone) at 30%-40% of the maximum depth of cut. Later, Ohlsson et al. [210] rediscover the phenomenon of multiple-step formation. Based on force measurements in the traverse direction as well as in the penetration direction, these authors develop a phenomenological model of striation formation based on variations in the geometry of the cutting front.

5.4 Macro-Mechanisms of Abrasive Water-Jet Material Removal   117

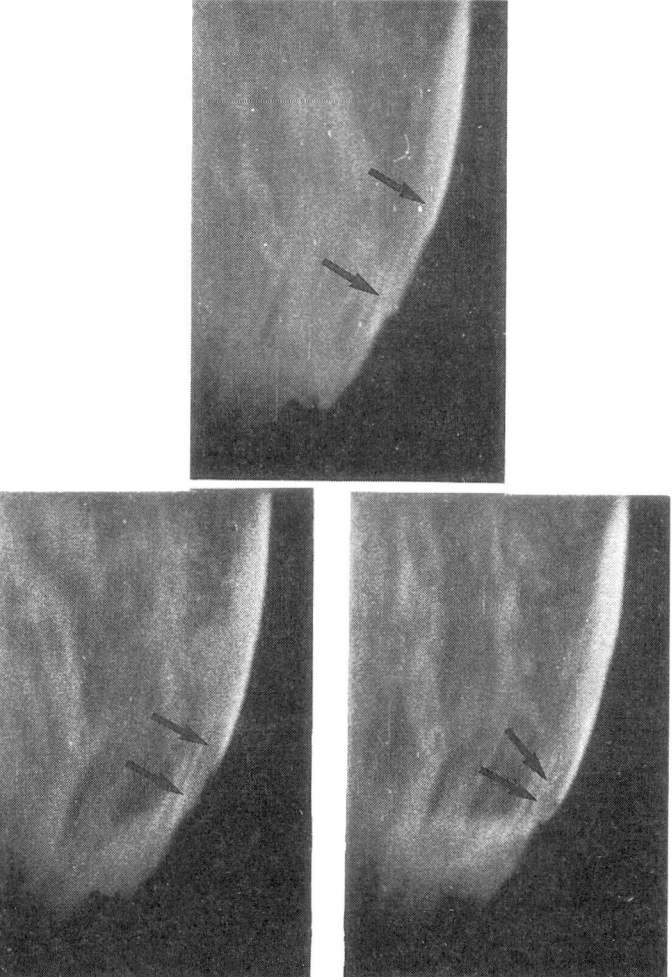

Figure 5.21 *Multiple step-formation (arrowed) in abrasive water-jet cutting of PMMA; distance between the pictures is ca. 50 ms (Univ. Hannover, IW)*

5.4.2.4 Step Formation at the Cutting Front

Mohaupt and Burns [211] first notice the step-formation phenomena during cutting plastics by a plain water jet. Based on the investigations of Hashish and Blickwedel, the general physics of the two-dimensional cut-generation process in abrasive water jetting is described as follows.

In the beginning of the material-removal process, the jet hits the surface parallel to its axis. With an increase in the depth of cut, the local impact angle increases, too.

118   5. Material-Removal Mechanisms in Abrasive Water-Jet Machining

Because of their high body-force, the abrasive particles can not follow the abrupt change in the flow direction, and a de-mixing process occurs. This process leads to a local accumulation of particle impacts. The removal process now restricts to a small portion of the cutting front. The result of this process is the formation of a local step. Using high-speed photography, the steps are observed in transparent materials (Figure 5.21). Typical step speeds are between 45 m/s and 200 m/s. The speed seems to be higher for brittle materials than for ductile materials [209]. In opaque materials, there is the possibility to suddenly interrupt the removal process. Then, the step is clearly seen. Additional experimental evidence for an unsteady removal process gives the impact force, measure after the abrasive water jet exits the workpiece. Figure 10.34 shows that the force signal of an exiting abrasive water jet exhibits a high-dynamic component compared to the signal of a non-working abrasive water jet.

The shape of the step changes as the material removal progresses until it reaches a critical shape. This critical shape is characterized by a 90°-impact angle. The material-removal process changes into a drilling process. The abrasive particles can not remove more material and the final stage of the cutting process is reached.

5.4.3 Three-Dimensional Model of the Integral Material Removal

5.4.3.1 Three-Dimensional Step Formation

Guo [180] developed a three-dimensional integral abrasive water jet cutting-model (see also [203]). Figure 5.22 presents the structure of the model .

Figure 5.23 illustrates the three-dimensional aspect of step formation: the movement of the step occurs in the direction of the traverse as well as perpendicular to it. This phenomena is observed in different materials, such as aluminum alloy, titanium alloy, steel, and ceramics [203]. Figure 5.23 additionally shows traces of material removal in a steel sample. These traces are caused by an abrasive water jet after cutting another sample. The traces appear as a series of dents that demonstrates an unsteady oscillation. This oscillation of the traces is observed only when striation formation is involved.

Thus, four major mechanisms play a significant role in the material removal process:

- individual abrasive-particle impact   → roughness formation,
- step formation in cutting direction   → striation formation,
- jet oscillation perpendicular to the   → groove formation,
  cutting direction
- oscillations of the guiding system   → external effects.

5.4 Macro-Mechanisms of Abrasive Water-Jet Material Removal 119

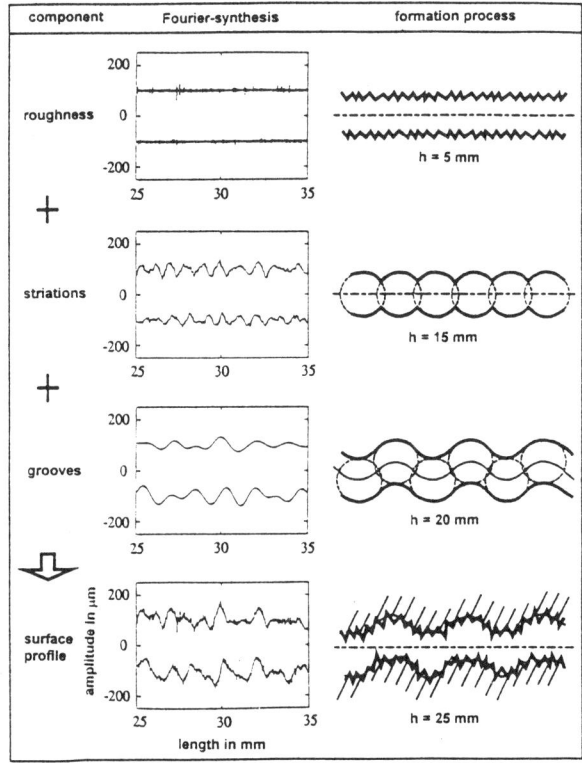

Figure 5.22 *Three-dimensional model of the surface formation [180]*

The interaction between the individual abrasive particles generates the microstructure of the cut surface (surface roughness) that dominates the surface profile in the upper region but can be detected in lower regions, too. The macroscopic geometry of the cut surface results from the accumulation of individual material removal actions that lead to the formation of the three-dimensional steps. As the step formation introduces, the properties of the steps and the hydrodynamics of the jet dominate the generation of the cut surface. First, the step formation and the step overlapping in the traverse direction generate a striation component of the surface profile. Second, the inclination and unsteady jet oscillation yields a three-dimensional step movement during cutting that is responsible for groove formation as an essential feature of the surface profile in the 'rough cutting' region. In order to separate the different physical phenomena contributing to the material-removal process, Guo [180] performs an extensive surface analysis.

120   5. Material-Removal Mechanisms in Abrasive Water-Jet Machining

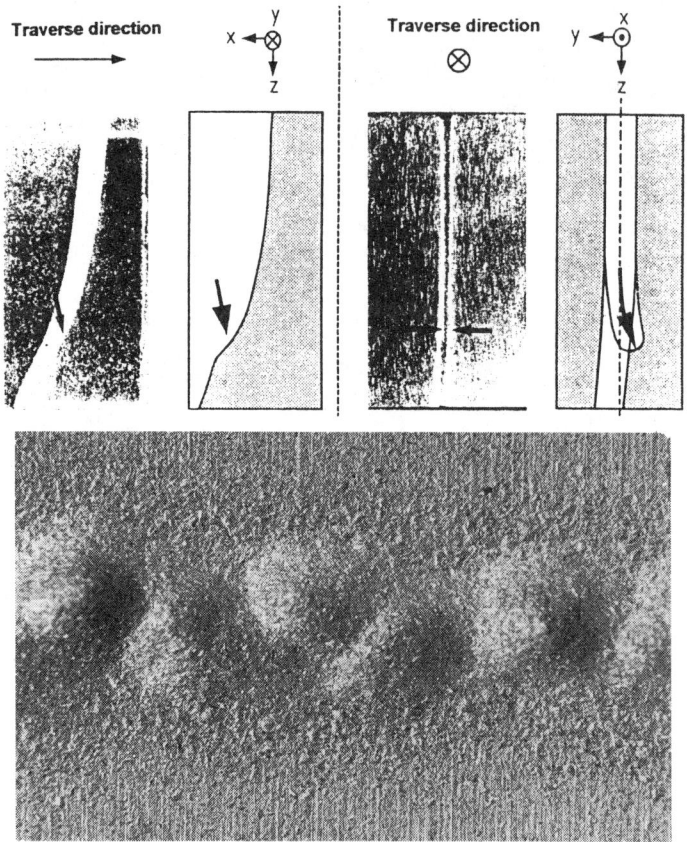

p=240 MPa, d_O=0.25 mm, v=6.67 mm/s, d_F=0.9 mm, l_F=40 mm, m_A=8 g/s, minersiv

Figure 5.23 *Two-dimensional step formation and traverse oscillation of an abrasive water jet during the cutting process (Univ. Hannover, IW)*

5.4.3.2 Influence of Machine Vibrations

Chao et al. [212] and Chao and Geskin [213] who investigate the correlation between the striations and instabilities of the operational conditions develop an alternative proposal to explain the formation of striations. In metallic materials, they notice significant differences in the structure of striations cut in the x-direction and the y-direction, respectively [212]. They conclude that the nozzle vibrations as well as by the vibrations of the traverse system determine the cutting-front structure to a great extent. Another finding is the existence of striation marks with separations. The distance between these separations is found to be a constant value and does not depend on the traverse rate [213]. This distance shows a correlation to the circular pitch of the moving system. From these results, the authors conclude that the smoothness of the traverse system is an important factor of the striation-mark formation during the abrasive water jet machining. The source of the striation formation is assumed to be the vibration of the cutting nozzle perpendicular to the cutting direction. If the cut in the material is deep enough, the spent water flows back along the already-cut channel. In case of vibrations, there is a radial velocity component added to the moving jet. Thus, the spent water has a sideways component that cuts into the wall and generates the typical striation structure. Nevertheless, this suggestion does not really explain the striation formation in the case of cutting through where no back flow of the spent water exists. Moreover, Guo [180] does not find any relation between the spectrum of surface-profiles and the spectrum of machine-oscillations, and concludes that machine oscillations influence the surface structure secondarily but are not a physical source of striation formation (Figure 5.24). Also, Hashish [214] shows that the regularity of the striations depends on the machinability of the materials. For hard-to-machine materials, such as ceramics, the striations are very regular; whereas, easy-to-cut materials show non-regular striation structures. As Hashish suggests, external effects, such as pressure fluctuations and vibrations, are excluded in the hard materials. These materials show the result of the plain, high-precise cutting cycles. In soft materials, the ideal removal process is randomly interrupted by external effects.

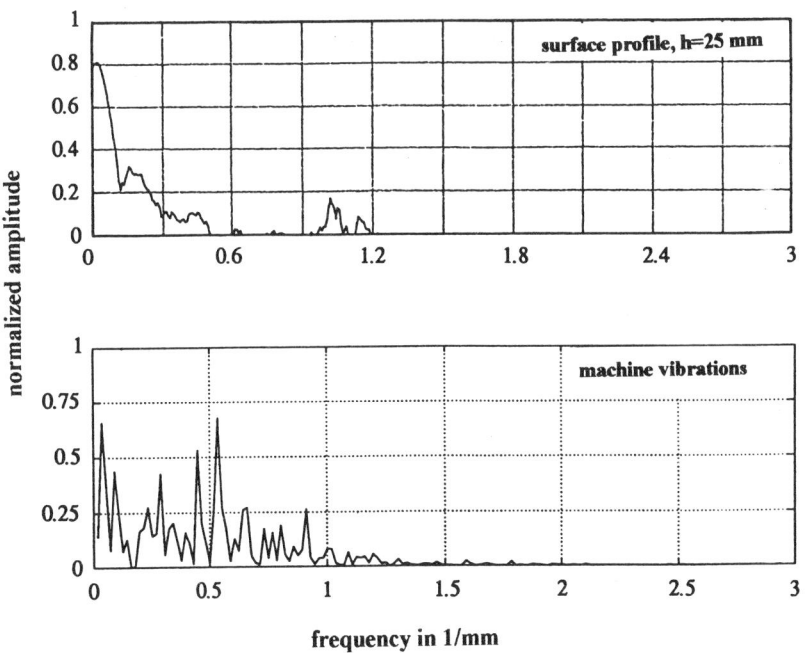

Figure 5.24 *Surface profiles and machine-vibration signals [180]*

5.4.4 Alternative Models of the Integral Material Removal

5.4.4.1 General Comments

The suggestion that the cutting front is divided into the different material-removal mechanisms is controversially discussed. As already pointed out in section 5.2, several investigators observe a constant material-removal mechanism on the entire cutting front which is in contrast to the two different material-removal modes assumed by Hashish and Blickwedel. Another related problem is that a curved cutting front is an essential feature of all stream-like cutting tools, such as lasers and electron beams [215]. For this reason, it is possible to assume a more general source of striation formation.

5.4.4.2 Two-Stage Impact Zone Model

Zeng and Kim [158] critisize the underlying concept in Hashish's model, that the particle impact angle changes with an increase in the depth of cut, and developed a two-stage model illustrated in Figure 5.25a. In a first stage, called the 'direct impact zone', the abrasive particles directly impact the entire cutting front. As the jet moves further into the material, the area behind the back side of the abrasive water jet is exposed to a 'secondary impact' by deflected abrasive particles that results in a sudden change of the cutting-front curvature. The entire cutting front consists of several cycles of step formation. The most important conclusion from the model of Zeng and Kim [158] is that the entire cutting process is associated with abrasive particle impacts at glancing angles, regardless of the type of target material. This idea gets some support from Ohlssen et al. [209] who measures cyclic fluctuations in the cutting forces exerted on the cutting front in both the vertical and the horizontal directions. From these measurements, they conclude a change in the geometry of the step on the cutting front as the step progresses downward. As the source of the frequencies, they assume the step generation process as well as mechanical vibrations set up in the cutting equipment (5 Hz - 20 Hz).

5.4.4.3 'Three-Zone' Cutting Front Model

Arola and Ramulu [181, 185] and Ramulu and Arola [187] develop an alternative 'three-zone' cutting-front model as illustrated in Figure 5.25b.

These authors introduce a 'top-entrance zone', or 'damage zone' that is characterized by a plastic deformation due to the almost perpendicular impacting abrasive particles in ductile materials. Earlier, Kovacevic et al. [174] discover and describe, but not name this zone. Guo [180] also discusses this zone. The size of this zone mainly depends on the standoff distance (chapter 8).

Arola and Ramulu [181, 185] call the following ranges the 'smooth-cutting zone' and the 'rough-cutting zone', respectively. These names indicate that rather the structure of the generated surface changes with the depth of cut but not the micro-

5.4 Macro-Mechanisms of Abrasive Water-Jet Material Removal

removal regime. Arola and Ramulu [181] observe that the wear tracks on the individual impact sites and their severity decrease with an increase in the depth of cut. The authors discuss this observations in terms of a loss of kinetic energy of the single abrasive particles involved in the material-removal process at greater depths. Section 5.5 addresses this aspect.

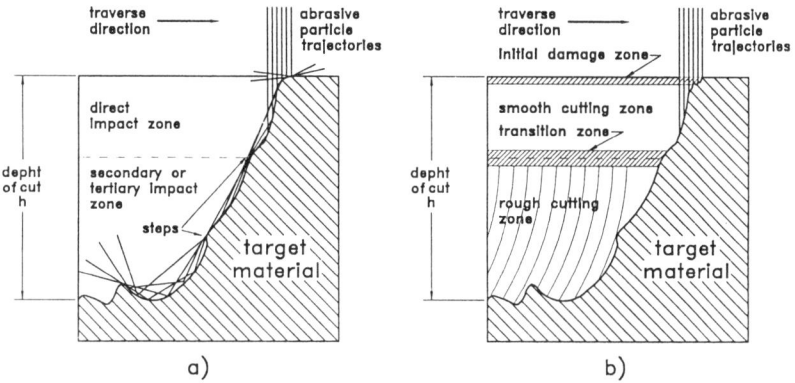

Figure 5.25 *Alternative cutting-front models for the abrasive water-jet cutting*
a - 'Two-stage' impact zone model [158]
b - 'Three-zone' cutting front model [181]

As discussed in chapter 10, several authors find an increase in the absolute value as well as in the dynamics of the workpiece reaction force with an increase in the depth of cut (Figure 10.32). Following the ideas introduced by Arola and Ramulu [181], these measurements are an indication of the decelerated material removal in the lower part of the cut. Due to energy losses in the upper part of the cut, the loading intensity of the single abrasive particles reduces. In brittle materials, the reduced stress intensity leads to less crack initiation and to slower crack growth. In ductile materials, the smaller kinetic energy of the abrasive particles leads to an accumulation of particle impacts at a given area that introduces higher cutting forces. Raju and Ramulu [216, 217] extend this phenomenological idea to a mechanistic model of the cutting-front formation process. Assuming steady cutting conditions, these authors develop governing equations for the depth of the 'smooth-cutting' zone as well as for the 'rough-cutting' zone. Section 6.2 discusses the basic relations of this model in detail.

5.4.4.4 'Energetic' Cutting Front Model

Raju and Ramulu [217] suggest that the kinetic energy of the abrasive-water slurry at the transition point between the 'smooth-cutting' zone and the 'rough-cutting' zone ($h_{SC} \rightarrow h_{RC}$) is a more-or-less constant value for a given material (Figure 5.26). Table 5.4 lists typical values for this transition-energy. These results suggest that the kinetic energy of the abrasive water jet rather than a change in the material removal mode is a critical parameter that dictates the formation of striation marks after the transition point.

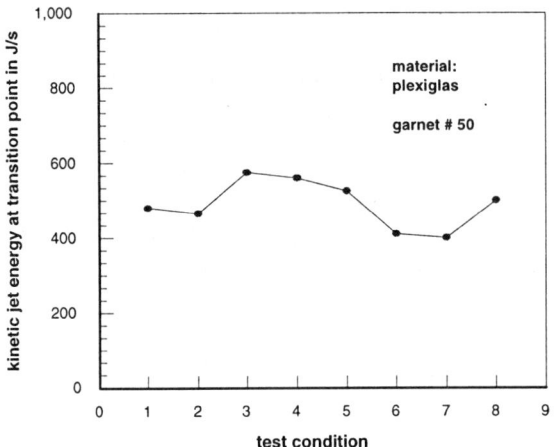

Figure 5.26 *Transition energies of abrasive water jets at the transition point [217]*

The trend in Table 5.4 correlates well with the machinability numbers of the materials (Figure 5.48). The lower the machinability number, the larger the amount of kinetic energy in the transition point.

Table 5.4 *Transition-point energies in different materials [217]*

Material	Transition energy in J/s	
	Garnet # 50	Garnet # 80
Aluminum	600	800
Plexiglass	500	700
Steel	1,700	4,500

5.4.4.5 Numerical Simulation of the Cutting Front

Fukunishi et al. [218] develop a computer-simulation model for the cutting-front formation in the abrasive water jet-cutting. The model bases on Bitter's [160, 161] formulas for the 'cutting wear' and the 'deformation wear'. A very interesting result of this simulation model is that step formation only happens if the input abrasive particles are randomly distributed (Figure 5.27). If, in contrast, the abrasive particles hit the target surface with a regular distribution, no step formation occurs.

5.5 Energy Balance of Abrasive Water-Jet Material Removal

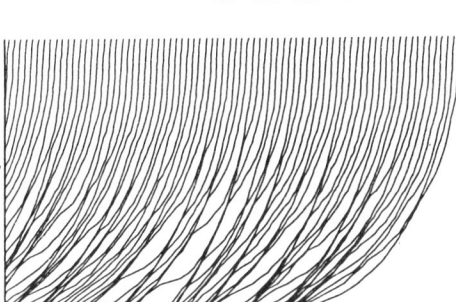

Figure 5.27 *Numerically calculated cutting front for abrasive water-jet cutting [218]*

Although this simulation is an interesting alternative aspect of the cutting-front generation process that links the cutting-front geometry to the abrasive particle distribution, the physical background as well as the entire simulation process are not sufficiently discussed in this contribution.

5.5 Energy Balance of Abrasive Water-Jet Material Removal

5.5.1 General Energy Situation

5.5.1.1 Dissipated Energy

Figure 5.28 illustrates the general energy situation in a material cut by an abrasive water jet. The energy dissipated in the workpiece during the cutting process is

$$E_{Diss} = E_A - E_{EX} . \tag{5.9}$$

In the equation, E_A is the kinetic energy of the abrasive water jet,

$$E_A = (\dot{m}_A + \dot{m}_W) \cdot v_P^2 \cdot \frac{d_F}{2 \cdot v} , \tag{5.10}$$

and E_{EX} is the kinetic energy of the high-speed slurry when it leaves the workpiece. Many investigations in abrasive water jet cutting show that a threshold energy (critical pump pressure, critical traverse rate) is necessary to destroy a material and to initiate the cutting process. Section 7.2 gives a review of this problem. This critical situation is also found at the bottom of the cut in a workpiece ($h=h_{max}$), when the jet does not have enough energy to penetrate the material further,

$$E_{EX} = E_{thr} . \tag{5.11}$$

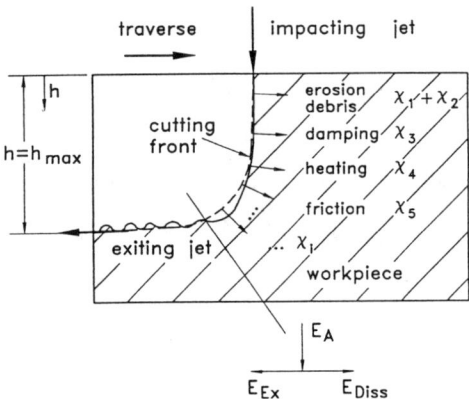

Figure 5.28 *Global energy situation on the cutting front [219]*

Eqs. (5.9) and (5.11) give

$$E_{Diss}(h = h_{max}) = E_A - E_{thr}. \qquad (5.12a)$$

Another critical condition exists for h=0. In this case, the workpiece does not dissipate energy, and the total amount of the abrasive water-jet kinetic energy goes into the exiting high-speed slurry. Therefore,

$$E_{Diss}(h = 0) = 0, \qquad (5.12b)$$
$$E_{EX} = E_A.$$

5.5.1.2 Energy-Dissipation Function

Introducing a parameter $\chi(h)$ that depends on the depth, the situation for all cases between these extremes is [219, 220]

$$0 < h < h_{max}: \quad E_{Diss}(h) = \chi(h) \cdot [E_A - E_{thr}]. \qquad (5.13)$$

For Eq. (5.12a), $\chi(h=h_{max})=1$ and for Eq. (5.12b), $\chi(h=0)=0$. The function $\chi(h)$ considers the energy-dissipation processes in the material during cutting.

Eq. (5.13) gives the amount of energy that is dissipated in the workpiece at any particular depth of cut. The parameter χ includes different mechanisms of energy dissipation (Figure 5.28), such as

- energy dissipation due to erosion debris formation, χ_1
- energy for accelerating the removed erosion debris, χ_2
- energy dissipation due to film damping on the cutting front, χ_3

- energy dissipated in heating the workpiece, χ_4
- energy dissipation due to friction on the cutting front, χ_5
- other energy dissipative processes, χ_i

Thus, $\chi(h)$ writes

$$\chi(h) = \sum_{i=1}^{N} \chi_i(h), \text{ or} \tag{5.14a}$$

$$\chi(\Phi) = \sum_{i=1}^{N} \chi_i(\Phi). \tag{5.14b}$$

In Eq. (5.14b), Φ is the relative depth of cut, $\Phi=h/h_{max}$ that is useful to unify the results.

5.5.2 Geometrical Energy-Dissipation Model

5.5.2.1 Special Solutions of the Energy-Dissipation Function

Mohan et al. [221, 222] derive a solution of Eq. (5.13) for the special case $h=h_{max}$ ($\chi=1$, $\Phi=1$),

$$E_{Diss} = \frac{\alpha^2 \cdot \mu^2 \cdot L_h \cdot (\dot{m}_W + \dot{m}_A)}{v \cdot \rho_W \cdot \left(1 + \frac{\dot{m}_A}{\dot{m}_W}\right)^2} \cdot [p - p_{thr}]. \tag{5.15}$$

Figure 5.30 shows results of Eq. (5.15). The energy dissipated in the material is between 60% and 90% of the abrasive water jet input-energy. These values are valid for kerfing a material. For a cutting-through process, the ratio between the dissipated energy and the abrasive water-jet kinetic energy is considerably lower. Also, these values depend on the depth of cut and available abrasive water jet input-energy, respectively. For lower abrasive water jet-energies and shallower depths, the relative amount of dissipated energy decreases.

The function in Figure 5.29 reasonably fits into a second-order equation,

$$E_{Diss}(h_{max}) = a_1 \cdot h_{max}^2 + b_1 \cdot h_{max} + c_1, \tag{5.16}$$
$c_1=0$.

A further special solution is $E_{diss}=0$ for $h=0$ ($\chi=0$, $\Phi=0$).

128   5. Material-Removal Mechanisms in Abrasive Water-Jet Machining

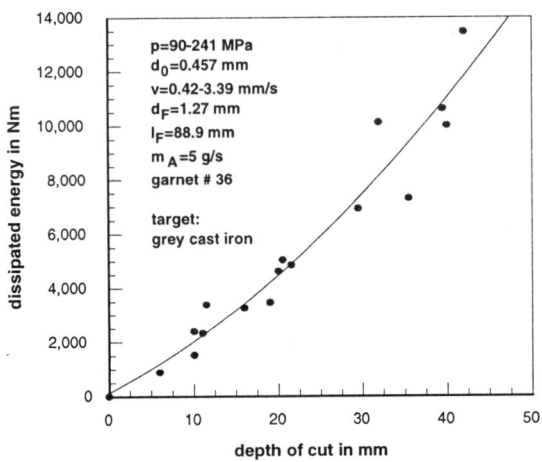

Figure 5.29 *Relation between the dissipated jet energy and depth of cut in a gray cast iron [222]*

5.5.2.2 Basics for a General Solution

In order to solve Eq. (5.13) for all cases $0<\chi<1$, Momber and Kovacevic [219, 220] and Momber [223] develop a mathematical model. Figure 5.30 shows the simplified assumptions of an abrasive water jet cutting-process without energy losses and with energy losses due to energy dissipation. As Figure 5.30a suggests, a rectangular area A_1 is generated as the abrasive water jet moves in the traverse direction without energy loss. Realistically, stream-like tools exhibit a curved cutting front (Figure 5.30b) that occupies an area

$$A_0 = h_{max} \cdot x - \int_0^x h(x)dx \ . \tag{5.17}$$

This area characterizes the energy loss in real cutting.

5.5.2.3 Striation Geometry

To describe the shape of an abrasive water jet generated cutting front, $h(x)$, Zeng et al. [224] suppose a parabolic curve. Chao and Geskin [213] find a similar relation - a second-order polynomial - for the relation between striation peak-amplitude and the depth of cut. Therefore,

5.5 Energy Balance of Abrasive Water-Jet Material Removal

$$h(x) = a \cdot (x-b)^2 + c. \tag{5.18}$$

In the equation, a, b, and c are constants that are evaluated by measurements of the cutting front. Approximations for $c = h_{max}$ and $b = x_{max}$ are

$$a = \frac{-c}{b^2} = \frac{-h_{max}}{x_{max}^2}. \tag{5.19}$$

The parameters x_{max} and h_{max} are easily estimated for every full-cut surface. Figure 5.31b gives some modeled shapes of cutting fronts generated by abrasive water jets. Table 5.5 lists the corresponding regression parameters.

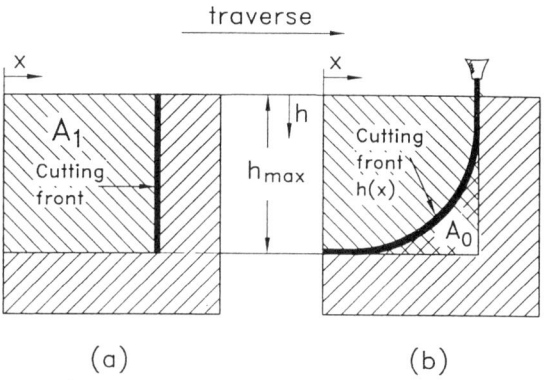

Figure 5.30 *Comparative schematics of an ideal (a) and a real (b) abrasive water jet cutting- process [219]*

Table 5.5 *Regression parameters of the parabola cutting-front [223]*

Specimen	Material	Parameter		
		a	b	c
# 1	aluminum	-0.337	8.3	22.0
# 2	aluminum	-0.336	8.0	20.8
# 3	ceramic	-0.200	15.0	45.0
# 4	concrete	-0.221	25.0	75.0
# 5	cast iron	-0173	13.4	41.6

The height of the curve is identical to the depth of cut. If the progress of the parabola function, dh/dx, is zero, the maximum possible depth of cut is reached for the given cutting conditions.

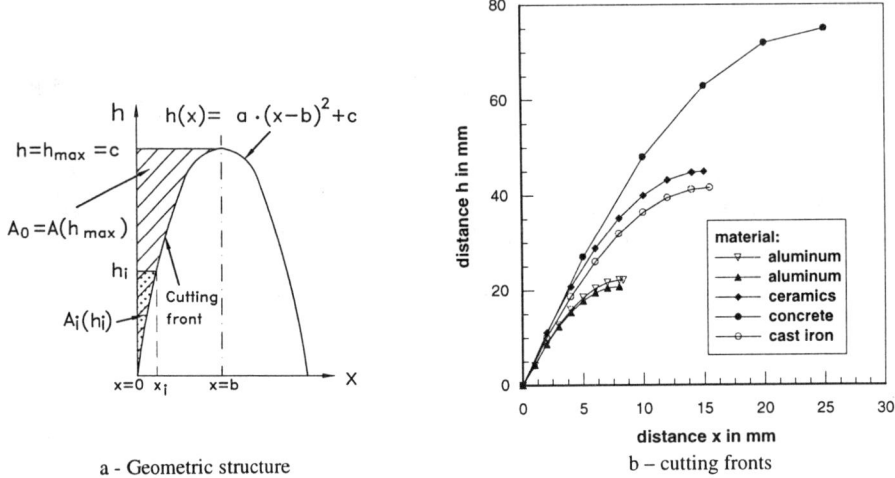

a - Geometric structure   b – cutting fronts

Figure 5.31 *Parabola model of the cutting front in the abrasive water-jet cutting [220]*

5.5.2.4 General Solution of the Energy-Dissipation Function

From Figure 5.31a, for x=b,

$$A_0 = b \cdot c - \int_0^b h(x)dx = -\frac{a \cdot b^3}{3}. \tag{5.20}$$

The area A_0 includes the energy dissipated during cutting up to the maximum depth of cut. A change of the regression parameters a, b and c with the depth of cut is neglected. A_0 is not considered to have the unit of energy, but it represents the energy dissipation during cutting. Relate this parameter is to other area values from different depths of cut can describe the relationship between energy dissipation and depth of cut. Figure 5.31a shows that certain areas A(h) exist on certain depth levels. Therefore,

$$A(h) = x \cdot [h(x)] - \int_0^x h(x)dx = a \cdot \left(\tfrac{2}{3} \cdot x^3 - b \cdot x^2\right). \tag{5.21}$$

From Eq. (5.18), x is

$$x = \sqrt{\frac{h-c}{a}} + b. \tag{5.22}$$

The dissipation function, $\chi(h)$, is

$$\chi(h) = \frac{A(h)}{A_0} = \frac{A(h)}{A(h_{max})} = \frac{2 \cdot x^3 - 3 \cdot b \cdot x^2}{(-b)^3}. \tag{5.23}$$

For $x=b$ and $h=h_{max}$, $\chi=1$, and Eqs. (5.21) and (5.20) are equal. Also, $\chi=0$ for $x=0$ ($h=0$).

Figure 5.32 shows calculations based on Eq. (5.23). The dissipation parameter is $\chi=0$ for non-cutting, and $\chi=1$ for the maximum depth of cut. Between these extremes, the function is non-linear and very reasonably approximates a second-order equation,

$$\chi(h) = A_1 \cdot h^2 + B_1 \cdot h + C_1. \tag{5.24}$$

Chao and Geskin [213] find a similar relation between the depth of cut and striation peak-amplitude. Also, Eq. (5.16) is a second-order polynomial, too. These results support the validity of Eq. (5.24).

Table 5.6 gives some regression parameters of Eq. (5.24). Notice that the functions for the concrete and the ceramic material are relatively straight lines. Because both these materials are brittle and multiphase, there is a relation between the material structure and mode of energy dissipation. A systematic study of this effect is still missing.

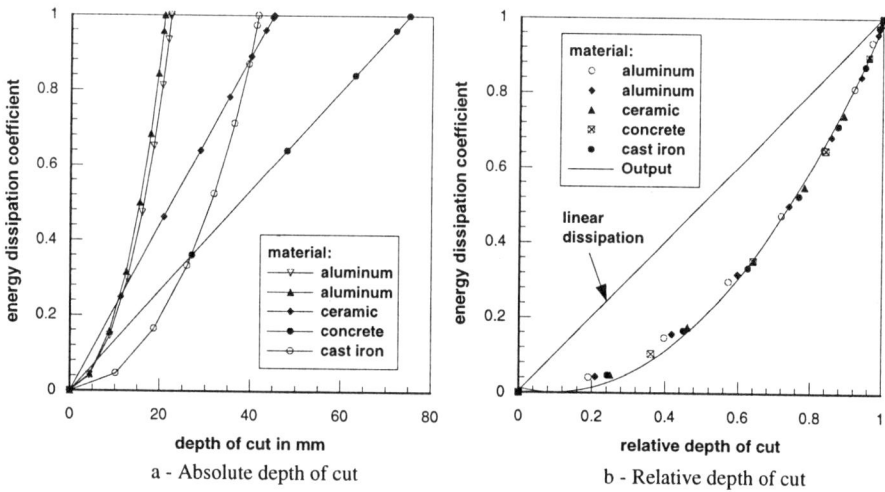

Figure 5.32 *Energy-dissipation functions $\chi(h)$ and $\chi(\Phi)$ [223]*

From Eqs. (5.13) and (5.24), the absolute value of the dissipated energy is

$$E_{Diss}(h) = [A_1 \cdot h^2 + B_1 \cdot h + C_1] \cdot [E_A - E_{thr}]. \qquad (5.25)$$

For $h=h_{max}$, Eq. (5.25) and Eq. (5.16) deliver identical results.

Table 5.6 *Regression parameters for the dissipation functions [223]*

Specimen	A_1	B_1	C_1	R^2
# 1	0.0023	-0.0069	0.0138	
# 2	0.0026	-0.0076	0.0121	
# 3	0.0060	-0.0050	0.0118	0.999
# 4	0.0020	-0.0028	0.0063	
# 5	0.0070	-0.0050	0.0125	
Average	$\overline{A}_2$	$\overline{B}_2$	$\overline{C}_2$	0.999
Relative depth of cut $\chi(\Phi)$	1.226	-0.277	0.037	0.999

5.5.2.5 Solution for the Relative Depth of Cut

Figure 5.32b plots the energy dissipation against the relative depth of cut for different cutting conditions. All plotted points fit into the same regression line

$$\chi(\Phi) = A_2 \cdot \Phi^2 + B_2 \cdot \Phi + C_2. \qquad (5.26)$$

Table 5.6 gives the regression parameters. In contrast to Figure 5.32a, the regression parameters are independent of the cutting conditions as well as the investigated materials. Although Capello and Gropetti [225] obtain a similar result in their cutting model (Table 6.7), this phenomena is not yet clarified.

5.5.2.6 Local Energy-Dissipation Intensity

Figure 5.32b exhibits a second line that characterizes how the intensity of the energy absorption is independent on the cutting depth. This function is simply a straight line, $\chi_S(\Phi)=\Phi$. Momber and Kovacevic [220] show that the difference $\xi=(\chi_S-\chi)$ characterizes the local energy-dissipation intensity

$$\xi(\Phi) = -A_2 \cdot \Phi^2 + (1-B_2) \cdot \Phi - C_2. \qquad (5.27)$$

The point of maximum energy-dissipation intensity is calculated by differentiating Eq. (5.27). For $d\xi/d\Phi=0$,

$$(\Phi)_{\xi=max} = \frac{B_2 - 1}{-2 \cdot A_2} .\qquad(5.28)$$

From the regression parameters listed in Table 5.6, a point of maximum energy-dissipation intensity exists at a depth of cut of $h=0.52 \cdot h_{max}$. At this relative depth of cut, Ohadi et al. [226] find a maximum in the cutting front temperature (section 5.8.2). This point is related to a critical striation angle

$$\alpha_{cr} = \arctan\left\{\frac{d[-a \cdot (x-b)^2 + c]}{dx}\right\} .\qquad(5.29)$$

For $h=0.52 \cdot h_{max}$, the critical striation angle varies in a very narrow range between 73° and 78° for all materials and all cutting conditions [220]. This result needs further discussion.

5.6 Erosion-Debris Generation and Acceleration

5.6.1 Properties of Generated Erosion Debris

5.6.1.1 Structure, Size and Shape of Erosion Debris

Louis et al. [227], Momber et al. [228] and Ohlsen [97] carry out experimental studies into the generation of erosion debris due to the abrasive water jet-erosion. Based on sieve analysis of debris samples, these authors draw several conclusions related to the erosion-debris generation process. Figure 5.33a gives a typical erosion-debris size-distribution diagram. The average diameter of the debris collection, d_{DSt}, is estimated by Eq. (2.10). Louis et al. [227] find that the amount of removed material debris is between 1-mass-% and 5-mass-% of the abrasive material involved in cutting. Ohlsen [97] reports values of about 3.5-mass-%. The size of the debris is a little bit smaller than the size of the impacting abrasive particles, sometimes even larger. For chip-like debris, Ohlsen finds that the chip width is about 50 % of the abrasive particle diameter. Typical average debris diameters are $d_{DSt}=42$ µm for steel and $d_{DSt}=62$ µm for aluminum. The higher value for aluminum suggests a more efficient material-removal performance due to a higher amount of micro-cutting.

Figure 5.33b shows the influence of the cut quality on the debris size. High-quality cutting generates smaller erosion debris than rough cutting. This effect appears especially if large abrasive particles are used. Nevertheless, a workpiece thickness variation of 800 % leads to a variation in the average debris diameter of only about 20 %.

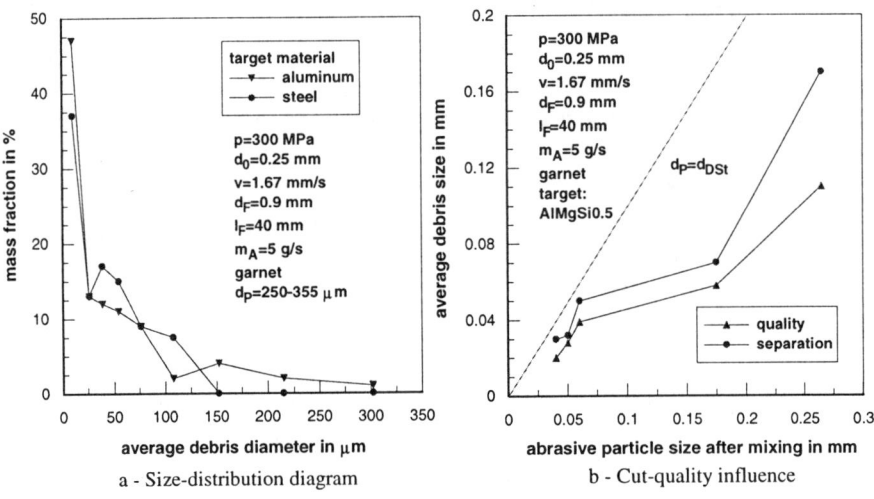

a - Size-distribution diagram   b - Cut-quality influence

Figure 5.33 *Aspects of the erosion-debris size [97]*

Figure 5.34 plots debris diameters estimated by Momber et al. [228] against the pump pressure. The diameters are in a narrow range between $d_{DSt}=60$ μm and $d_{DSt}=70$ μm, that suggest that the general removal mechanism does not depend significantly neither on pump pressure nor on depth of cut. Nevertheless, the average debris size drops with an increase in the pump pressure. SEM-photographs support these results from the sieve analysis [228].

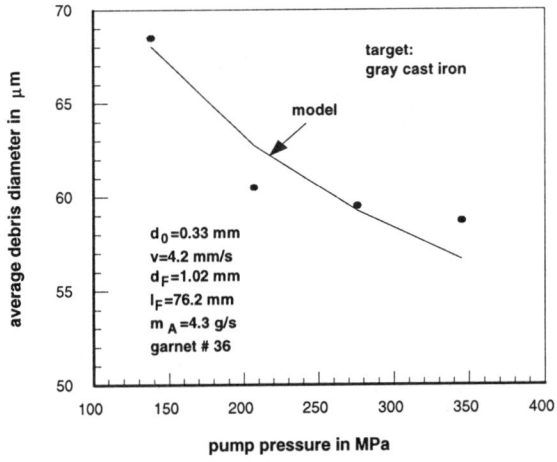

Figure 5.34 *Relation between the pump pressure and average erosion-debris diameter [228]*

Momber et al. [228] combine a model for dynamic fragmentation of brittle solids [101] with elements of contact mechanics [229] and derive a semi-empirical relation between the average debris diameter and pump pressure,

$$d_{DSt}(p) = a_1 \cdot p^{-\frac{1}{5}} \qquad (5.30)$$

The constant a_1 includes material parameters from the specimens as well as from the abrasives. Figure 5.34 shows that the results of Eq. (5.30) are in good agreement with experimentally-estimated values.

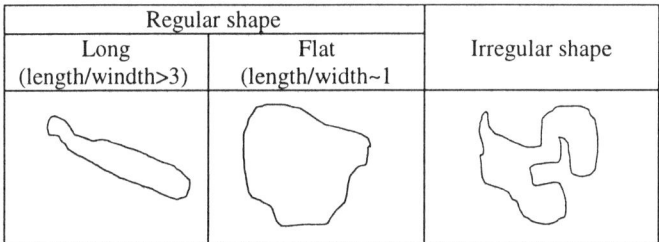

Figure 5.35 *Systematization of the erosion-debris shape [227]*

Louis et al. [227] and Ohlsen [97] perform a preliminary classification of debris shapes (Figure 5.35). They find for steel as well as for aluminum that about 40 % to 50 % of all debris show regular shapes, among them 15 % to 30 % a long-structure. The amount of long-structure debris increases as the abrasive-particle diameter increases. This result indicates a more pronounced micro-cutting mode for the larger abrasives.

5.6.1.2 Contact-Number Estimation

Based on the average debris-particle diameter and on known process parameters, Momber et al. [199] introduce a contact-number that gives the ratio between the number of removed target-material debris to the number of abrasive particles involved in the material-removal process

$$C_N = \frac{N_M}{N_P}. \qquad (5.31)$$

Figure 5.15 shows results of Eq. (5.31). The range of the estimated contact-numbers is between $C_N=3$ and $C_N=11$, which is surprisingly high and exceeds values reported from dry solid-particle erosion experiments (that are between $C_N=0.1$ and $C_N=1$). Momber et al. explain this phenomenon by the contribution of the high-speed water flow to the material-removal process at high pump pressures (section 5.3). The estimated contact-numbers as well as the assumed water-flow action are restricted to the material (brittle gray cast iron) and the process conditions applied by these authors. Nevertheless, Figure 6.10, published by Zeng and Kim [230] to illustrate their abrasive water jet cutting-model for brittle materials (section 6.2), suggests contact-number higher than $C_N = 1$ at least for this material group.

5.6.1.3 Erosion-Debris Size Distribution Function

Momber et al. [228] find that a Rosin-Rammler-Sperling (RRSB) grain-size distribution according to Eq. (2.9b) describes the distribution of brittle debris separated by an abrasive water jet (section 2.3). Figure 5.36a shows an example. Interestingly, the experimental results obtained from aluminum and steel (taken from [97]) do not fit into a straight line. This result indicates a different material-removal mode for these conventionally ductile-behaving materials. The values for the reference diameter for the cast iron are between $d^*=99$ μm and $d^*=110$ μm and decrease with an increase in the pump pressure (Figure 5.37b). These results agree with Figure 5.34.

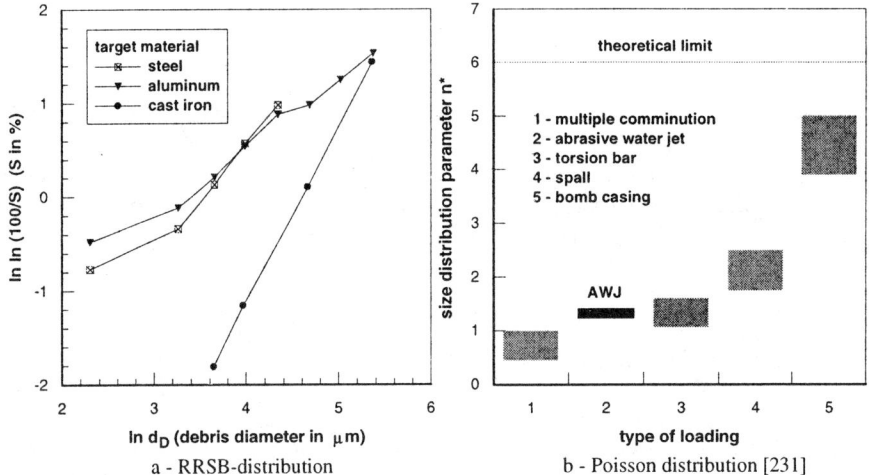

Figure 5.36 *Grain-size distributions for erosion debris*

The value for the distribution modulus of a grain-size distribution is infinite if the grain sample consists of grains with identical diameters (section 2.3). Related to the abrasive water jet material-removal process, this result is valid in an idealized, homogeneous removal process. Therefore, the distribution modulus characterizes the machining regime. Figure 5.37a shows that the values of n depend on the pump pressure. The values lie between $n=1.9$ and $n=2.6$, and exhibit a minimum at a pump pressure of about $p=200$ MPa. These values are slightly higher than the regularity numbers for water jet cutting [232]. This result indicates that abrasive water jet erosion is a more controlled and localized process. The narrow range of the values of n suggests that the general material-removal mechanism does not depend mainly on the pump pressure. Nevertheless, an increase in the distribution modulus at very high pressures is an indicator that the influence of material instabilities, such as graphite flakes in the investigated gray cast iron, reduces at higher abrasive water jet energy. Recently, Mohan et al. [222] verify this aspect by acoustic-emission measurements. Momber and Kovacevic [195] make a similar observation for concrete cut by a water jet. These authors suggest, that the material-removal process in a multiphase material becomes regular if the jet kinetic energy of the jet exceeds the maximum possible local material resistance-energy.

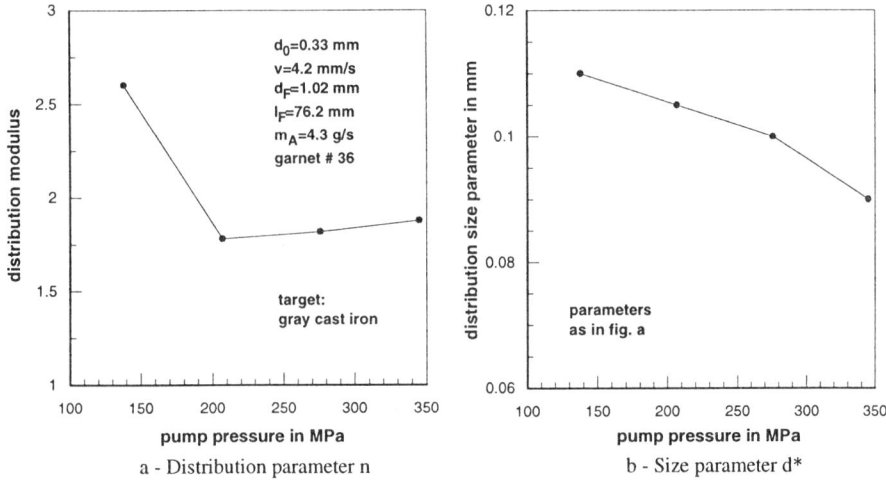

Figure 5.37 *Debris-size distribution parameters [228]*

Figure 5.36b shows a size-distribution parameter n* for several types of loading. This parameter is calculated according to Grady and Kipp [231] based on grain-size measurements from Momber et al. [228]. For abrasive water jet cutting, this parameter is between 1.3<n*<1.4. These values are in the range of values estimated for multiple comminution and torsion-bar failure.

5.6.2 Efficiency of Erosion-Debris Generation

5.6.2.1 Surface-Based Efficiency Estimation-Model

From the results of the erosion-debris sieve analysis, Momber et al. [228] calculate the surfaces of the erosion debris samples,

$$S_M = f(d^*, n) \cdot V_M .\tag{5.32}$$

Figure 5.38 plots the surface values against the pump pressure. The surface increases with a rise in the pump pressure. This result explains due to the smaller average particle-diameter and the higher fineness of the debris samples, and the larger number of removed grains for higher pump pressures. Interestingly, the function drops at a pump pressure of about p=300 MPa.

A certain amount of abrasive water jet-energy contributes to the generation of a certain number of erosion debris

$$E_D = \chi_1 \cdot E_{Diss}. \tag{5.33}$$

With a known surface of the erosion-debris sample is, this energy approximates

$$E_D = 2 \cdot \Gamma_M \cdot S_M. \tag{5.34}$$

Therefore,

$$\chi_1 = \frac{2 \cdot \Gamma_M \cdot S_M}{E_A - E_{thr}}. \tag{5.35}$$

Momber et al. [228] apply Eq. (5.35) to calculate the relation between the pump pressure and χ_1. The estimated values are between χ_1=0.017 and χ_1=0.024 for gray cast iron. The results indicate that about 2.0 % of the abrasive water jet input-energy dissipate during the generation of wear-particle surfaces. These dissipation values are in the range of efficiency values for the impact comminution [43] and single particle-abrasion [233]. As Figure 5.39a shows, the dissipation parameter exhibits a maximum at a pump pressure of about p=180 MPa. This result agrees with Eq. (7.6).

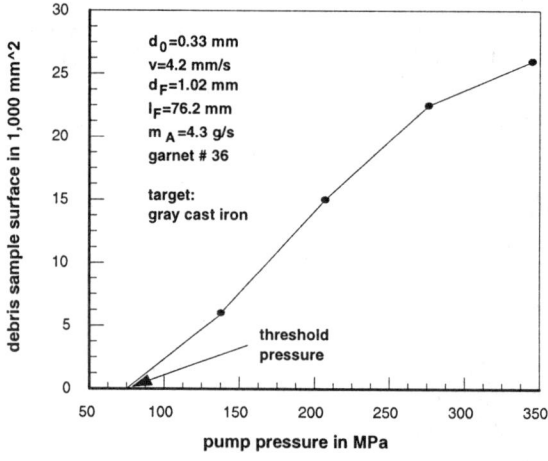

Figure 5.38 *Pump-pressure influence on the erosion-debris surface [228]*

5.6.2.2 Fracture-Based Efficiency Estimation-Model

An alternative experimental procedure to solve Eq. (5.30) bases on the idea that the energy dissipated during the removal of a certain material volume is identical to the dynamic energy, E_{AbsD}, that is absorbed during the compression test of the material (section 6.2). The energy E_{AbsD} is the area under the stress-strain curve of the given material under compression, multiplied by a dynamic parameter (Figure 5.53c). In the engineering literature, a similar parameter is known as the strain energy, that is the area under the tensile stress-strain curve. The strain energy is related to solid-particle erosion processes [234, 235]. Thus,

$$E_D = E_{AbsD} \cdot V_M .\tag{5.36}$$

A combination with Eq. (5.33) gives

$$\chi_1 = \frac{E_{AbsD} \cdot V_M}{E_A - E_{thr}} .\tag{5.37}$$

The stress-strain curves of engineering materials even under compression are complex. Abrasive water jet-experiments on rocks show, that a simplification of the curves yields unsatisfactory results [236]. Momber and Kovacevic [237] perform a study on concrete samples. The absorbed compression fracture energies of these materials, E_{AbsD}, is estimated by integrating the stress-strain curves up to the ultimate strains (Figure 5.53c). The parameter χ_1 is calculated by Eq. (5.37). Figure 5.39b shows some results. The estimated values are between $\chi_1=0.0005$ and $\chi_1=0.0045$, and depend on the material properties as well as on process parameters. These values are one order of magnitude lower than the values calculated with Eq. (5.31). This difference is attributed to the action of the high-speed water flow that contributes to the material removal and is not considered when the absorbed fracture energy is estimated.

5.6.2.3 Parameter Influence on the Efficiency

Figure 5.39b shows that the dissipation parameter χ_1 almost linearly increases as the material's compressive-strength increases. This result occurs especially for high-energy jets (high pressure, low traverse rate). For low-energy jets, the progress starts to drop at a certain strength value. To further illuminate the material influence on the efficiency of the erosion-debris generation, the estimated dissipation parameters are compared to the machinability-numbers of the investigated materials calculated from Eq. (5.55). A significant reduction of the abrasive water-jet energy involved in the erosion-debris formation is noticed for high machinability-numbers. The more resistant the machined material, the greater percent of the available jet energy is involved in the erosion-debris generation. Thus, from the point of view of the material-removal efficiency, abrasive water jet is more efficient in machine high-resistant materials. It is further noticed that χ_1 decreases with an increase in the pump pressure and a decrease in the traverse rate. Obviously, higher abrasive water jet-energy leads to worse erosion-debris formation efficiency. This trend agrees with the results in Figure 5.39a at least for very high pump pressures.

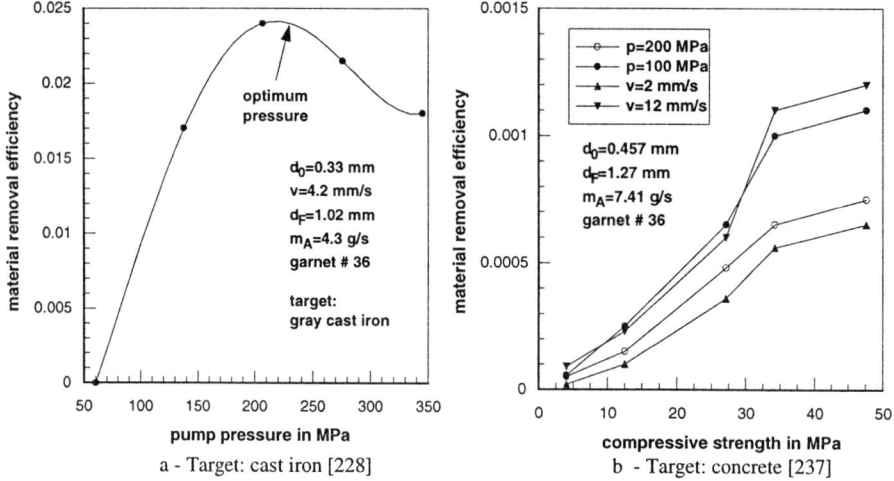

a - Target: cast iron [228]   b - Target: concrete [237]

Figure 5.39 *Efficiency of the erosion-debris generation*

5.6.3 Erosion-Debris Acceleration

After they are separated, the erosion debris undergo an acceleration by the high-speed abrasive-water slurry. This aspect is investigated by Momber [238]. The kinetic energy of the removed wear particles as they leave the cutting site is equal to a certain portion of the kinetic energy of the high-speed slurry

$$E_d = \xi \cdot E_{EX}. \qquad (5.38)$$

Also, the removed wear particles leave the workpiece with the same velocity as the abrasive-water mixture. Thus,

$$E_d = \tfrac{1}{2} \cdot m_M \cdot v_{EX}^2. \qquad (5.39)$$

For straight kerfs, the mass of the removed wear particles is

$$m_M = h \cdot b \cdot L_h \cdot \rho_M. \qquad (5.40)$$

To solve Eq. (5.37), the velocity of the removed wear particles, $v_M = v_{EX}$, is estimated. Physically, the velocity of a moving slurry is related to its impulse flow, that is given by Eq. (3.3). Since the mass of the removed wear particles is very small compared to the mass of the abrasive-water slurry (typical values are between 1 % and 5 % [227]).

$$\Omega_F = \frac{v_{EX}}{v_P} = \frac{F_{EX}}{F_A}.$$ (5.41)

Therefore,

$$E_d = \tfrac{1}{2} \cdot L_h \cdot b \cdot h \cdot \rho_M \cdot (\Omega_F \cdot v_P)^2.$$ (5.42)

Momber [238] shows that the impact force of the abrasive water jet significantly drops after the jet leaves the material. This result is due to the energy loss of the jet during the material-erosion process. Additionally, the dynamics of the force signal increases during cutting. This increase is a good indicator for the unsteady material-removal process in the abrasive water-jet cutting.

Figure 5.40a shows the relation between the workpiece thickness and velocity of the removed wear particles. The velocity decreases as the specimen thickness increases. This decrease is first due to the energy loss of the abrasive water jet during the penetration process. As the penetration depth increases, the amount of dissipated jet energy increases. Secondly, as the workpiece thickness increases, the mass of the wear particles that are accelerated by a given abrasive water-jet kinetic energy also increases. Figure 5.40b illustrates the influence of the workpiece thickness on the kinetic energy of the removed wear particles. The kinetic energy of the removed wear particles increases almost linearly as the specimen thickness increases.

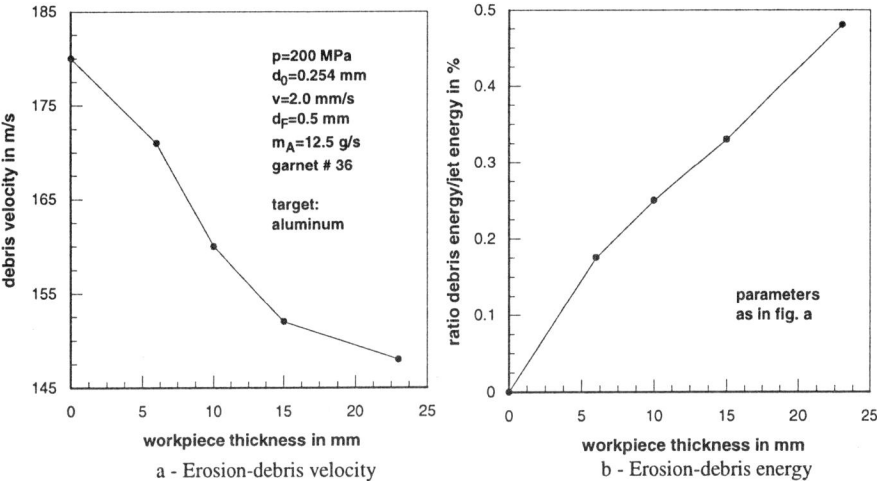

Figure 5.40 *Influence of the workpiece-thickness on the erosion-debris acceleration [238]*

This result indicates a much more significant influence of the mass of the removed wear particles compared to the influence of the reduction in the jet velocity. The higher mass of removed wear particles for thicker specimens is responsible for the increase in the kinetic energy of the particles with an increase in the workpiece thickness. The portion of abrasive water jet-energy involved in the wear particle acceleration process is [238]

$$\chi_2 = \frac{E_d}{E_A} = \frac{2 \cdot E_d}{(\dot{m}_A + \dot{m}_W) \cdot v_P^2 \cdot t}. \tag{5.43}$$

Figure 5.41b shows that the energy dissipated by the wear-particle acceleration is about 0.3 % of the abrasive water-jet energy before the jet cuts the material.

5.7 Damping Effects in Abrasive Water-Jet Material Removal

5.7.1 Damping During Single Particle-Impact

5.7.1.1 Observations in Solid-Particle Erosion

Part of the kinetic energy of an impacting abrasive particle dissipates by damping from layers of water, abrasive fragments, and fragments of the removed material.

This aspect usually does not play a role in dry, solid-particle erosion, Nevertheless, Zu et al. [239] show for low impact velocities ($v_P \cong 5$ m/s) that the presence of water improves the solid-particle erosion process because the water flow removes embedded abrasive fragments as well as removed target-material particles. On the other hand, Clark and Burmeister [240] find that the pressure required to exude a fluid from between an impinging solid particle and the target, is of sufficient magnitude to reduce the impact velocity of the abrasives. For glass and quartz particles, respectively, that are suspended in water, these authors estimate a loss in the kinetic energy due to a water-film damping of about 33 %. A water film also significantly influences the rebound behavior of impacting particles. Nevertheless, both references cited above only study the influence of a plain fluid-film on the impact characteristics, and do not consider the possible influence of layers of broken abrasive fragments and target-material debris.

5.7.1.2 Damping of Free-Falling Objects

Yong and Kovacevic [241] and Kwak [242] perform investigations into the influence of water-solid films on the impact behavior of free-falling objects. Based on an energy balance on the impact site, Yong and Kovacevic derive

5.7 Damping Effects in Abrasive Water-Jet Material Removal

$$F_P(l, h_F) = K_0^{\frac{2}{5+a_3}} \cdot \left[\frac{5+a_3}{2} \cdot \left(M - a_1 \cdot \sqrt{\frac{a_3 \cdot l}{h_F^{a_2}}} \right) \cdot h_F \cdot g \right]^{\frac{3+a_3}{5+a_3}} \tag{5.44}$$

In this equation, l is the film thickness. For uncovered surfaces, $a_3=0$. The accuracy of Eq. (5.44) decreases if notable plastic deformations occur on the impact site. This fact is a serious limitation because plastic deformation is a major feature of solid-particle erosion at high impact velocities even in very brittle materials. The unknown parameters a_1 to a_3 in Eq. (5.44) are approximated from the experimental results. Table 5.7 gives some values for different film mixtures. The force ratio between dry and damped impact is

$$R_F = \frac{F_P(l=0, a_3=0)}{F_P(l, h_F)} = \frac{K_0^{\frac{2}{5}}(0) \cdot \left[\frac{5}{2} \cdot M \cdot h_F \cdot g \right]^{\frac{3}{5}}}{K_0^{\frac{2}{5+a_3}} \cdot \left[\frac{5+a_3}{2} \cdot \left(M - a_1 \cdot \sqrt{\frac{a_3 \cdot l}{h_F^{a_2}}} \right) \cdot h_F \cdot g \right]^{\frac{3+a_3}{5+a_3}}}. \tag{5.45}$$

The larger the parameter R_F, the more serious the film effects become. Figure 5.41a plots the force ratio for the two different film mixtures that are given in Table 5.7, versus the abrasive velocity (calculated from Eq. (5.46)). Notice that the effect of water-mixture film varies greatly with the particle impact-velocity. However, the damping rapidly reaches a stable value and becomes less noticeable. Also, the damping is more serious in the case of a water/abrasive-film.

Table 5.7 *Experimentally estimated damping parameters [241]*

Film materials	Film thickness (l)	a_1	a_2	a_3
steel powder and water	356.5 μm	0.96	0.22	0.08
garnet and water		1.25	0.16	0.07

Kwak [242] carries out damping experiments with impact bodies of different shapes. Figure 5.41b shows some experimental results of these impact studies. The figure exhibits two remarkable features. First, the influence of the damping effect significantly reduces at high impact velocities. The functions show a saturation behavior similar to that shown in Figure 5.41a. Secondly, the intensity of film damping depends on the shape of the impacting particles.

5.7.1.3 Critical Particle Velocities for Damping

The transition point between the unstable and the stable damping range (Figure 5.41a) is used to calculate a critical abrasive-particle velocity that is required to overcome serious damping effects. This critical velocity is [241]

$$v_{PCr} = \sqrt{\frac{M}{m_P} \cdot 2 \cdot g \cdot h_{FCr}} \ . \tag{5.46}$$

This purely energetic balance does not consider the influence of the abrasive-particle size, the abrasive-particle shape, and the abrasive-particle velocity on the certain impact regime and material-removal process. Smaller and sharper particles bring plastic deformations in front. Neglecting this effect, for a spherical garnet abrasive particle with the size mesh # 60, the critical velocity is about v_{PCr}=370 m/s. Beyond this abrasive-particle velocity, damping effects on the material-removal mechanism become stable. In Figure 5.41b, a transition impact velocity is noticed, too. This velocity is $v_{PCr}\cong$350 m/s. Beyond this limit, conical particles are more sensitive to damping effects. A discussion of these effects has not happened.

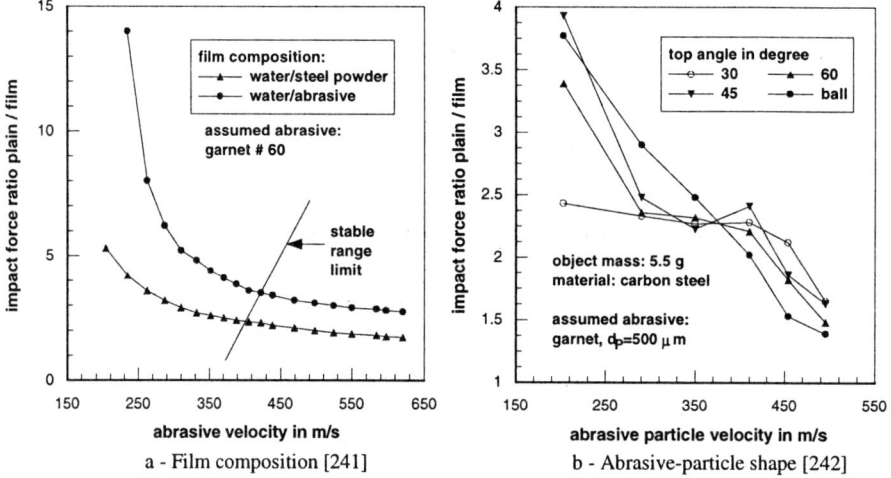

Figure 5.41 *Influences on the damping-force ratio*

The single-particle impact experiments discussed above enable the estimation of the damping parameter, χ_3, at a given depth of cut that is characterized by a certain film composition and thickness. Globally, the film thickness increases with an increase in the depth of cut. However, the complete damping function, $\chi_3(h)$, cannot be evaluated by these experiments.

5.7.2 Damping During Abrasive Water-Jet Penetration

5.7.2.1 Concept of Force Measurement for Damping Estimation

In the case of cutting-through (characterized here by the symbol 'T') of a specimen by an abrasive water jet, damping effects due to a water-solid film are neglected because the abrasive water jet can leave the workpiece at the bottom exit. The ratio between the energy of the exiting suspension and the energy of the incoming abrasive water jet is the summary of the energies dissipated due to wall friction, heating, and erosion-debris generation and acceleration. Therefore,

$$\chi_1(h) + \chi_2(h) + \chi_4(h) + \chi_5(h) = \frac{E_{EX,T}(h)}{E_A} = \chi(h) - \chi_3(h). \tag{5.47}$$

Thus,

$$\chi_3(h) = \chi(h) - \frac{E_{EX,T}(h)}{E_A}. \tag{5.48}$$

Consider that the abrasive-mass flow rate is constant during the cutting process (the mass flow of the removed material is neglected). In this case, the ratio between the exit energy and the input energy is the ratio of the square velocities and the ratio of the squares of the measured impact forces, respectively. Thus,

$$\chi_3(h) = \chi(h) - \left[\frac{F_{EX,T}(h)}{F_A}\right]^2. \tag{5.49}$$

5.7.2.2 Results of Force Measurements

Momber et al. [237] and Momber and Kovacevic [243] perform experiments to solve Eq. (5.49). Figure 10.40 shows a typical plot of a measured impact force generated by an abrasive water jet during the cutting-through process. The maximum possible depth of cut is estimated by an additional kerfing test under the given process conditions. Figure 5.42 exhibits some results of the experiments. The results approximate a second-order polynomial,

$$\chi_3(h) = (A_1 - A_3) \cdot h^2 + (B_1 - B_3) \cdot h + (C_1 - C_3). \tag{5.50}$$

The value for $\chi_3(h=h_{max})$ is obtained by using the regression parameters A_3 to C_3.

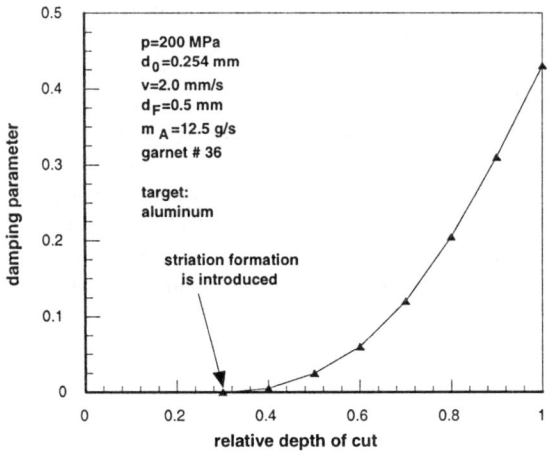

Figure 5.42 *Damping effects in abrasive water jet-cutting [237]*

5.7.2.3 Efficiency Losses due to Damping

Because $\chi(h)$ delivers $\chi=1$ for $h=h_{max}$ (Figure 5.32), about 50% of the dissipated abrasive water jet energy is involved in damping. This value is higher than that estimated by Clark and Burmeister [240] for solid-particles hitting water-covered plain surfaces. This discrepancy is probably due to the additional damping by finely-broken solid particles distributed over the erosion site. The value of 50% is also slightly higher than the estimated force ratios. This result indicates that particle-particle interaction contributes to the damping process during the abrasive water jet cutting. The damping function shows only a small non-linearity but has a progressive increase as the depth of cut increases.

5.8 Heat Generation During Abrasive Water-Jet Material Removal

5.8.1 Sources of Heat Generation

Heat generation during abrasive water-jet cutting occurs due to friction between the impinging abrasive particles and the cutting front, as well as due to plastic deformation during the material-removal process. Presently, there is a limited knowledge about the generated heat and thermal energy distribution through the workpiece during cutting with an abrasive water jet.

5.8 Heat Generation During Abrasive Water-Jet Material Removal 147

5.8.2 Results from Thermocouple Measurements

5.8.2.1 General Results

Ohadi et al. [226] and Ohadi and Whipple [244] perform preliminary experimental studies into the thermal energy distribution in a specimen cut by an abrasive water jet. In these investigations, the authors use thermocouples for measuring the temperature at several locations in the workpiece. The maximum developed temperature is $T-T_0=75°C$, where T_0 is the room temperature.

5.8.2.2 Process-Parameter Influence

Figure 5.43a shows that the temperature, especially directly on the cutting front, remarkably depends on the pump pressure. The higher the pump pressure, the higher the specimen temperature. This result is probably due to the higher abrasive particle impact-velocities that increase frictional forces and the amount of plastic deformation in the ductile material. Nevertheless, at locations far from the cutting front, the pump pressure influence is not that significant.

Ohadi and Cheng [245] investigate the influence of the traverse rate on the heat development. Figure 5.44 illustrates that a lower traverse rate results in a higher heat input. This increase is attributed to the larger number of abrasive particle-cut surface contacts [see Eq. (2.13)] per unit traverse length at a lower traverse rate. A second striking feature of Figure 5.44 is that at the immediate vicinity of the abrasive water jet, the temperature contours approach the contour of the heat wave for a stationary jet that is a family of circles. With an increase in the distance from the abrasive water jet, the temperature contours approach a family of ellipses.

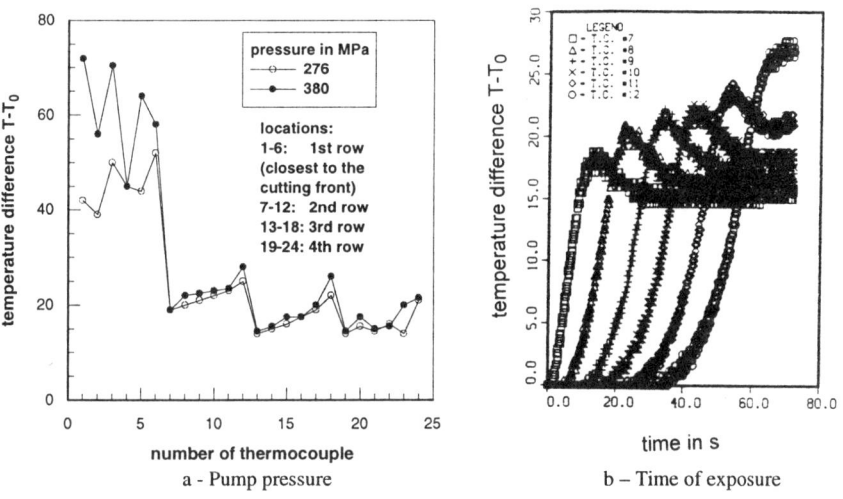

Figure 5.43 *Parameter influence on the workpiece temperature [226]*

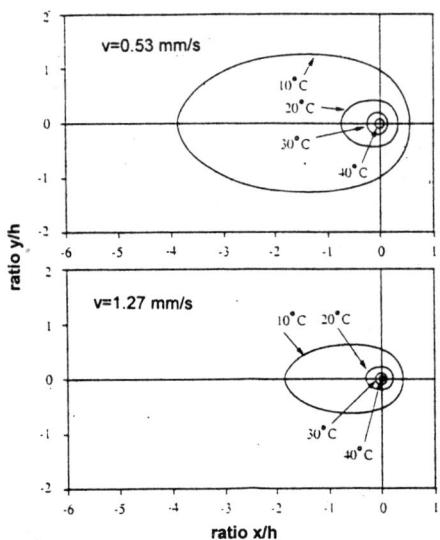

Figure 5.44 *Influence of the traverse rate on the temperature distribution [245]*

5.8.2.3 Local Temperature Distribution

From the saw-tooth structure of the curve, the conduction of the heat along the cutting path causes a gradual rise in temperature. Figure 5.43b plots an exemplary local temperature-distribution in respect to the time. A review of these graphs indicates again the different behavior of points in the immediate neighborhood and those further from the cutting interface. Not only are much higher peaks in the temperature observed (as also shown in Figure 5.43a) for those close to the interface, but also the way in which the temporal temperature variations take place is different from those further away from the interface. For locations very close to the cutting interface as the abrasive water jet approaches the individual thermocouples, a spontaneous response by the sharp rise in the temperature to the peak value is observed. The peak point represents the time instant when the abrasive water jet is in the immediate vicinity of the thermocouple, and the subsequent drop-off corresponds to the time when the jet has passed the individual thermocouple site. Maximum temperature occurs at a depth of $h=0.5 \cdot h_{max}$. Although this result is explained by the reduction of the cooling effects, attention is drawn to section 5.5.2.6 and to the work of Momber and Kovacevic [220] who find a maximum in the energy-dissipation intensity function at exactly the same relative depth of cut.

5.8.3 Results from Infrared-Thermography Measurements

5.8.3.1 General Remarks

Using infrared-thermography, Kovacevic et al. [208] and Mohan et al. [246] conduct an alternative systematic study in the temperature generation in workpieces cut by an abrasive water jet. The results of the infrared measurements are presented by two methods, namely isotherms and linescans. Isotherms are the loci of points of equivalent temperatures; whereas, linescans are thermal profiles along selected lines. Several process parameters, such as pump pressure, traverse rate, and the distance between the coated and cut surface, are varied in order to analyze their influence on the generated thermal energy.

5.8.3.2 Process-Parameter Influence on Linescans

Figure 5.45a plots the linescans along the middle of the coated surface with a change in the pump pressure. These linescans indicate that with an increase in the pump pressure, the workpiece temperature increases. This result is in agreement with Figure 5.43a and is explained by similar arguments as for this figure. Figure 5.46a shows the variation of the estimated heat flux with change in the pump pressure. From this figure, with an increase in the pump pressure, the heat flux at the kerf increases linearly and causes the workpiece temperature to increase.

Figure 5.45b shows the linescans for different traverse rates. With the increase in the traverse rate, there is a marginal increase in peak temperatures. This result is consistent with Figure 5.44. The heat-flux graph in Figure 5.46b reflects the same trend.

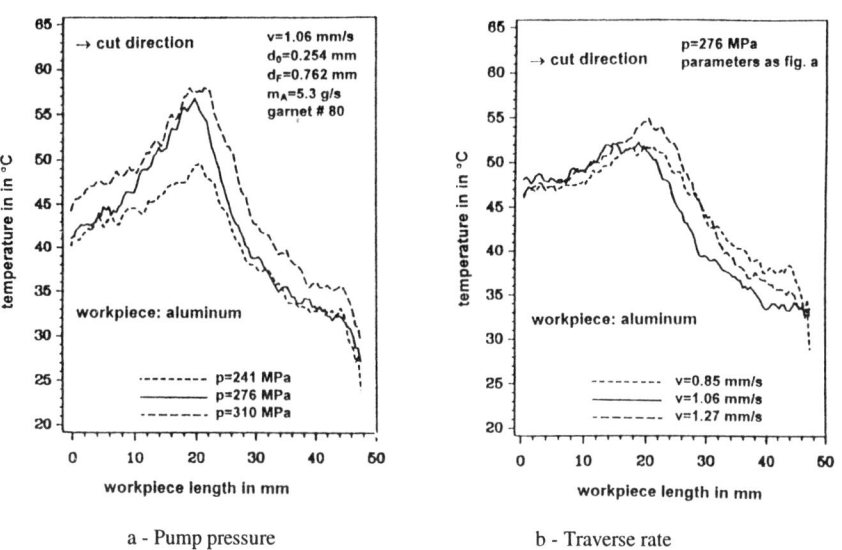

a - Pump pressure   b - Traverse rate

Figure 5.45 *Infrared-thermography linescans for different cutting conditions [208]*

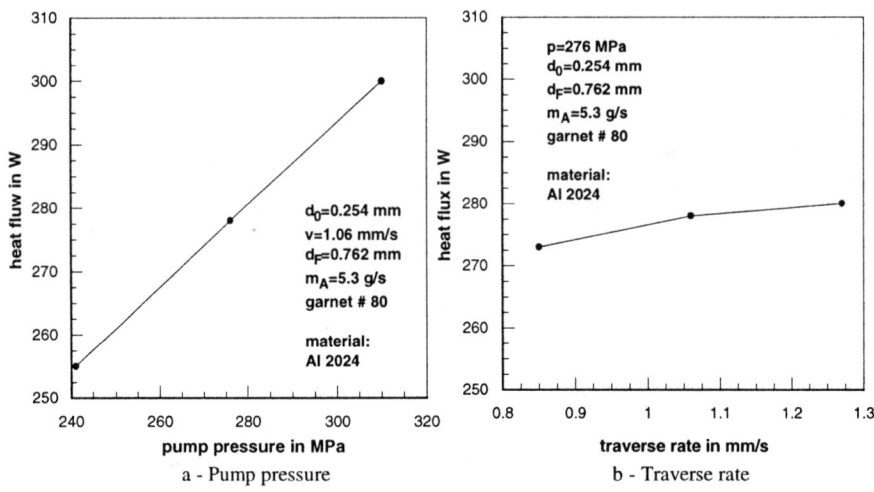

Figure 5.46 *Parameter influence on the heat flux [208]*

Linescans along the middle of the coated surface which are at different distances from the cut kerf verify that higher temperatures are present on the planes closer to the cutting zone (Figure 5.43).

5.8.3.3 Material Isotherms

A change in the thermal conductivity of the material yields a different temperature-distribution pattern in the workpiece. Hence, Kovacevic et al. [208] use workpieces of similar heat capacity and contrasting thermal conductivity (an aluminum alloy as a good conductor of heat and a titanium alloy as a poor conductor of heat) to investigate this problem. Figure 5.47a shows the isotherms for the aluminum, and Figure 5.47b shows the isotherms for the titanium when the abrasive water jet has traversed approximately half the length of cut. In case of titanium, two peak temperatures are present, one in the upper half and one in the lower half. On the contrary in aluminum, only one peak temperature is observed.

5.8.4 Comparison Between Thermocouples and Infrared-Thermography

To compare different heat-measurement methods, Kovacevic et al. [208] measure temperatures at discrete points using thermocouples, and compare the results with infrared thermography results. For comparison, Figure 5.48 plots the time-temperature graphs from the infrared-thermography images and the thermocouples. The time-temperature graphs obtained from infrared images and the thermocouple match very closely with each other.

5.8 Heat Generation During Abrasive Water-Jet Material Removal   151

The overall workpiece temperature increases as time progresses, indicated by the slightly higher peak temperature towards the end of the cut. The temperature plots in Figure 5.48 show that the lower thermocouple (as well as the corresponding pixel) reaches its peak temperature slightly later than the upper thermocouple (corresponding pixel). This occurrence is due to the deflection of the abrasive water jet away from the direction of cut. Also, the peak temperature of the bottom thermocouple is marginally higher than the top one. This trend is due to the higher frictional forces in the lower cutting zone.

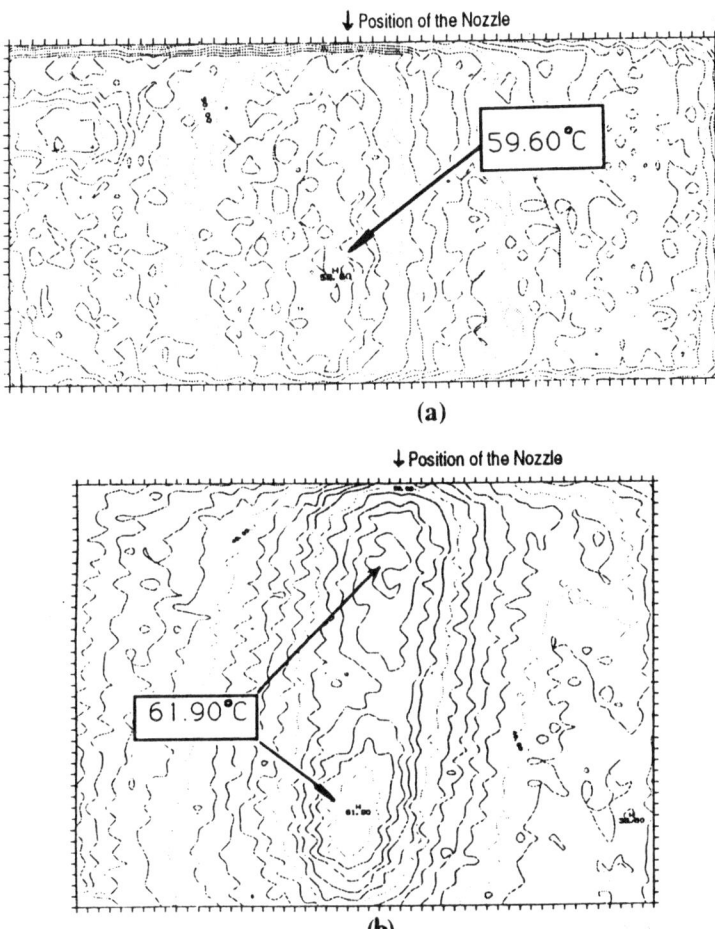

Figure 5.47 *Isotherms acquired during the abrasive wate- jet cutting [208]*
a - Aluminum, b - Titanium

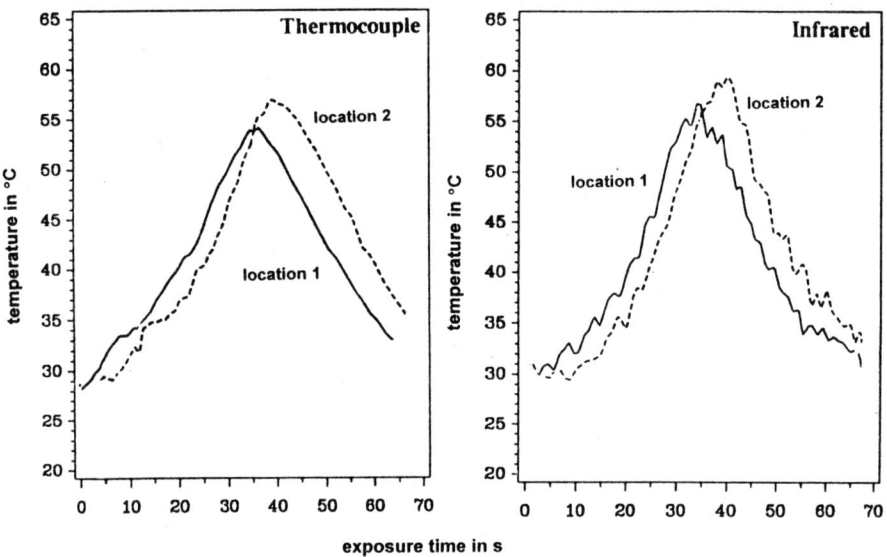

Figure 5.48 *Comparison between results obtained with thermocouple and infrared-thermography on titanium [208]*

5.8.5 Modeling of the Heat-Generation Process

5.8.5.1 Basic Equations

Ohadi and Cheng [245] and later Kovacevic et al. [208] develop a theoretical analysis for the workpiece temperature-distribution. From solutions of the direct heat-conduction problem they obtain

$$T - T_0 = \frac{q_H}{2 \cdot \pi \cdot K \cdot g} \cdot e^{-\eta_H \cdot \lambda_H \cdot v_H \cdot \psi_H}, \quad \eta_H = 2, \quad (5.51)$$

The heat flux, $\hat{q}_H$, is derived from solutions of the inverse heat conduction problem

5.8 Heat Generation During Abrasive Water-Jet Material Removal

$$\hat{q}_H = \tilde{q}_H + \frac{\sum_{i=1}^{m}\sum_{j=1}^{n}[T_m(i,j) - (T_0 + \hat{q}_H \cdot G(i,j)] \cdot G(i,j)}{\sum_{i=1}^{m}\sum_{j=1}^{n} G(i,j)^2}. \tag{5.52}$$

The complete deviations are given in the original papers.

5.8.5.2 Results of the Modeling

Ohadi and Cheng's model delivers satisfactory agreements with the experiments for the locations close to the interface. With an increase in the distance from the cutting interface, the model accuracy drops. Also, the model does not hold for the entry and the exit zone of the abrasive water jet in the workpiece.

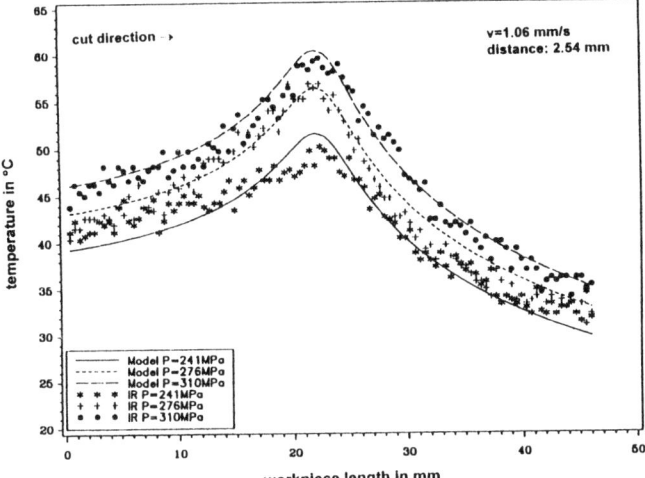

Figure 5.49 *Comparison between calculated workpiece temperatures and experimental results [208]*

Figure 5.49 plots temperature distributions calculated by Kovacevic et al. for various pump pressures. The model and the experimental data have a very close correlation with an RMS-error <1.5°C. This accuracy indicates that Eq. (5.51) is fairly accurate. The model and the infrared measurements results match very closely as the heat source approaches the point of observation and after it leaves. The predicted temperature of the model is higher than the infrared-thermography results when the heat source is farther behind the point of observation. This result is because the specimen has a finite length; whereas, the model is derived for the general case of a workpiece of infinite length. However, as the primary objective of a thermal analysis in abrasive water-jet cutting is to predict/measure the maximum temperature at the cutting zone so as to avoid material failure, these discrepancies can be ignored.

5.9 Target-Material Property Influence on Material Removal

5.9.1 Hardness and Modulus of Fracture

5.9.1.1 General Observations

Hunt et al. [247] perform an early investigation that relates the mechanical material properties to the abrasive water-jet cutting process. Using an abrasive water jet for piercing, these authors investigate the influence of the material's Brinell-hardness as well as the modulus of fracture on the piercing time. The modulus of fracture is (Figure 5.50)

$$M_F = f(\sigma_t, E_M, \varepsilon_{cr})$$ (5.53)

Figure 5.51 shows results of this study. The general trend in Figure 5.51a is that the piercing time increases as the material hardness increases. However, two materials do not exactly fit into the linear relationship. For ceramics, high hardness is often combined with a low modulus of fracture. Figure 5.51b shows that the relation between this property and the piercing time is approximately linear for all materials, except the ceramics.

Matsui et al. [236] conduct very similar experiments and find that the hardness (Vickers) alone is not a suitable property to describe the resistance of a ductile-behaving as well as a brittle-behaving material against abrasive water-jet cutting (Figure 5.52a).

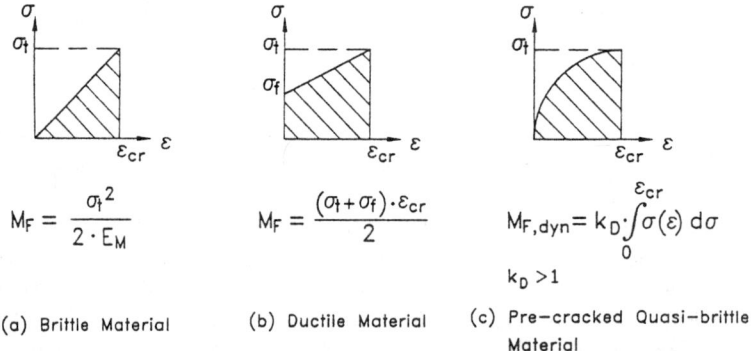

Figure 5.50 *Modulus of fracture of different types of materials*

Matsui et al. show that the modulus of fracture characterizes the behavior of brittle materials insofar as they are not pre-cracked, such as rocks. Figure 5.52b illustrates this effect. In contrast, the resistance of ductile-behaving materials is characterized by the plastic-deformation energy that is absorbed during the tensile test or, alternatively, by the product of Vickers hardness and elongation. For isotropic materials, Thikomirov et al. [56] find an almost linear decrease in the material-removal rate as the target-material hardness increases.

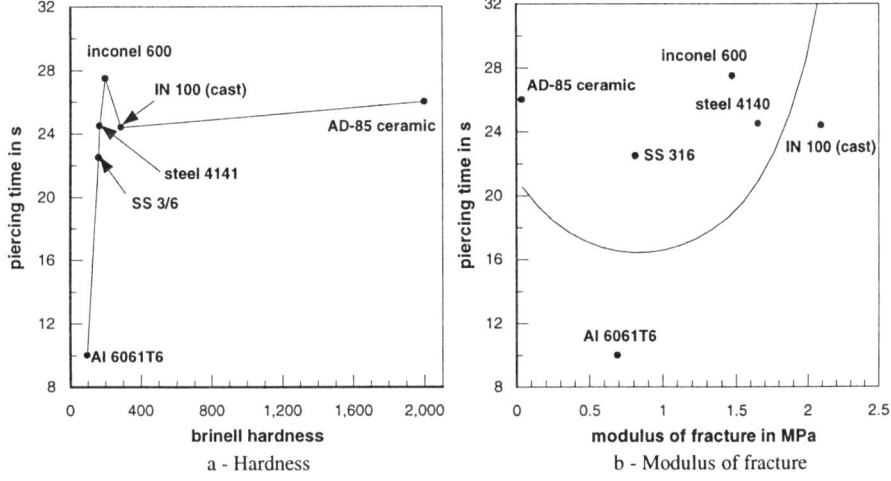

Figure 5.51 *Influence of material properties on the piercing time [247]*

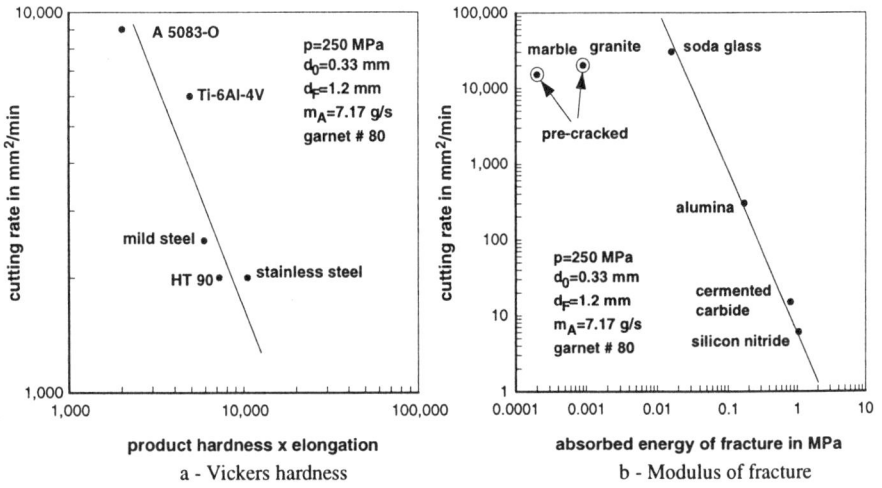

Figure 5.52 *Influence of material properties on the cutting rate [236]*

Miranda et al. [248] find during rock cutting experiments that the exit angle of an abrasive water jet increases with an increase in the micro-hardness of the rocks. Because the exit angle expresses the energy absorbed in the material during abrasive water-jet cutting [224], harder rocks dissipate more abrasive water-jet energy during the cutting process.

5.9.1.2 'Two-Stage' Resistance Approach

Hashish [249], who does not consider these early investigations suggests, based on his two-zone cutting model, a combination of mechanical properties to describe the materials resistance against the 'cutting wear' and the 'deformation wear'. According to this model (chapter 6), the entire depth of cut is

$$h = h_C + h_d. \qquad (5.54)$$

Hashish assumes that the 'cutting-wear' depth depends on the Vickers hardness of the material; whereas, the 'deformation-wear' depth is a function of the Young's modulus. Thus,

$$h = \frac{A}{\sqrt{H_V}} + \frac{B}{E_M}. \qquad (5.55)$$

These relations are well known from solid-particle erosion. For example, Eyres [250] points out that the material hardness determines the erosion resistance under shallow angles; whereas, the volume removal due to large angles depends on the Young's modulus.

5.9.2 Concepts of Material Machinability

5.9.2.1 The 'Machinability Number'

Zeng et al. [251] develop the machinability number (section 6.2) to define the material resistance in the abrasive water-jet cutting. Table 5.8 lists machinability numbers of several engineering materials. Considering only 'classical' mechanical parameters, this number is

$$N_m = A_1 \cdot \frac{D_M \cdot \sigma_f}{\gamma_M \cdot E_M} + \frac{B_1}{\sigma_f}. \qquad (5.56)$$

Because the flow stress is often proportional to the hardness for ductile-behaving materials [32] and $N_m \propto h$,

$$h = A_2 \cdot \frac{D_M \cdot H_M}{\gamma_M \cdot E_M} + \frac{B_2}{H_M}. \tag{5.57}$$

This equation has a very similar structure as Eq. (5.55). However, Eq. (5.57) is partly in contrast to experimental results obtained on rocks and concretes.

Table 5.8 *Machinability numbers of engineering materials*

Material	Machinability number	
	Absolute	Relative (%)
Alumina Ceramic AD 85	17.3	8.1
Alumina Ceramic AD 90	10.3	4.8
Alumina Ceramic AD 94	17.3	8.1
Alumina Ceramic AD 99.5	13.1	6.2
Alumina Ceramic AD 99.9	1.6	0.8
Aluminum, AL 6061-T6	213	100
Asphalt Concrete	461	216
B4C	4.2	2
Concrete (medium strength)	516	242
Concrete (high strength)	468	220
Copper	110	52
Dupont Corian	455	214
Glass	596	280
Granite	322	151
Graphite	875	411
Gray Cast Iron	121	57
Lead	490	230
Magnesia Chromite	430	202
Mortar	858	403
Nylon	538	252
Pine Wood	2,637	1,240
Plexiglas	690	324
Polypropylene	985	462
Refractory bauxite	106	50
Silica Carbide	12.6	6
Silica Ceramic Si3N4, hot pressed	1.1	0.5
Silica Ceramic SS 304	81.9	38
Silica Ceramic SS 316L	83.1	39
Sintered Magnesia	408	190
Stainless Steel 304	115	52
Steel, ASTM A34	87.6	41
Ti3B2	4.3	2
Titanium	115	54
Tool Steel 901	120	56
White Marble	535	251
Zinc-Alloy	136	64

158   5. Material-Removal Mechanisms in Abrasive Water-Jet Machining

Momber et al. [252], for example, find that the resistance of concrete against the abrasive water jet increases with an increase in the diameter of the inclusion grains (Figure 5.53a). Heβling [253] observes the same tendency during rock cutting. In this study, the rock with the largest grains is very difficult to cut (Figure 5.53b). Heβling find that the energy release rate, that depends on the surface energy, γ_M, is only weakly related to the depth of cut in brittle rocks. On the other hand, the reverse relation between material resistance and Young's modulus is confirmed (Figure 5.55).

Like all other material properties, a variation in the machinability number must be expected. For aluminum, for example, this number varies within ±10 % for 61 % of the data and within ±20 % for 90 % of the data. If a standardized set of process parameters is used, the erroris less than ±10 %. The concept of a machinability number is extended to other machining operations, such as turning and piercing (chapter 9).

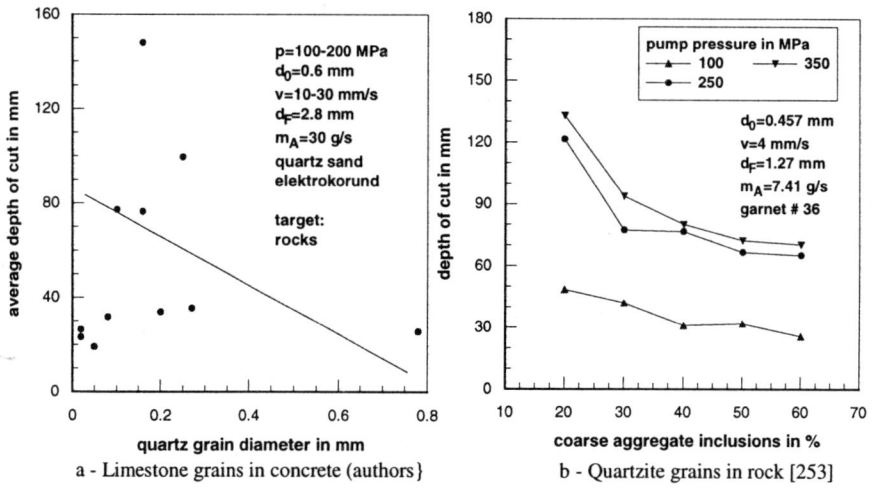

Figure 5.53 *Influence of the inclusion-grain size on the depth of cut*

5.9.2.2 Other 'Machinability' Conceptions

In their cutting model, Capello and Gropetti [225] introduce a 'machinability parameter' γ_1. Nevertheless, the authors report values for only three engineering materials (Table 6.8), and do not install relationships to conventional strength parameters. Also, a comparison of the γ_1-values for aluminum and glass show a difference of more than one order of magnitude. In Table 5.8, the difference in the corresponding machinability numbers is only about 250 %.

Gou et al. [254] develop a database that relates the specific cutting efficiency of material to a reference material. As the reference material, the authors use aluminum. Table 5.9 lists several materials. Good agreement with the order of materials obtained by using the machinability number is noticed.

5.9 Target-Material Property Influence on Material-Removal

Table 5.9 *Material-resistance listing [254]*

Material	Specific cutting efficiency in %
Alumina ceramic (Al$_2$O$_3$)	5 - 20
Aluminum	100
Concrete	310
Glass (lead crystal)	350
Ni-alloy (Inconel)	42
Sandstone	1,435
Silica ceramic (Si$_3$N$_4$)	0.1
Silica ceramic (SiSiC)	2
Steel	40
Steel, hardened	36
Titanium	55

5.9.3 Properties of Pre-Cracked Materials

5.9.3.1 Stress-Strain Behavior

Pre-cracked materials, such as rocks, concrete, and some ceramics, do not fit into the relations discussed in section 5.9.1. These materials typically show a non-linear stress-strain behavior under compression and tension (Figure 5.50c). Therefore, the fracture modulus can not be estimated by simplified methods as illustrated in Figure 5.50a. But if the energy absorbed during the stress-strain determination is properly estimated by integration [252], the modulus of fracture reasonably characterizes the behavior of pre-cracked materials (Figure 5.54a).

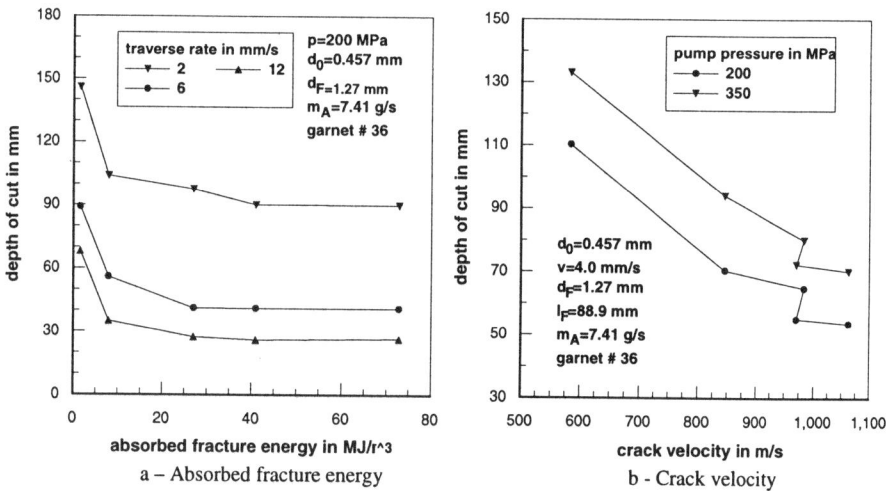

Figure 5.54 *Influence of material properties on the cutting resistance of concrete [252]*

5.9.3.2 Relations to Conventional Testing Procedures

Momber et al. [252], Heβling [253] and Momber and Kovacevic [255] carry out systematic investigations in the resistance of rocks and concrete against abrasive water-jet cutting. Using statistical methods, Momber et al. [252] find that the crack velocity of concrete shows the strongest relation to the depth of cut generated by an abrasive water jet (Figure 5.54b).

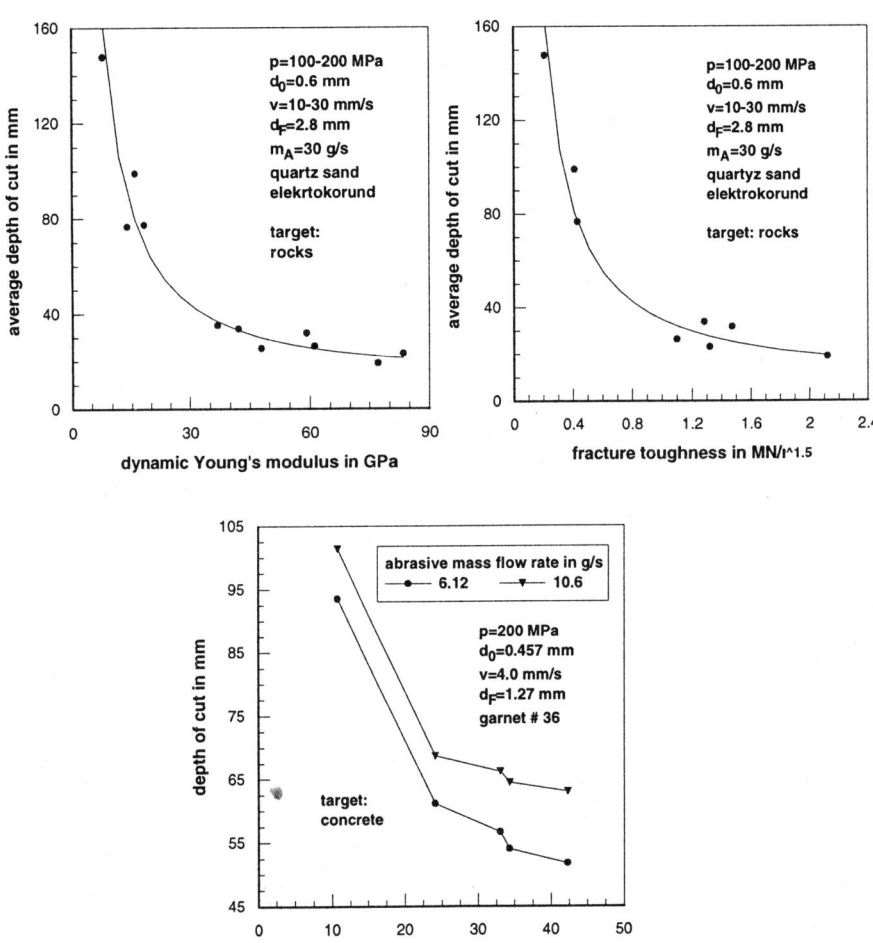

Figure 5.55 *Influence of strength parameters on the depth of cut in pre-cracked, multiphase materials [252, 253]*

5.9 Target-Material Property Influence on Material-Removal

Also, the threshold pump pressure, the specific material-removal energy, and the threshold traverse rate of a material are effectively related to the crack velocity [255].

Significant relationships also exist between the depth of cut and several other mechanical properties of rocks and concrete (Figure 5.55). Momber and Kovacevic [255] and Momber et al. 190, 191, 257] give an explanation for these surprisingly good trends. These authors suggest that the major failure-mechanism of this group of materials in compression as well as in abrasive water-jet erosion is the widening of pre-existing micro-cracks. Momber et al. [190] find that concrete samples, failed by spalling fracture during the compression test, are removed by trans-granular fracture in the abrasive water jet cutting. In contrast, materials that failes by a crumble mode in the compression test, are removed mainly by inter-granular erosion in the abrasive water-jet cutting. This specific result is independently obtained by acoustic-emission measurements on concrete samples (section 10.7).

5.9.4 Other Material Properties

5.9.4.1 Material Porosity

Zeng and Kim [258] investigate the influence of the porosity of materials on their machinability. Figure 5.56a shows two opposite effects. Whereas, the porosity has a beneficial effect on the material resistance at low-porosity values due to jet deflection, the effect is opposite at high-porosity values. Here, the large amount of pores makes the material removal easier. For comparatively highly porous materials, such as rocks, the resistance against the abrasive water jet almost linearly decreases with an increase in the porosity (Figure 5.56b).

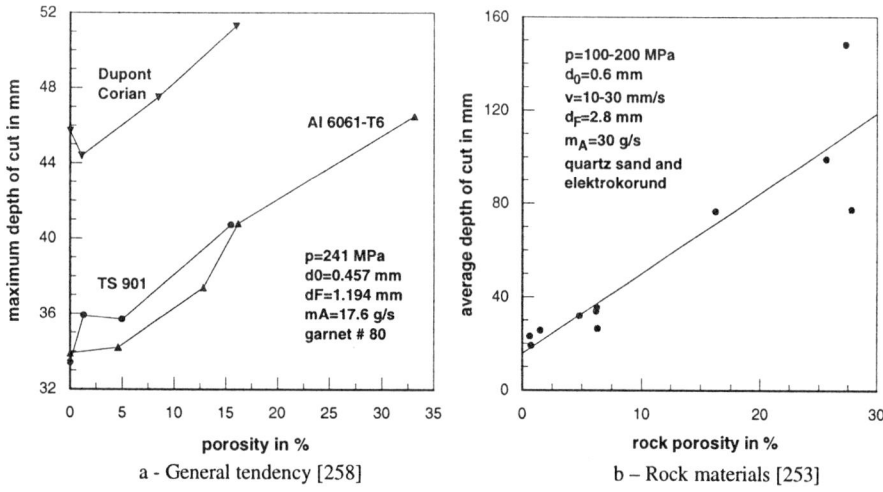

Figure 5.56 *Relation between the target-material porosity and depth of cut*

5.9.4.2 Thermal-Shock Factor

For ceramics, Kahlman et al. [193] suppose an alternative mechanism of thermal spalling. This spalling is the result of the creation of highly-localized temperatures (T≅1280°C). The main resistance parameter against this type of failure in brittle materials is a thermal-shock factor,

$$R^* = \frac{\sigma_B \cdot (1 - v_M)}{E_M \cdot \alpha_H} \cdot \lambda_H . \tag{5.58}$$

Figure 5.57 presents some results. The figure indicates a decrease in the removed volume with an increase in the thermal-shock resistance. The somewhat low cutting resistance of the aluminum whisker is explained by a reinforcement mechanisms of the whiskers that locally reduces fracture and holds the matrix together. This alternative material-resistance parameter needs some further support from additional experimental work. Eq. (5.58) shows the same Young's modulus dependence as all equations presented in this section.

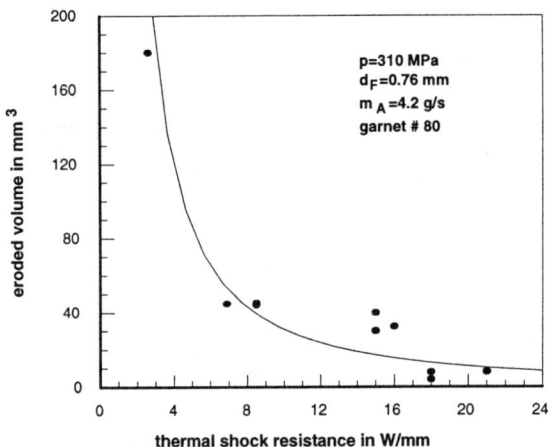

Figure 5.57 *Relation between the thermal shock resistance of ceramics and material-volume removal [193]*

6 Modeling of Abrasive Water-Jet Cutting Processes

6.1 Introduction

Over the years, several models are developed to describe the abrasive water-jet cutting process. Generally, these models are attempts to estimate the depth of cut achievable in different materials cut by abrasive water jets under certain process conditions. The reader will find other models, such as for material-removal processes and for abrasive water jet turning, milling, and drilling, in chapter 9. Chapter 8 separately discusses models for geometry and quality aspects, such as cut geometry and surface topography.

The present chapter focuses on models for the calculation of the depth of cut. A first rough view of the models developed for this purpose so far shows, that the models divides into at least four groups:

- volume-displacement models
- energy-conservation models
- parameter-regression models
- numerical simulations

The models are not completely derived in this chapter, and the reader may consult the corresponding original works. Nevertheless, the basic assumption as well as the general treatment of the problem are given. Also, the final equations are processed in a way that the reader can uses the models for calculations. This processing includes most of the empirical and regression parameters that are relevant to the models. Figure 6.1 gives a summary of the models discussed in this chapter.

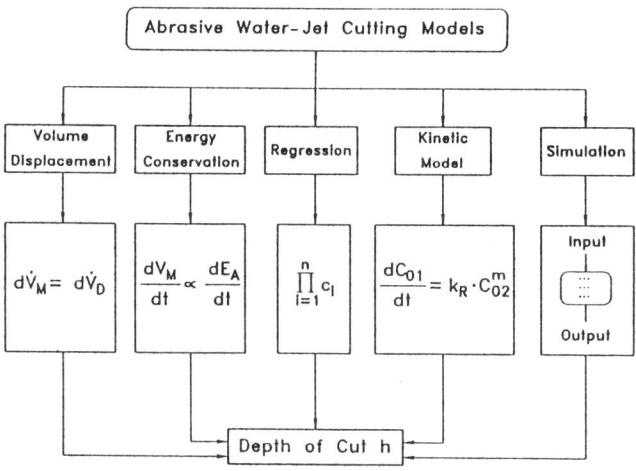

Figure 6.1 *An overview of models for abrasive water-jet cutting*

6.2 Volume-Displacement Models

6.2.1 Volume-Displacement Model for Ductile Materials

The basic assumption of the models contained in this group is the equivalence between the physically-estimated differential volume-removal rate and the geometrically-determined displacement-rate. This assumption is valid for a steady-state removal process. Thus,

$$d\dot{V}_M = d\dot{V}_D. \tag{6.1}$$

Hashish [259] develops an erosion model for the volume removal by an impacting abrasive particle at a shallow angle of impact that is based on Finnie's [155] micro-cutting analysis (section 5.1)

$$d\dot{V}_M = \frac{7}{\pi} \cdot \left[\frac{R_f^3 \cdot d\dot{m}_A}{\rho_P}\right] \cdot \left[\frac{v_P}{v_C}\right]^{2.5} \cdot \sin 2\varphi \cdot \sqrt{\sin \varphi}. \tag{6.2}$$

In the equation, v_C is a characteristic abrasive velocity

$$v_C = \sqrt{\frac{3 \cdot \sigma_f \cdot R_f^{3/5}}{\rho_p}}. \tag{6.3}$$

For small impact angles, Eq. (6.2) gives,

$$d\dot{V}_M = \frac{14 \cdot d\dot{m}_A}{\pi \cdot \rho_P} \cdot \left[\frac{v_P}{v_C}\right]^{2.5} \cdot \varphi^{1.5}. \tag{6.4}$$

From the geometry of the cutting process, the local volume-displacement rate is (Figure 6.2)

$$d\dot{V}_M = dh \cdot v \cdot d_{jet}. \tag{6.5}$$

Combining Eqs. (6.1), (6.4), and (6.5), setting $d\dot{m}_A = \dot{m}_A \cdot dx/d_{jet}$ (Figure 6.2), solving the final equation for dh, and integration yields [260]

$$h_C = \frac{c_1 \cdot d_{jet}}{2.5} \cdot \left[\frac{14 \cdot \dot{m}_A}{\pi \cdot \rho_P \cdot v \cdot d_{jet}^2}\right]^{0.4} \cdot \left[\frac{v_P}{v_C}\right]. \tag{6.6}$$

In Eq. (6.6), $c_1=1-(\varphi_0/\varphi_{Cr})$ is a parameter that depends on the local abrasive-particle impact angle (Figure 6.2). For simplification, d_{jet} is replaced by d_F. Eq. (6.6) gives the depth of cut generated by abrasive grains impacting under low angles in the so-called 'cutting-wear' zone (section 5.4).

The abrasive-particle velocity, v_P, depends on the depth of cut. Due to wall friction and damping, the particle velocity reduces at deeper locations on the cut.

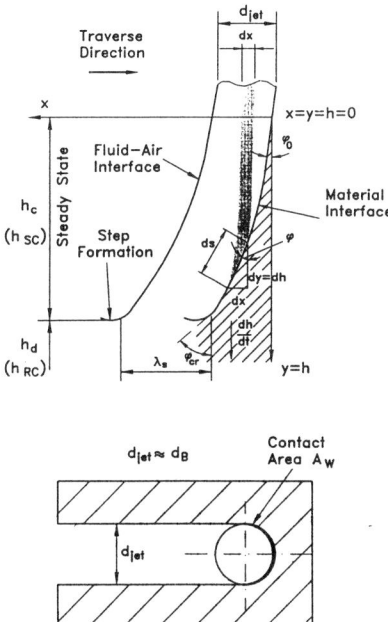

Figure 6.2 *Geometrical situation during the abrasive water-jet cutting*

A simple momentum balance in a control volume that neglects the influence of the entrained air gives

$$(\dot{m}_W + \dot{m}_A) \cdot v_P(h) = (\dot{m}_W + \dot{m}_A) \cdot v_P(h=0) - F_f . \tag{6.7}$$

The friction force is [261]

$$F_f = C_f \cdot A_W \cdot \frac{\rho_{jet}}{2} \cdot v_P^2 . \tag{6.8}$$

In the equation, C_f is the coefficient of friction, A_W is the area of contact between the abrasive water jet and the kerf, and ρ_{jet} is the abrasive water-jet density. The contact area is (Figure 6.2)

$$A_W = \frac{\pi \cdot d_{jet} \cdot h}{2}. \tag{6.9}$$

A combination of Eqs. (6.7) to (6.9), delivers the local abrasive-particle velocity

$$v_P(h) = v_P - \left[\frac{\pi \cdot C_f}{4} \cdot \frac{d_{jet} \cdot \rho_{jet} \cdot v_P^2}{(\dot{m}_A + \dot{m}_W)} \cdot h\right]. \tag{6.10}$$

As large abrasive-particle impact angles in deep cuts are observed by Hashish [207] and Blickwedel [69], Bitter's [160, 161] particle-impact model is used to estimate the volume that is removed by an abrasive grain (section 5.1). A substitution of v_P in Bitter's equation by Eq. (6.10) yields

$$d\dot{V}_M = \frac{\dot{m}_A}{2 \cdot \varepsilon_M^{def}} \cdot \left[v_P \cdot (1 - \frac{C_f}{d_{jet}} \cdot h) - v_{thr}\right]^2. \tag{6.11}$$

From the geometry of the step formation on the cutting front (Figure 6.2),

$$d\dot{V}_D = A_C \cdot \frac{dh}{dt} = \frac{\pi}{4} \cdot d_{jet}^2 \cdot \frac{dh}{dt}. \tag{6.12}$$

Hashish [260] solves Eqs. (6.11) and (6.12) for dh, replaces ε_{Mdef} by σ_f, rearranges some terms, and integrates the final expression. Finally, he obtains

$$h_d = \frac{1}{\frac{\pi \cdot d_{jet} \cdot \sigma_f \cdot v}{2 \cdot c_2 \cdot \dot{m}_A \cdot (v_P - v_{thr})^2} + \frac{C_f}{d_{jet}} \cdot \frac{v_P}{v_P - v_{thr}}}. \tag{6.13}$$

Eq. (6.13) delivers the depth of cut for the material removal by abrasive particles impacting at large angles in the so-called 'deformation-wear' zone (section 5.4).

Hashish [260] expresses Eqs. (6.6) and (6.13) in terms of non-dimensional numbers (Table 6.1). Thus, the entire depth of cut is

$$h = h_C + h_d = d_F \cdot (N_C + N_d). \tag{6.14}$$

In this group of equations, the strength parameter, σ_f, and the threshold velocity, v_{thr}, need to be determined. Hashish [260] assumes $\sigma_f \cong E_M/14$. The threshold particle velocity, in particular, is a more complex value that depends on the abrasive shape and type.

Table 6.2 lists several experimentally-estimated values. Notice that σ_f and v_{thr} are not independent on each other. Materials with high flow-stress show high threshold velocities. In contrast, materials with low flow-stress are characterized by low threshold values. For the drag coefficient, a value of $c_f=0.002$ is found adequate for depths up to 50 mm.

Table 6.1 *Non-dimensional numbers of Hashish's model [260]*

Non-dimensional number	Physical meaning
$N_C = \dfrac{h_C}{d_{jet}}$	Cutting-wear depth number
$N_d = \dfrac{h_d}{d_{jet}}$	Deformation-wear depth number
$N_1 = \dfrac{\rho_P \cdot v \cdot d_{jet}^2}{\dot{m}_A}$	Traverse number
$N_2 = \dfrac{\rho_P \cdot v_P^2}{\sigma_f}$	Relative strength number
$N_3 = \dfrac{\rho_P \cdot v_{thr}^2}{\sigma_f}$	Minimum relative strength number
$N_4 = C_f$	Coefficient of wall drag
$N_5 = 3 \cdot R_f^{3/5}$	Particle-shape number
$N_6 = \dfrac{v_{thr}}{v_P} = \dfrac{N_3}{N_2}$	Threshold-velocity number

Figure 6.3 shows a flow chart of a simplified calculation procedure as Hashish [260] recommends. The correlation between the model and experiment, as expressed by the correlation coefficients in Table 6.2, is quite good.

From results published by Hashish, the model delivers values that are too low for deep cuts (h>30 mm), and too high values for shallow cuts (h<15 mm). The analysis neglects the influence of the high-speed water flow. Also, the model is restricted to ductile-behaving metals. A further problem with this model is that it contains the abrasive-particle velocity that is difficult to measure (section 10.5) and is calculated by approximate methods (section 3.4), respectively.

6. Modeling of Abrasive Water-Jet Cutting Processes

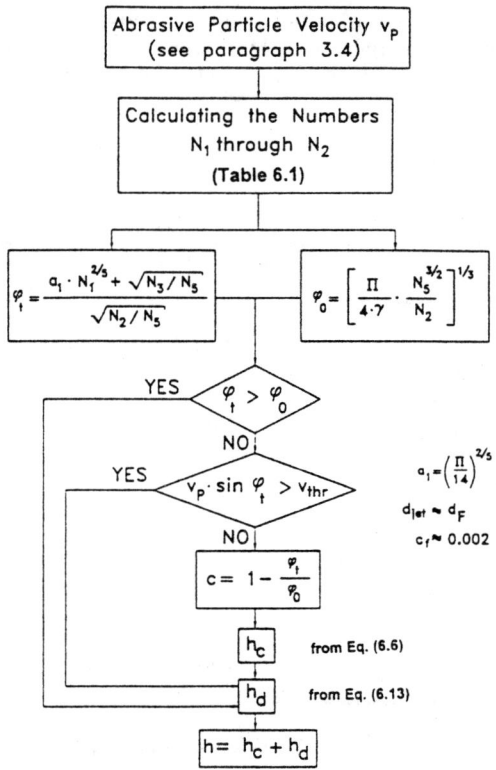

Figure 6.3 *Flow chart of a simplified calculation procedure [260]*

Table 6.2 *Comparison between the predicted depth of cut and measured depth of cut [260]*

Material	Parameter		
	v_{thr} [m/s]	σ_f [GPa]	R^2
Alloy steel 4340	90	-	0.927
Aluminum	40	5.0	0.931
Carbon steel 1018	90	-	0.930
Cast iron A48.C40	75	7.9	0.939
Inconel	60	15.0	0.944
Mild steel A36	90	14.8	0.927
Stainless steel 319	90	14.1	0.917
Stainless steel 304	90	13.8	0.894
Stainless steel 17-4PH	60	14.1	0.988
Stainless steel 15-5PH	90	14.1	0.915
Titanium 6AL-AV	60	8.1	0.906
Tool steel A2-DCF	110	14.8	0.925

6.2.2 Volume-Displacement Model for Brittle Materials

Zeng and Kim [158] and Zeng et al. [224] derive a cutting model that relates the macroscopic material-removal rate on the cutting front to the accumulated effect of micro-cutting by individual abrasive particles. The cutting front, h(x), follows a parabolic function (Figure 6.4 and section 5.5). Thus,

$$\frac{db_S}{dt} = \frac{dx}{dt} \cdot \cos\varphi. \tag{6.15}$$

In the equation, $dx/dt=v$, and db_S/dt is the one-dimensional, linear material-removal rate. With $d_{jet}=\cos\varphi \cdot d_F$, the total material-removal rate along the cutting front is

$$\dot{V}_M = \int \dot{M} \cdot d_F \cdot \cos\varphi \cdot ds. \tag{6.16}$$

Zeng and Kim [158] give the depth of cut further treatment,

$$h = \frac{f(\varphi_c) \cdot \left(\frac{\dot{m}_A}{\dot{m}_w}\right) \cdot \dot{V}_M}{d_F \cdot v}. \tag{6.17}$$

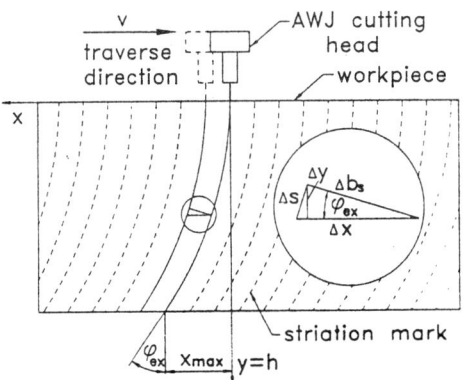

Figure 6.4 *Geometry of the cutting front in abrasive water-jet cutting [224]*

In this equation, φ_C is the abrasive water jet exit-angle as the maximum depth of cut is reached (Fig. 6.7). Table 6.3 gives the function $f(\varphi_C)$. Substitute the volume, V_M, by Eqs. (6.48) and (6.49) as derived in the next paragraph. This substitution yields

6. Modeling of Abrasive Water-Jet Cutting Processes

$$h = \left[\frac{\eta_T \cdot \alpha \cdot \mu^3 \cdot f_{comp}}{1 + \frac{\dot{m}_A}{\dot{m}_W}}\right]^2 \cdot \frac{C_{col} \cdot \dot{m}_A \cdot p}{\rho_W \cdot d_F \cdot v} \cdot \left[\frac{2 \cdot f_w \cdot f(v_M) \cdot D_M \cdot \sigma_f \cdot \varphi^2}{3 \cdot \gamma_M \cdot E_M} + \frac{\varphi}{\sigma_f}\right]. \quad (6.18)$$

In the equation, C_{Col} is a collision factor (C_{Col}=0.104), and $f(v_M)$ is a function of the material's Poisson's-ratio (Figure 6.5a). The factor f_W ($f_W \cong 6.65 \cdot 10^{-4}$) is the amount of stress-wave energy that is involved in fracturing the material. Summarize the target material properties into one parameter, R_E, and summarize all constants in Σ const=2,670. This treatment finally gives

$$h = \frac{\dot{m}_A \cdot \dot{m}_w^2 \cdot p}{2{,}670 \cdot (\dot{m}_A + \dot{m}_w)^2 \cdot d_F \cdot v \cdot R_E}. \quad (6.19)$$

In the equation, R_E is the 'erosion resistance'. The reverse of the erosion resistance, $N_m = 1/R_E$, called 'machinability number' by Zeng et al. [251], represents the behavior of materials in abrasive water-jet cutting. Table 5.8 lists a database of estimated machinability numbers for a wide range of engineering materials.

Table 6.3 *Function $f(\varphi_C)$ for the abrasive water-jet exit angle [158]*

φ_C [°]	0	10	20	30	40	50	60	70	80	90
$f(\varphi_C)$	1	1	1	0.998	0.993	0.981	0.953	0.891	0.752	0.053

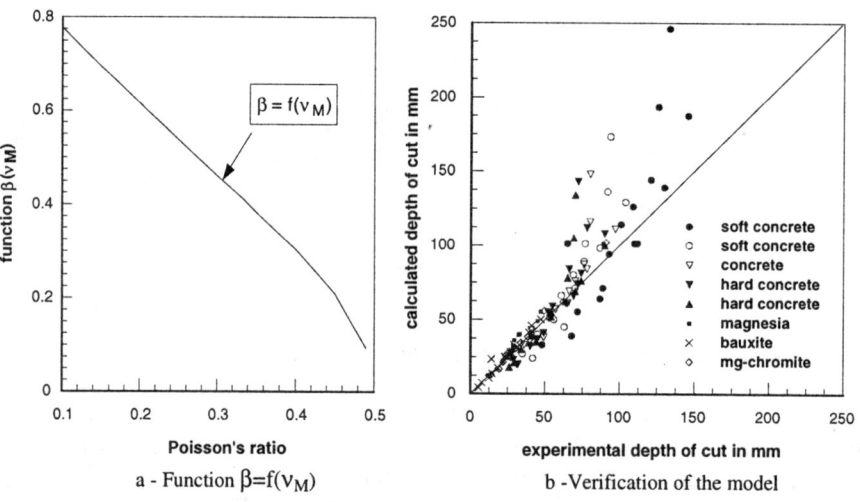

a - Function $\beta = f(v_M)$ b - Verification of the model

Fig. 6.5 *Zeng/Kim's elasto-plastic model for the cutting of brittle materials*

Zeng and Kim [262] simplify Eq. (6.19) to a group of power expressions

$$h = \frac{N_m \cdot p^{1.25} \cdot \dot{m}_A^{0.343} \cdot \dot{m}_W^{0.687}}{8,800 \cdot v^{0.866} \cdot d_F^{0.618}}. \tag{6.20}$$

Figure 6.5b shows a comparative plot between experimentally-estimated data and data calculated by the model. In tendency, the model computes values higher than experimentally estimated. Especially for comparatively soft materials, such as several concrete mixtures, the difference between the model and experiment is sometimes 80 %. Also, the model does not fit very well for deep cuts. Nevertheless, the accuracy is sufficient for the ceramic materials. This model expresses the first rigorous deviation for the abrasive water-jet cutting of brittle material.

6.2.3 Generalized Volume-Displacement Model

Raju and Ramulu [216, 217] perform a similar procedure as described in section 6.2.1 for the estimation of the depth of the 'smooth-cutting' zone, h_{SC}, and the depth of the 'rough-cutting' zone, h_{RC} (section 5.4). These references substitute the abrasive-particle impact angle by trigonometric functions (Figure 6.2),

$$\sin \varphi = \frac{dx}{ds}, \quad \text{and} \tag{6.21}$$

$$\cos \varphi = \frac{dy}{ds} = \sqrt{1 - \left(\frac{dx}{ds}\right)^2}. \tag{6.22}$$

In the equations, s is the arclength of the cutting front. After exploiting a momentum balance similar to Eq. (6.7), Raju and Ramulu [216] derive for the 'smooth-cutting' zone

$$\frac{dx}{ds} = K_3 \cdot [K_1 + K_2 \cdot s], \tag{6.23}$$

$$K_1 = \frac{1}{v_{jet}}, \quad K_2 = \frac{\pi \cdot d_{jet} \cdot \rho_{jet} \cdot C_{fl}}{4 \cdot (\dot{m}_A + \dot{m}_W)}, \quad K_3 = \frac{v_C}{\left[\frac{2}{v \cdot d_{jet}} \cdot \frac{7}{\pi} \cdot \frac{\dot{m}_A}{\rho_P \cdot d_{jet}}\right]^{2/5}}.$$

This ordinary differential equation is solved for x=0 and s=0. Simplify the right term in Eq. (6.23) to $K_3 \cdot [K_1 + K_2 \cdot s] = \Theta$, and $K_3 \cdot K_1 = \Theta_0$, replace dx/ds in Eq. (6.21) by Eq. (6.23), and assume y=0 for s=0. After integration,

$$y = h_{SC} = \frac{1}{K_2 \cdot K_3} \cdot \left[\frac{\Theta \cdot \sqrt{1-\Theta^2}}{2} - \frac{\Theta_0 \cdot \sqrt{1-\Theta_0^2}}{2} + \frac{\arcsin \Theta}{2} - \frac{\arcsin \Theta_0}{2} \right]. \quad (6.24)$$

Eqs. (6.23) and (6.24) definitely give the depth of the 'smooth-cutting' zone.

Based on the geometry of the step formation (section 5.4 and Figure 6.2), Raju and Ramulu [216] perform a differential analysis for the 'rough-cutting' zone. In the analysis, the authors use Finnie's cutting-wear model as well as Bitter's deformation-wear concept (section 5.1). Raju and Ramulu derive the following set of differential equations:

$$\frac{dx}{ds} = K_4 \cdot \frac{b}{v_{jet}^2}, \quad (6.25a)$$

$$\frac{dv_{jet}}{ds} = -b \cdot K_7 \cdot v_{jet}^2, \quad (6.25b)$$

$$\frac{db}{ds} = K_7 \cdot K_6^2 \cdot \left[\frac{b}{v_{jet}}\right]^5 \cdot \left[\frac{5 \cdot v_{jet}^{10} - 9 \cdot K_4^2 \cdot b^2 \cdot v_{jet}^6}{2 \cdot K_4^2 \cdot K_6^2 \cdot b^5 - 3 \cdot v_{jet}^9}\right]. \quad (6.25c)$$

In these equations,

$$K_4 = \frac{4 \cdot \lambda_S \cdot v \cdot \sigma_f}{\dot{m}_A},$$

$$K_6 = \frac{7}{\pi} \cdot \left[\frac{\dot{m}_A}{\rho_P \cdot v \cdot \lambda_S}\right] \cdot \left[\frac{1}{v_C}\right]^{2.5} \cdot K_4^{2.5},$$

$$K_7 = \frac{\pi \cdot \rho_{jet} \cdot C_{f2}}{2 \cdot (\dot{m}_A + \dot{m}_W)}.$$

These equations are closed, and involve one argument, s, three dependent variables, x, b, v_{jet}, and three empirical constants, λ_S, σ_f, $C_{f1,2}$. To solve this group of equations and to obtain the depth of cut of the 'rough-cutting' zone, h_{RC}, Raju and Ramulu [216] suggest a numerical procedure as shown in Figure 6.6a.

6.2 Volume-Displacement Models

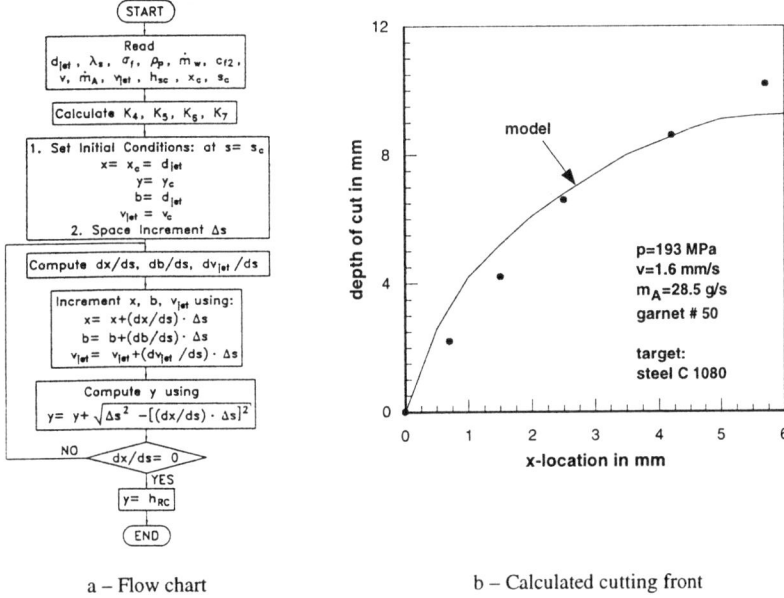

a – Flow chart   b – Calculated cutting front

Figure 6.6 *Flow chart of the numerical solution of Eqs. (6.25a-c) and a calculated cutting front [217]*

Table 6.4 contains typical values for the unknown parameters. The values for σ_f are between $E_M/2$ and $E_M/30$. The friction coefficients are one magnitude higher than those estimated by Hashish [260].

Table 6.4 *Empirical constants for Raju-Ramulu's model [217] abrasive: garnet # 80*

Material	Parameter		
	σ_f [GPa]	C_{f1}	C_{f2}
Aluminum 6061-T6	2.14	0.03	0.04
Plexiglass	1.43	0.03	0.03
Steel C 1080	28.57	0.01	0.02

Figure 6.6b gives the cutting front that is calculated from the simulation model. This figure agrees qualitatively with the real cutting-front geometry insofar as the general shape of the cut is concerned. At the point of transition ($h=h_{SC}$), a discontinuity in the function is noticed. Also, realistically, the progress of the function tends to zero as the final depth of cut, $h=h_{SC}+h_{RC}$, is obtained. A quantitative comparison between the model and experimental results gives very good agreement for the 'smooth-cutting' zone [217]. However, the experimental values for the entire depth of cut seriously deviate from the model, sometimes up to 200 %. However, the model represents a significant progress since it is not restricted to ductile metals, but is suitable for brittle materials, too.

6.3 Energy-Conservation Models

6.3.1 Two-Parameter Energy-Conservation Model

The basis of the energy conservation models is either the proportionality between the abrasive water-jet input energy and the material-removal rate,

$$\frac{dV_M}{dt} \propto \frac{dE_A}{dt},\qquad(6.26)$$

or the conservation of the abrasive water-jet input-energy during the material-removal process. In Eq. (6.26), a proportionality parameter is introduced that physically expresses a specific removal energy. Also geometrically, the material removal, dV_M, is

$$dV_M = h \cdot b \cdot dx.\qquad(6.27)$$

Assume that all particles have identical impact velocities without any temporal fluctuations. Also, dx/dt=v [69]. Thus,

$$h \cdot b \cdot v \propto \tfrac{1}{2} \cdot \dot{m}_A \cdot v_P^2.\qquad(6.28)$$

Consider the square-root relation between the abrasive-particle velocity and pump pressure (Eqs. (3.2) and (3.25)), assume the abrasive-mass flow rate and the width of cut as constant, and introduce a factor of proportionality, C_0. Thus,

$$h = C_0 \cdot \frac{p}{v}.\qquad(6.29)$$

Experimental evidence shows that a critical threshold pressure, p_{thr}, has to be exceeded to introduce the material-removal process (section 7.2). Blickwedel [69] experimentally shows that the relation between the depth of cut and the traverse rate is not exactly inversely proportional. Therefore, Blickwedel adds a traverse exponent, f(v) that expresses the energy loss of the jet flowing through the cut. This loss increases with an increase in the depth of cut. These modifications, including a regression analysis for f(v), give

$$h = C_0 \cdot \frac{p - p_{thr}}{v^{0.86 + \frac{2.09}{v}}}.\qquad(6.30)$$

Table 6.5 lists typical values for C_0 and p_{thr}. Figure 6.7 gives a comparison between Eq. (6.30) and experimental values.

6.3 Energy-Conservation Models

Table 6.5 *Parameters for Blickwedel's model [263]*

Material	Parameter	
	C_0	p_{thr} [MPa]
Aluminum alloy	0.99	55.6
Austenitic steel	0.42	55.0
Glass	2.34	16.0
Titanium	0.53	52.5

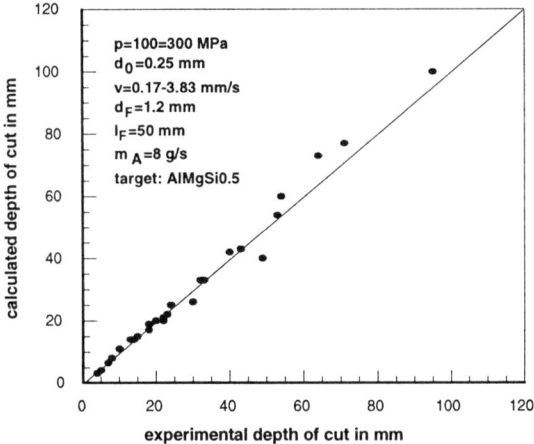

Figure 6.7 *Verification of Blickwedel's model [69]*

6.3.2 Regression Energy-Conservation Model

Oweinah [264] uses Eq. (6.28) in a modified form and writes

$$d\dot{V}_M \cdot \varepsilon_M = \frac{dE_A}{dt}. \tag{6.31}$$

With $\dot{V}_M = h \cdot v \cdot b$, Oweinah yields after some manipulations

$$h = \eta_h \cdot \frac{\dot{m}_A \cdot v_P^2}{2 \cdot v \cdot b \cdot \varepsilon_M}. \tag{6.32}$$

In this equation, the efficiency parameter, η_h, depends on several process and material properties,

$$\eta_h = f_1(\dot{m}_A) \cdot f_2(A) \cdot f_3(\varphi) \cdot f_4(M). \tag{6.33}$$

Oweinah [264] developed regression models for the solution of Eq. (6.33), (Table 6.6). The parameter ε_M is estimated by reference cuts. A typical value for aluminum cut with quartzite abrasives is ε_M=3.4 GPa [264]. This value is between Hashish's [260] and Raju and Ramulu's [217] values for aluminum (Tables 6.2 and 6.4).

Table 6.6 *Regressions for Oweinah's model [264]*

Function	Regression
$f_1(\dot{m}_A)$	$4.823 - 0.04733 \cdot \dot{m}_A + 1.658 \cdot \left(\frac{1}{\dot{m}_A} - 2\right)$
$f_2(A)$	$0.79 \cdot \frac{h_{HV} \cdot h_{(KG)}}{h_{(HV100)} \cdot h_{(130[\mu m])}}$
$f_3(\varphi)$	$51.86 \cdot \varphi^3 - 174.6 \cdot \varphi^2 + 162.6 \cdot \varphi - 0.55$
$f_4(M)$	$HB - \beta \cdot \frac{HB - MT}{3.5343}$

6.3.3. Semi-Empirical Energy-Conservation Model

Capello and Groppetti [225, 265] use Eq. (6.26) in a modified form

$$dV_M = \eta_h \cdot dE_A. \tag{6.34}$$

In the equation, η_h is related to the removal characteristics, abrasive characteristics, and material parameters, similar to Oweinah's [264] model. Introduce the exposure time. The introduction gives,

$$\frac{dV_M}{dt} = \eta_H \cdot \frac{dE_A}{dt}, \tag{6.35}$$

which is identical to Eq. (6.26). As an extension of the energy models presented above, Capello and Groppetti [265] assume that the abrasive water-jet kinetic energy depends on the depth of cut

$$E_A(h) = f(h) \cdot E_A(h = 0). \tag{6.36}$$

For f(h), the authors choose

$$f(h) = \frac{1}{(1+h)^{\gamma_1}}. \tag{6.37}$$

More recently, Momber and Kovacevic [220] and Momber [223] show that the energy dissipation in a workpiece cut by an abrasive water jet is realistically expressed by a second-order polynomial (section 5.5). Eq. (6.35) and $dV_M = dx \cdot db \cdot dh$ yield

$$\frac{dh(x,b,h)}{dt} = \frac{\gamma_2}{(1+h)^{\gamma_1}} \cdot \frac{dE_A(x,b)}{dxdbdt} . \qquad (6.38)$$

The left term of this equation is the penetration-rate of the abrasive water jet inside the workpiece. The right term of the equation is the kinetic energy of the abrasive water jet per area and time

$$\frac{dE_A}{dxdbdh} = \frac{2 \cdot \dot{m}_A \cdot v_P^2}{\pi \cdot d_F^2} . \qquad (6.39)$$

Divide Eq. (6.39) by dx, introduce dx/dt=v, and keep $2/\pi$ into γ_2. The treatment finally gives

$$(1+h)^{\gamma_1} dh = \gamma_2 \cdot \frac{\dot{m}_A \cdot v_P^2}{v \cdot d_F^2} dx , \qquad (6.40)$$

An integration of Eq. (6.40) between 0 and h, and 0 and d_F, respectively, yields

$$h = \left[1 + \gamma_2 \cdot (1+\gamma_1) \cdot \frac{\dot{m}_A \cdot v_P^2}{v \cdot d_F} \right]^{\frac{1}{1+\gamma_1}} - 1 . \qquad (6.41)$$

The parameter γ_2 is related to the proportionality factor in Eq. (6.34). The parameter depends on the material as well as on abrasive characteristics and is very sensitive to the depth of cut. The parameter γ_1 represents the energy losses and the global efficiency inside the cut. As this parameter depends on the target material, it is considered as a 'machinability' parameter. Section 5.9.2.2 discusses this aspect. Both parameters are estimated by non-linear analysis of reference cuts [225]. As Table 6.7 shows, the energy-dissipation parameter, γ_1, is less sensitive to the machined materials. Momber and Kovacevic [220] and Momber [223] who find a unique energy-dissipation characteristics for a set of different materials cut by an abrasive water jets (Figure 5.32), confirm this result.

Table 6.7 *Empirical constants for the model of Capello and Groppetti (1993)*

Material	Parameter	
	γ_1	γ_2
Aluminum		$8.752 \cdot 10^{-11}$
Fiber-reinforced plastics	22.8	$2.595 \cdot 10^{-10}$
Glass		$3.115 \cdot 10^{-10}$

6.3.4 Elasto-Plastic Model for Brittle Materials

Zeng and Kim [158] use an energy approach for the derivation of their plastic-elastic model for the abrasive water-jet cutting of brittle materials. They assume the total material removal of a single impact as the sum of the removal due to a plastic flow and brittle, intergranular fracture

$$V_M = V_{PL} + V_{FR}. \tag{6.42}$$

The volume removed by plastic flow and micro-cutting is assumed to be identical to Bitter's [160, 161] formula for nominal incidence and to Finnie and McFadden's [162] model for low incidence angles (section 5.1).

The energy balance for the brittle fracture is the equality between the fracture energy required to form a crack network by inter-granular fracture to remove a certain volume (Figure 6.8),

$$E_{FR} = \frac{6 \cdot \gamma_M \cdot V_{FR}}{D_M}, \tag{6.43}$$

and the energy that is absorbed by stress waves during the impact [266],

$$E_{STR} = f_e \cdot f(v_M) \cdot \left(\frac{\rho_M}{\rho_P}\right)^{0.5} \cdot \left(\frac{H_M}{E_M}\right)^{1.5} \cdot E_p. \tag{6.44}$$

For low incidence angles, the authors finally derive

$$V_{M1} = \underbrace{\frac{f_w \cdot f_e \cdot D_M \cdot \sigma_f \cdot f(v_M) \cdot m_P \cdot v_P^2 \cdot \sin^2\varphi}{3 \cdot \gamma_M \cdot E_M}}_{\text{Fracture}} + \underbrace{\frac{m_P \cdot v_P^2}{4 \cdot \sigma_f} \cdot \left[\sin 2\varphi - 4 \cdot \sin^2\varphi + 38.12 \cdot v_P \cdot \sin^3\varphi \cdot \sqrt{\frac{\rho_P}{\sigma_f}}\right]}_{\text{Plastic Flow}} \tag{6.45}$$

For normal incidence, the authors derive

$$V_{M2} = \underbrace{\frac{f_w \cdot f_e \cdot \varphi \cdot f(v_M) \cdot m_P \cdot v_P^2}{2 \cdot \gamma_M} \cdot \left(\frac{\rho_M}{\rho_P}\right)^{0.5} \cdot \left(\frac{H_M}{E_M}\right)^{1.5}}_{\text{Fracture}} + \underbrace{\frac{m_P \cdot (v_P - v_{el})^2}{2 \cdot \varepsilon_M}}_{\text{Plastic Flow}}. \tag{6.46}$$

Eqs. (6.45) and (6.46) are used for the derivation of the depth of cut in the previous section (Eq. (6.18)).

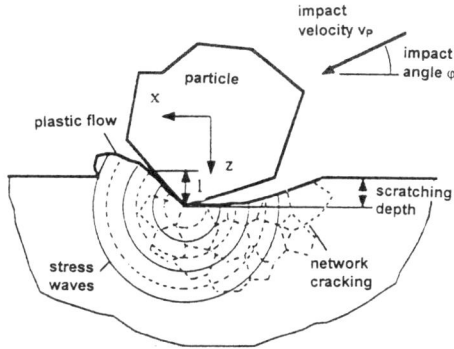

Figure 6.8 *Intergranular-network model for the removal of brittle materials by an abrasive water jet [230]*

But certainly, the assumption of a general intergranular fracture is questionable. As discussed in Sections 5.2 and 5.9, this assumption is valid only for certain combinations of material properties and process parameters.

6.3.5 Energy-Conservation Models for Pre-Cracked Materials

For multiphase pre-cracked materials, Momber and Kovacevic [267] construct an energy balance between the specific material-removal energy and the dynamic energy that is absorbed during a standard compression-test (section 5.9)

$$E_{AbsD} = \chi_M \cdot \frac{P_H}{\dot{V}_M}. \tag{6.47}$$

Assume the material-removal rate as $\dot{V}_M = h \cdot b \cdot v$ and rearrange Eq. (6.47). These manipulations give

$$h = \frac{c_1 \cdot (c_2 \cdot E_{AbsD}) \cdot p \cdot \dot{Q}_W}{E_{AbsD} \cdot b \cdot v}. \tag{6.48}$$

In this equation, the function in the bracket is a regression for the relation between the material-removal efficiency (section 5.6) and absorbed fracture energy [267]. Figure 5.39 shows this relation. The parameter c_2 is sensitive to the traverse rate and pump pressure. Figure 6.9 plots a comparison between results from Eq. (6.48) and measurements. The model holds for the concrete materials but delivers too low values for the low-strength mortar materials. The reason is that, in the mortars, the abrasive water jet does not produce straight kerf walls as assumed in the model.

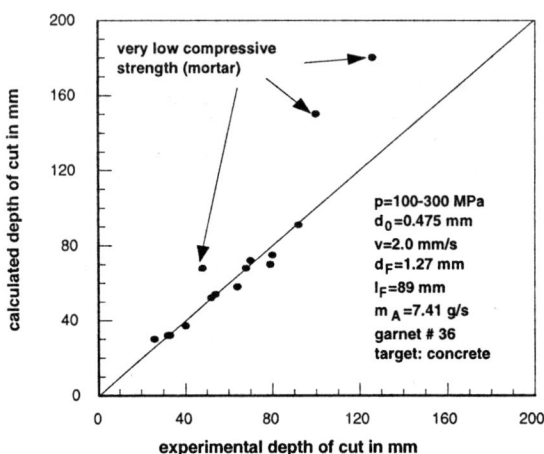

Figure 6.9 *Verification of Momber-Kovacevic's model [267]*

Iihoshi et al. [268] develop a model for cutting rocks by abrasive water jets. The general idea is the direct relation between the water-jet power, P_W, and the depth of cut. A special feature of this model is the consideration of the cutting action of the plain water as well as the standoff distance. Thus,

$$P_W \propto h_A + h_W. \qquad (6.49)$$

As a general relation, the authors find,

$$h = h_A + h_W \propto \frac{1}{v \cdot x}. \qquad (6.50)$$

Finally, the depth of cut is

$$h = \left[c_{01} + c_{02} \cdot \left(\frac{\dot{m}_A}{\dot{m}_W} \right)^{c_2} \right] \cdot \frac{P_W}{v \cdot x}. \qquad (6.51)$$

The parameters c_{01} and c_{02} are estimated by multiple regression. The parameter c_{02} is especially sensitive to the selected abrasive material. Table 6.8 lists the regression results.

Table 6.8 *Regression parameters for Iihoshi's model [268]*

Abrasive type	Parameters		
	c_{01}	c_{02}	c_2
Garnet	0.44	5.55	0.762
Silica sand		2.93	0.762

p=180-340 MPa, v=0.17-2.5 cm/s

Hlavac [269] presents a model for abrasive water-jet cutting without a detailed deviation. The author points out that the model bases on an energy balance during the cutting performance. The final formula is

$$h = \frac{\sqrt{2} \cdot R_f \cdot \pi \cdot d_P^2 \cdot d_0 \cdot \rho_P \cdot p \cdot H_P \cdot \dot{m}_W^2 \cdot \sin\varphi}{12 \cdot N_P \cdot v \cdot \rho_W \cdot (\sigma_C + \sigma_S) \cdot H_M \cdot (\dot{m}_W + \dot{m}_A)^2}. \tag{6.52}$$

6.4 Regression Models

6.4.1 Multi-Factorial Regression Models

Chung et al. [270] investigate the correlation between the operational parameters of abrasive water-jet cutting of ductile materials and the cut geometry. The authors carry out a series of factorial experiments and use the experimental results to construct regression equations. The final model is

$$h = c_{01} \cdot \frac{\dot{m}_A^{c_3} \cdot (p - p_{thr})}{v \cdot b} + c_{02}. \tag{6.53}$$

Table 6.9 lists the process parameter ranges as well as values for the regression constants.

Kovacevic [271] and Kovacevic et al. [272] also adapt a factorial approach to the experimental design in order to mathematically model the depth of cut. The general structure of the functional relationships is [271]

$$h = c_0 \cdot p^{c_1} \cdot \dot{m}_A^{c_2} \cdot v^{c_3} \cdot d_F^{c_4} \cdot x^{c_6}. \tag{6.54}$$

Data from cutting experiments based on a 2^n-composite factorial design are fitted with Eq. (6.54). Table 6.10 lists the final regression coefficients. Figure 6.10 shows an application of the model for calculating the depth of cut in a concrete material.

Table 6.9 *Regression parameters for Chung's model [270]*

Material	d_P [μm]	d_0 [mm]	c_{01}	c_3	c_{02}	p_{thr} [MPa]
Steel	300	0.254	0.0365	0.6	-0.638	70
		0.305	0.0397		-0.085	
		0.178	0.0383		-2.415	
		0.229	0.0431		-1.536	
	177	0.254	0.0447		-0.207	
		0.305	0.0513		-1.605	
		0.356	0.0490		1.294	
	125	0.254	0.0501		0.049	
	65	0.254	0.0285		1.547	
		0.356	0.0401		0.468	
Titanium	177	0.178	0.0416	0.7	-0.827	60
		0.254	0.0217		-0.334	
Aluminum	300	0.203	0.0542	0.65	-0.374	63
		0.254	0.0728		-0.616	
		0.305	0.0855		-0.337	
	177	0.203	0.0602		-0.716	
		0.254	0.0898		-0.520	
		0.305	0.1093		-0.181	
	65	0.254	0.0882		-0.987	
		0.305	0.0934		-0.987	

Table 6.10 *Parameter range and regression parameters for Eq. (6.54) [271, 272]*

Parameter	Parameter range		c_i	Regression coefficient	
	Concrete	Mild steel		Concrete	Mild steel
			c_0	1.3545	0.0014
p [MPa]	103 - 241	170 - 275	c_1	0.7903	1.47
m_A [g/s]	4.54 - 9.07	3.02 - 18.87	c_2	0.1844	0.211
v [mm/s]	0.85 - 6.77	0.42 - 0.95	c_3	-0.5671	-0.74
d_F [mm]		1.2 - 2.6	c_4	-	0.756
x [mm]	6.35 - 25.4	3.1 - 12.7	c_6	-0.0068	-0.139

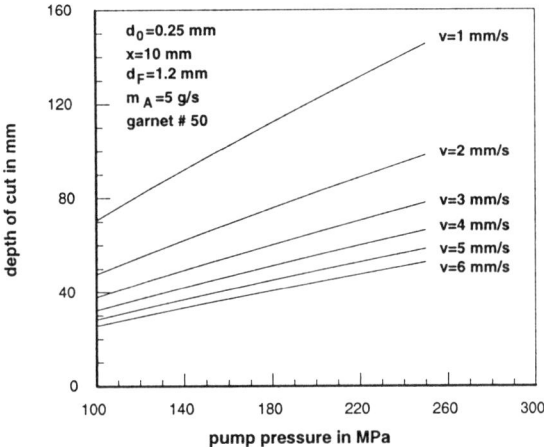

Figure 6.10 *The estimation of the depth of cut for concrete by a multiple-regression model [272]*

6.4.2 Further Regression Models

Matsui et al. [236] develop some regression models to describe the relation between the cutting process and several material properties. To characterize the cutting process, they use the cutting rate as given by Eq. (7.2). The target materials are characterized by idealized (linear) stress-strain curves (Figure 5.50) that are questionable for pre-cracked materials. Based on SEM-observations, the authors suggest different material-removal mechanisms for ductile-behaving materials and brittle-behaving materials. Therefore, the authors use the modulus of fracture to evaluate brittle-behaving materials. In contrast, the ductile-behaving materials are characterized by the product of hardness and elongation as well as by the plastic-deformation energy. Section 5.9 discusses these aspects.

Due to the regression of experimental results, the authors find for ductile materials

$$A_h = 10^{4.74} \cdot [H_M \cdot \varepsilon]^{-0.67} \text{ , and} \tag{6.55a}$$

$$A_h = 10^{4.98} \cdot \left[\frac{(\sigma_u + \sigma_y) \cdot \varepsilon}{2} \right]^{-0.64} . \tag{6.55b}$$

For brittle materials, Matsui et al. [236] derive

$$A_h = 10^{0.91} \cdot \left[\frac{\sigma_t^2}{2 \cdot E_M} \right]^{-1.97} .$$ (6.56)

Eq. (6.56) does not hold for rocks, such as granite and marble. This result indicates that pre-existing cracks influence the erosion behavior of this group of materials. Momber and Kovacevic [267] further investigate this aspect (section 5.9.3 and Eq. (6.48)).

6.4.3 Regression Model for Cutting with Suspension-Abrasive Water Jets

Yazici [273] develop a regression model for the estimation of the depth of cut in rocks that are cut by suspension-abrasive water jets

$$h = c_0 \cdot p^{c_1} \cdot \dot{m}_A^{c_2} \cdot d_F^{c_4} .$$ (6.57)

Table 6.11 lists the regression parameters as well as the process-parameter ranges for which Eq. (6.57) is valid. This model is the only available for cutting with suspension-abrasive water jets.

Table 6.11 *Parameter range and regression parameters for Yazici's model [273]*

Parameter	Parameter range	Regression coefficient	
		c_0	3.799
p [MPa]	20.7 - 34.0	c_1	1.012
m_A* [g/s]	46.0 - 149.0	c_2	0.318
d_F [mm]	1.98 - 2.77	c_4	0.308
v [cm/min]	15.2	-	-

*abrasive type: garnet, d_P=296 μm

6.5 Kinetic Model of the Abrasive Water-Jet Cutting Process

Momber [148] develops a kinetic model of the material-removal by abrasive water jets. The model bases on the equation for reaction kinetics [274]

$$\frac{dCo_1}{dt} = k_R \cdot Co_2^m .$$ (6.58)

In the equation, Co_1 is the concentration of the product per unit surface, Co_2 is the concentration of the reactant per unit surface, k_R is the reaction velocity, and m is

6.5 Kinetic Model of the Abrasive Water-Jet Cutting Process

the order of reaction. The general process of material removal during abrasive water jets is the generation of a certain number of wear particles in a given time period (section 5.6). Thus,

$$\frac{dCo_1}{dt} = \frac{dN_M}{dt} \propto h. \tag{6.59}$$

The depth of cut increases as the number of material particles removed in the time period increases (Figure 6.11). During the cutting process, the target material is subjected to a number of impinging abrasive particles

$$Co_2 \propto \dot{m}_A. \tag{6.60}$$

Eqs. (6.58) to (6.60) give

$$h = k_A \cdot \dot{m}_A^m. \tag{6.61}$$

In this equation, k_A is the reaction velocity for abrasive water-jet cutting, and m is the order of reaction for the abrasive water-jet cutting. The models presented in Table 6.12 can be generalized by Eq. (6.61). This result suggests the possibility of discussing the abrasive water-jet cutting using the structure of reaction kinetics.

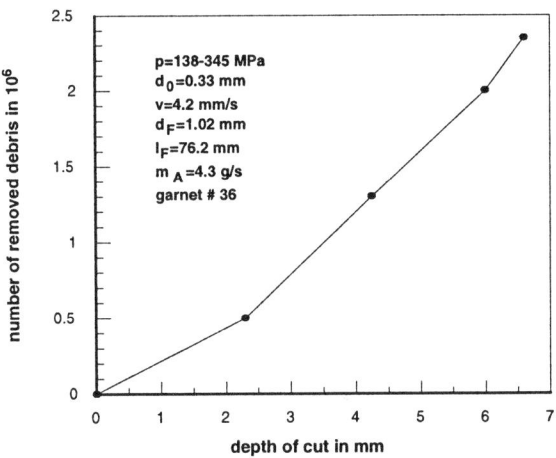

Figure 6.11 *Relation between the depth of cut and number of removed wear particles (authors)*

Table 6.12 *Generalization of abrasive water-jet cutting models*

Reference	Power exponents		
	p	$\dot{m}_A$	v
Matsui et al.	-	0	1
Kovacevic et al.	0.79	0.1844	0.51
Kovacevic	1.47	0.211	0.74
Zeng and Kim	1.25	0.343	0.86
Hashish, 'cutting'	0.5	0.4	1
Chung et al.	1	0.6	1
Ihoshi et al.	1.5	0.762	1
Blickwedel	1	1	f(v)
Oweinah	1	1	1
Hashish, 'deformation'	1	1	1
Tikhomirov et al.	1.5	1	-

From Figure 6.12,

$$m = f(\dot{m}_A). \quad (6.62)$$

Momber [148] divides the following cases:

$$m=1 \qquad h = k_A \cdot \dot{m}_A, \qquad \frac{dh}{d\dot{m}_A} = k_A = \text{const.} \quad (6.63a)$$

$$0<m<1 \qquad h = k_A \cdot \dot{m}_A^m, \qquad \frac{dh}{d\dot{m}_A} > 0 = f(\dot{m}_A) \quad (6.63b)$$

$$m=0 \qquad h = k_A, \qquad \frac{dh}{d\dot{m}_A} = 0 \quad (6.63c)$$

$$m<0 \qquad h = k_A \cdot \dot{m}_A^m, \qquad \frac{dh}{d\dot{m}_A} < 0 = f(\dot{m}_A) \quad (6.63d)$$

Momber [148] discusses these equations in detail.

According to the transition state theory of chemical reactions one can define a probability, P_E, at which the reaction (cutting process) starts to occur

$$k_A = A \cdot P_E. \quad (6.64)$$

The reaction probability is

$$P_E = f(E_{Act}, E_P). \quad (6.65)$$

6.5 Kinetic Model of the Abrasive Water-Jet Cutting Process

In this equation, E_{Act} is an energy barrier that must be overcome by the impacting abrasive particles. The energy potential of these particles is their kinetic energy. The higher the kinetic energy, the higher the probability of forming a reaction product. The rate (or velocity) of product formation depends on the value A in Eq. (6.64).

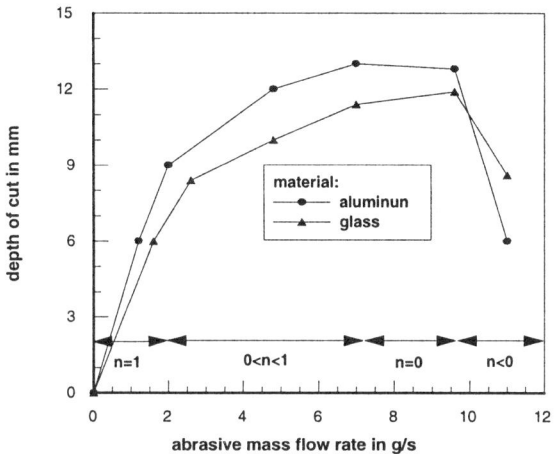

Figure 6.12 *Relations between the depth of cut and abrasive-mass flow rate, indicating different orders of reaction [148]*

The kinetic energy of an abrasive particle is calculated by Eq. (2.14). The range of the individual abrasive-particle energies in an abrasive water jet is limited by

$$E_P^{min} \leq E_P \leq E_P^{max} \tag{6.66}$$

From Figure 4.16, a Gaussian-distribution is applied for the distribution function of the abrasive-particle kinetic energy. Assume a narrow diameter range for the impacting abrasive grains, and a constant density for a given abrasive material. In this case, the particle energy distribution is

$$g(E_P) = \frac{1}{\sqrt{2 \cdot \pi} \cdot \sigma_{E_P}} \cdot e^{\frac{-(E_P - \overline{E}_P)^2}{2 \cdot \sigma_{E_P}^2}} . \tag{6.67}$$

Momber [148] defines a critical energy value of the target material for the energy barrier in Eqs. (6.65). This value expresses the energy that is required to remove a certain minimum number of target-material debris. This energy is

$$E_{Act} = f(\gamma_M). \tag{6.68}$$

6. Modeling of Abrasive Water-Jet Cutting Processes

Figure 6.13 shows that the reaction probability, or the probability that the activation energy will be exceeded by the energy potential of an abrasive grain, is the area enclosed by $g(E_P)$ and E_{Act}

$$P_E = \frac{1}{\sqrt{2 \cdot \pi} \cdot \sigma_{E_P}} \cdot \int_{E_{act}}^{E_P} e^{\frac{-(E_P - \overline{E}_P)^2}{2 \cdot \sigma_{E_P}^2}} dE_P . \qquad (6.69)$$

For $P_E = P_{EMax}$, the maximum possible depth of cut is reached. For $P_E = P_{EMin}$, the minimum depth of cut is obtained. Momber (1995) distinguishes between the three conditions (a): $E_{PMin} > E_{Act}$, (b): $E_{PMin} > E_{Act} > E_{PMax}$, and (c): $E_{PMax} < E_{Act}$. For (a), every impacting abrasive particle contributes to the erosion process. The result $P_E = 0$ is invalid under these conditions. For (b), $P_E = 0$ is possible for some values of E_P. For (c), no cutting occurs, and $P_E = 0$. Figure 6.13 presents an exemplary calculation.

Finally, the depth of cut is [148]

$$h = \frac{p \cdot d_F \cdot \dot{m}_A^m}{\sqrt{2 \cdot \pi} \cdot v \cdot \sigma_{E_P}} \cdot \int_{E_{Act}}^{E_P} \exp\left[\frac{-(E_P - \overline{E}_P)^2}{2 \cdot \sigma_{E_P}^2}\right] dE_P . \qquad (6.70)$$

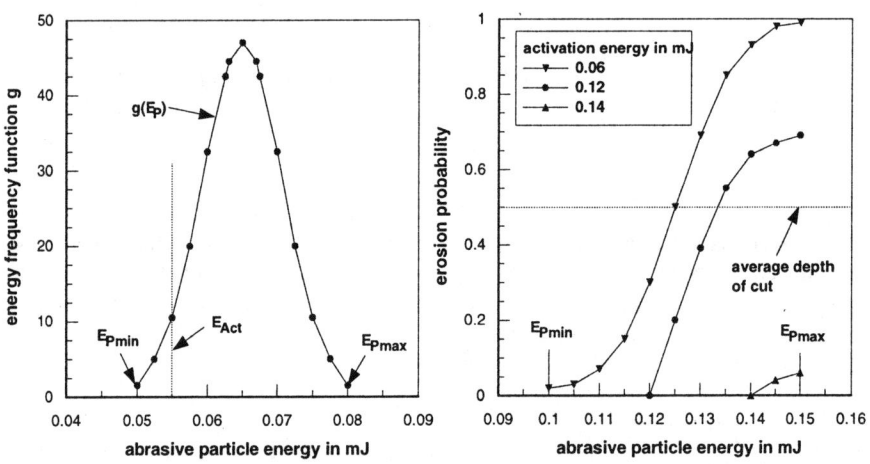

Figure 6.13 *Relations between the distribution of the abrasive-particle kinetic energy, material-activation energy, and erosion probability [148]*

6.6 Fuzzy Rule-Based Model of the Abrasive Water-Jet Cutting Process

The models for the abrasive water-jet cutting discussed in the previous sections generally suffer from the inherent complexities and a less then full understanding of the physics of the cutting process. The fuzzy logic in such cases is very useful as it has proved to be successful in analyzing uncertain and complex systems that can not be described mathematically.

Kovacevic and Fang [275] and Kovacevic et al. [276] conduct investigations in order to determine the value of the depth of cut for selected cutting conditions by fuzzy rules. Namely, while the experiments are performed, one of several selected process variables is varied amongst three levels (minimum, medium and maximum) while the other variables are kept constant. The effect of the cutting parameters on the depth of cut is expressed with respect to a reference depth value. The changes in the depth of cut with respect to the reference depth value caused by the change in the process parameters are selected as the universe of the input, and the corresponding cutting variables are selected as the universe of the output.
The total depth of cut for the given cutting variables is

$$h = h_{ref} \cdot \prod_{i=1}^{N} \Delta h_i . \qquad (6.71)$$

In the equation, h_{Ref} is the reference depth that is obtained under standard cutting conditions, $\Delta h_i = h_i / h_{Ref}$ are the changes in the depth of cut with respect to the reference depth caused by the changes in the cutting conditions, and N is the number of process parameters. Universes of discourse for the selected process parameters and the change of the depth of cut are discretized into a number of levels with five terms (primary fuzzy sets). The relationship between the input and the output is found using Cartesian product expressions of the two sets

$$\Delta h = input \bullet output. \qquad (6.72)$$

In this equation, • represents the Cartesian product. A membership function of this relation is,

$$\mu_{\Delta h} = \min\{\mu_{Input}, \mu_{Output}\}. \qquad (6.73)$$

The input to the system is the change in the depth of cut with respect to the reference depth and the output is the magnitude of the focus diameter. The fuzzy variables in the form as negative big, negative small, and positive small and positive big are specified with respect to the reference points that are described as zero points.

6. Modeling of Abrasive Water-Jet Cutting Processes

Table 6.13 *Results of case studies of fuzzy-logic modeling of the abrasive water-jet cutting [275, 276]*

No*	h_{Ref} [mm]	h_{Req} [mm]	d_F [mm]	Δh_1	p [MPa]	Δh_2	m_A [g/s]	Δh_3	v [mm/s]	Δh_4
1		10	1.65	1.00	224	0.85	12.85	0.99	0.87	0.87
2	13.7	15	1.73	0.83	231	0.89	16.34	1.07	0.47	1.37
3		20	1.47	0.91	280	1.20	15.47	1.05	0.42	1.27
4		30			104	0.96	4.50	0.93	4.65	0.67
5	69.8	60			172	1.00	4.50	0.93	2.90	0.92
6		90			241	1.28	9.07	1.06	2.74	0.85

*1-3: steel, 4-6: concrete

Figure 6.14a illustrates the procedure of selecting the process parameters of the abrasive water-jet cutting for the required depth of cut. The first stage of the procedure is to determine the required depth of cut and then through the iterations procedure select the most suitable abrasive water-jet parameters whose combination produces the required depth of cut.

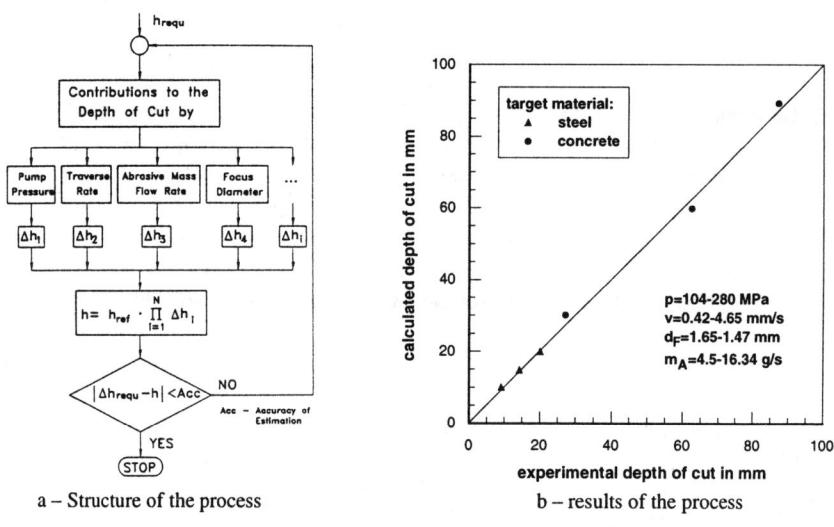

a – Structure of the process   b – results of the process

Figure 6.14 *Structure and results of the parameter estimation by fuzzy rules [275, 276]*

Examples are presented by Kovacevic and Fang [275] for steel, and by Kovacevcic et al. [276] for concrete cutting. Table 6.13 contains results for the estimation of the depth of cut by using fuzzy rules.

6.7 Numerical Models

6.7.1 Numerical Simulations

Mazurkiewicz [159, 277] develops a step-by-step calculation method to estimate the target material volume removed by a single abrasive particle. Table 6.14 illustrates the method. The abrasive flow-stream is divided into several columns, and the material is cut in a column-column mode. The results of this simulation agree well with Finnie's micro-cutting erosion model for ductile-behaving materials (section 5.1).

Corcoran et al. [278] develop a time based, three-dimensional computer model to simulate the abrasive water jet cutting mechanism. The water jet velocity is assumed to follow a sixth-degree polynomial over the jet diameter. The abrasive mass flow rate is assumed to be constant. The algorithm allows for variations in particle size due to mixing. The model gives choices for the particle position distribution within the jet stream, such as uniform, annular, and random. The target material is established within a three-dimensional Cartesian-frame that is linked to the cylindrical abrasive trajectory frame. The authors use Finnie's erosion model. The impact area over the target is divided into four quadrants, and a number of rows and columns. The model defines three primary strike types, such as row wall, column wall, and element surface. Figure 6.15 gives the construction of the FORTRAN-program. The program reflects, as least qualitatively, the influence of the abrasive-mass flow rate, the piercing time, and traverse rate. Nevertheless, the computation time is very high even for a low number of elements.

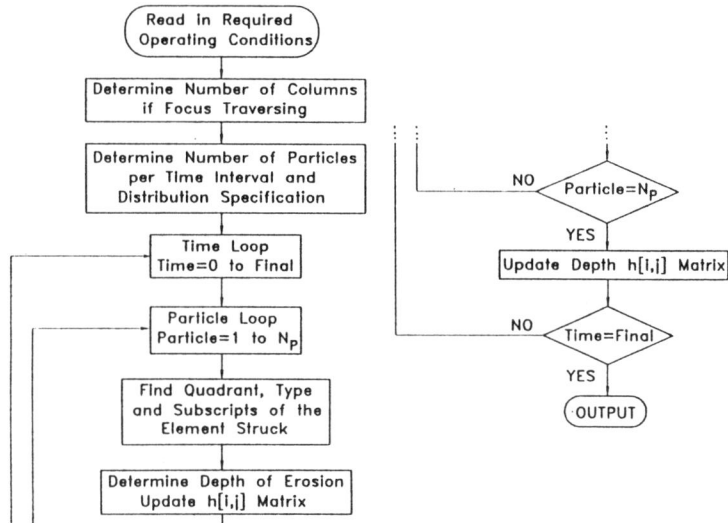

Figure 6.15 *FORTRAN flow-chart for the finite-element simulation of the abrasive water-jet cutting [278]*

Table 6.14 *Procedure of the mass-loss calculation for a single abrasive particle [277]*

Step	Procedure
1	Input conditions: $d_{jet} = 1.5$ mm $\quad l_A = 315$ mm $v = 5$ m/s $\quad d_0 = 0.35$ mm $h = 10$ mm $\quad$ material: aluminum $v_P = 200$ m/s $\quad \sigma_f = 1.38$ MPa $d_P = 425$ μm $\quad \rho_M = 2{,}700$ kg/m^3
2	Abrasive-particle spacing: $x_P = d_P + l_A = 0.425 + 0.315 = 0.74$ mm
3	Number of columns formed by abrasive particles on the half of the cutting front - jet periphery: $N_{col} = \dfrac{\pi \cdot d_{jet}}{2 \cdot x_P} = \dfrac{\pi \cdot 1.5}{2 \cdot 0.74} = 3.18$ columns
4	Numbers of abrasive particles in a column for one second (t = 1 sec) cut: $N_P = \dfrac{v_P}{x_P} = \dfrac{200 \cdot 10^3}{0.74} = 270{,}270$ grains
5	Volume of material eroded during one second (t = 1 sec) action: $V_M = \left[(d_{jet} \cdot s_t) + \dfrac{\pi \cdot d_{jet}^2}{8} \right] \cdot h = \left[(1.5 \cdot 5) + \dfrac{\pi \cdot 1.5^2}{8} \right] \cdot 10 = 83.83$ mm^3
6	Abrasive grains involved in the volume (V_M)-removal process: $N_{PV} = N_{col} \cdot N_P = 3.18 \cdot 270{,}270 = 859{,}458$ grains
7	Volume removed by a single abrasive grain: $V_{MA} = \dfrac{V_M}{N_{VP}} = \dfrac{88.83}{859{,}458} = 9.754 \cdot 10^{-5}$ mm^3/grain
8	Mass removed by a single abrasive grain: $M_A = V_{AM} \cdot \rho_M = 9.754 \cdot 10^{-5} \cdot 2{,}700 = 0.263$ g/grain
9	Mass removed by a single abrasive grain calculated according to Finnie [155]: $M_A = \left[\dfrac{\rho_M}{\psi \cdot \sigma_f} \right] \cdot \dfrac{m_P \cdot v_P^2}{2} \cdot [\sin(2 \cdot \varphi) - 3 \cdot \sin^2 \varphi]$ g/grain $M_A = \left[\dfrac{2.7}{2 \cdot 13{,}800 \cdot 10^5} \right] \cdot \dfrac{6.11 \cdot 10^4 \cdot 200^2 \cdot 10^6}{2} \cdot 0.0312 = 0.373$

6.7.2 Numerical Process Model

Yong and Kovacevic [279, 280] develop a numerical process-model for abrasive water-jet machining that covers several major subjects of the process, such as the simulation of multiphase pipe-flow, tracer record of abrasive particles, and energy transformation in a defined material 'memory cell'. Three coordinate systems that are linked by transform equations fix the kinematic behavior of the abrasive particles. Figure 6.16 shows a network that divides the surface area to be cut on the workpiece. The figure also illustrates how each element of this network is assumed as a 'memory cell'. The model utilizes the information recorded in such a cell about the particle history to predict the depth of cut at the point (x_i, y_i, z_i).

If a number of N_P abrasive particles are trapped by a cell, the total kinetic energy E_K contained is

$$E_K = \frac{1}{2} \cdot \sum_{i=1}^{N_P} \left[m_{Pi}(t_i) \cdot v_{Pi}^2(t_i, x_i, y_i, z_i) \right], \tag{6.74}$$

where $m_{Pi}(t_i)$ and $v_{Pi}(t_i,x_i,y_i,z_i)$ are the mass and the velocity, respectively, of an abrasive particle at instant time t_i. The z-ordinate is the depth of cut direction.

The next step is the modeling of the dynamic behavior of the abrasive particles in a cell and the calculation of the depth of cut for each cell as output of the model. The expression of the depth of cut at a given instance is

$$h = f(c_1, c_2, ...c_j) \cdot v_{Pz}^k(h) \geq 0,$$
$$f(c_1, c_2, ...c_j) > 0. \tag{6.75}$$

In this equation, k is a constant (usually between 2 and 3 for erosion models), and $c_j = c_j(t, x_j, y_j, z_j)$ (j=1,1,...) stands for any other general parameter, such as abrasive type and material resistance. If v_{Pz} is treated as a constant, independent on the depth of cut, Eq. (6.86) is general in agreement with abrasive water jet cutting-formulas (see previous sections). Nevertheless, for deep cutting or drilling with abrasive waterjet, the influence of the depth of cut is considered. After a small time increment, Δt, one obtains

$$\Delta h \frac{v_{Pz}^k \cdot \sum_{i=1}^{j} \left[\frac{\partial f}{\partial c_i} \cdot \Delta c_i \right]}{1 - k \cdot f \cdot v_{Pz}^{k-1} \cdot \frac{\partial v_{Pz}}{\partial h}}. \tag{6.76}$$

In the equation, $\partial v_{Pz}/\partial h$ characterizes the loss in the abrasive-particle velocity due to friction, damping and collision in deep cavities. Eqs. (6.75) and (6.76) are evaluated by a numerical algorithm to ensure its adaptability for different conditions.

194   6. Modeling of Abrasive Water-Jet Cutting Processes

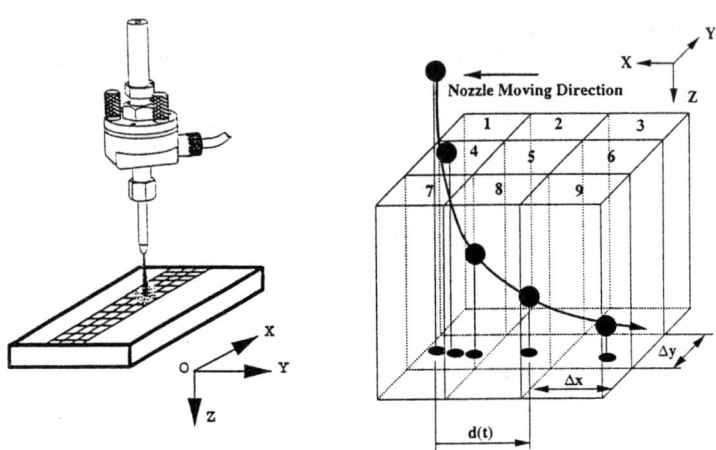

Figure 6.16 *Basics of Yong-Kovacevic's process model [279, 280]*

Figure 6.17 shows a comparative plot between the experiment and simulation. This model is applied to model other machining tasks, such as milling and drilling (chapter 9).

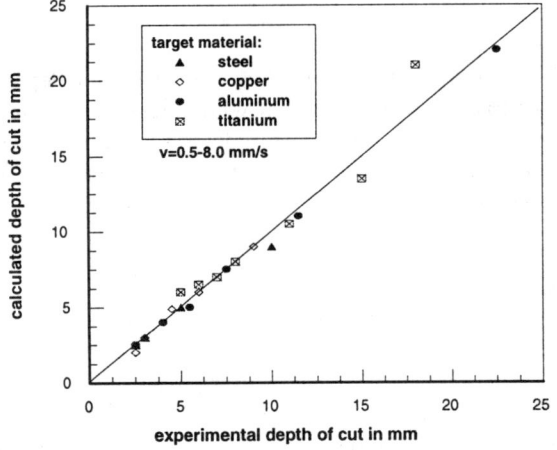

Figure 6.17 *Verification of Yong-Kovacevic's model [279, 280]*

7 Process-Parameter Optimization

7.1 Definition of Process and Target Parameters

7.1.1 Process Parameters

The abrasive water jet-cutting process is characterized by a large number of process parameters that determine efficiency, economy, and quality of the whole process. Therefore, optimization of the process is a primary requirement for a successful application.

Generally, the process parameters in the abrasive water-jet cutting divide as follows (Figure 7.1):

1 - hydraulic parameters:

- pump pressure (p)
- water-orifice diameter (d_0)

2 - cutting parameters:

- traverse rate (v)
- number of passes (n_P)
- standoff distance (x)
- impact angle (φ)

3 – mixing-and-acceleration parameters:

- focus diameter (d_F)
- focus length (l_F)

4 - abrasive parameters:

- abrasive-mass flow rate (m_A)
- abrasive-particle diameter (d_P)
- abrasive-particle size distribution ($f(d_P)$)
- abrasive-particle shape,
- abrasive-particle hardness (H_P)
- abrasive-recycling capacity

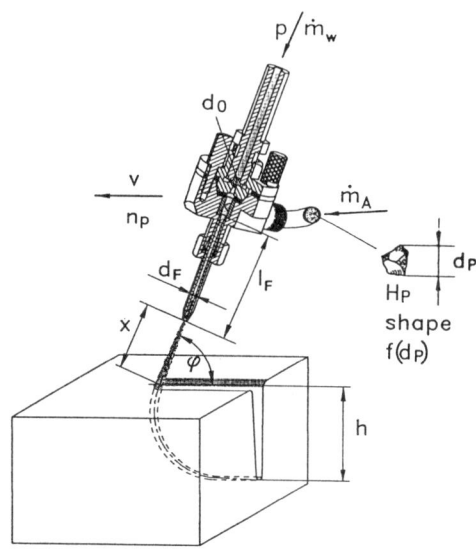

Figure 7.1 *Optimization parameters in the abrasive water-jet cutting*

7.1.2 Target Parameters

The most important target parameter in cutting applications is the depth of cut, h, or the workpiece thickness, h_S, that can be separated under certain conditions. Guo [180] finds

$$h_S = 0.8 \cdot h .\tag{7.1}$$

Figure 7.2 illustrates this relation.

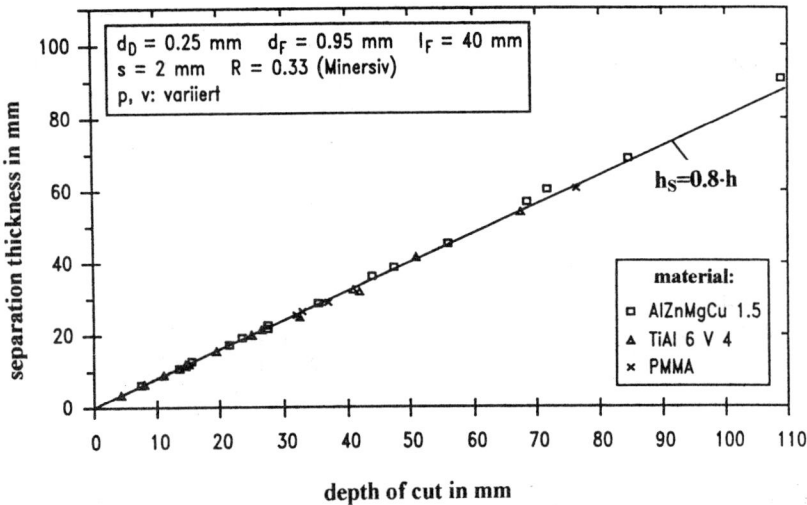

Figure 7.2 *Relation between the depth of cut and separation thickness [180]*

Another parameter in cutting is the cutting rate

$$A_h = h \cdot v ,\tag{7.2}$$

that is the area generated during a certain period of time. For identical thickness, the cutting rate can be different because it depends also on the traverse rate.

In material removal applications, such as milling and turning, the removed material volume is important. For some conditions, this volume is

$$V_M = h \cdot b_h \cdot L_h .\tag{7.3}$$

Usually, the removed volume is measured by filling the generated cavities with a reference material. The volume-removal rate is

$$\dot{V}_M = h \cdot b_n h \cdot v = \frac{V_M}{t_E}. \tag{7.4}$$

For optimization purposes, consider the effects of water-volume flow rate and abrasive-mass flow rate because they depend on the pump pressure and orifice diameter. In these cases, relate conventional target parameters to these process parameters. The specific cutting rate for optimization of the abrasive-mass flow is, for example,

$$A_{hA} = \frac{A_h}{\dot{m}_A} = \frac{h \cdot v}{\dot{m}_A}. \tag{7.5}$$

7.2 Influence of Hydraulic Process Parameters

7.2.1 Influence of Pump Pressure

7.2.1.1 General Trends

Fig. 7.3 shows the general relation between the pump pressure and depth of cut. A mathematical formulation for this function is

$$h(p) = C_1 \cdot (p - p_{thr})^{C_2}, \tag{7.6}$$
$$C_1 = \frac{\Delta h}{\Delta p}.$$

The general structure of the function is the result of complex relations between the pump pressure and the processes of jet formation, abrasive acceleration and mixing, and material removal.

From Eq. (3.2) is evident that the water jet velocity increases with rising pump pressure. Himmelreich and Rieß [281] and Neusen et al. [48] experimentally verify this trend. Similar to the trend shown in Figure 7.3a, the progress of the velocity-function decreases with an increase in the pressure. The efficiency parameter μ and the compressibility factor κ nearly linearly decrease with rising pump pressure [45, 49] (chapter 3). Thus,

$$v_0(p) = \sqrt{2} \cdot \mu(p) \cdot \sqrt{\frac{p}{\rho_W(p)}}. \tag{7.7}$$

The influence of the pump pressure on the mixing process shows a similar complexity. All parameters that influence the average velocity of the abrasive particles depend on the pump pressure.

7. Process-Parameter Optimization

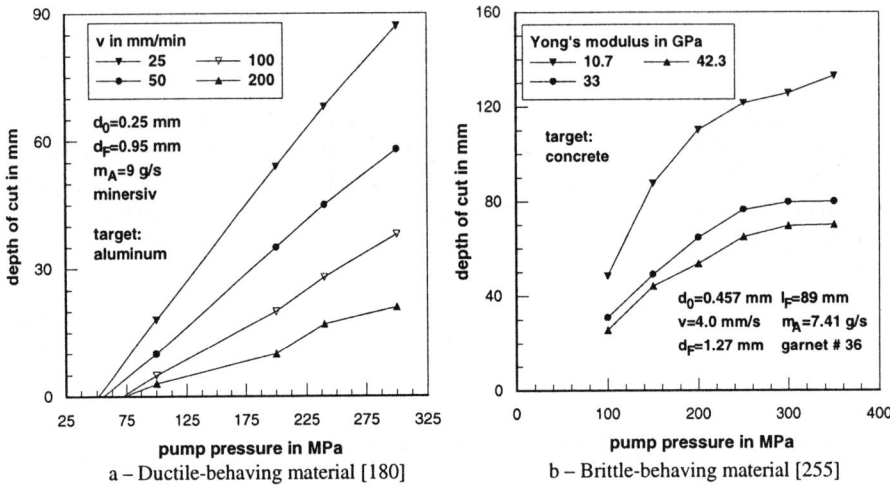

Figure 7.3 *Relation between the pump pressure and depth of cut*

Figure 3.17 shows that the momentum-transfer coefficient has an optimum and decreases at high pump pressures if all other process parameters, especially the orifice diameter, remain constant. The pump pressure influences the actual abrasive-mass flow rate. The maximum possible abrasive-mass flow rate is determined by calibrating the abrasive-delivery unit. But the amount of abrasives that is sucked into the mixing chamber depends on the velocity of the air flow that is a function of the suction pressure in the delivery hose. Louis and Meier [76] show that the suction pressure increases with an increase in the pump pressure. Also, the air-flow rate in the mixing chamber increases with a rise in the pump pressure (Figure 3.13). From the point of view of momentum transfer, a certain water-jet momentum is required to effectively accelerate the abrasive-mass flow rate. This relation is illustrated by plotting the depth of cut against the abrasive-mass flow rate for different pressure levels. The optimum abrasive-mass flow rate increases as the pump pressure increases. These results give

$$v_p(p) = \eta_T(p) \cdot \frac{v_0(p)}{1 + \frac{\dot{m}_A(p)}{\dot{m}_W(p)}} \ . \tag{7.8}$$

Neusen et al. [145] observe that the pump pressure influences the distribution of water as well as of abrasives in the abrasive water jet (Figure 4.7). A summary of all these effects gives the relation plotted in Figure 7.3 and approximated by Eq. (7.6).

The general relation supposed by Eq. (7.6) is valid also for material-removal processes (section 9.2) as well as for the cutting rate [254].

Figure 7.3 distinguishes between three process stages.

7.2.1.2 Incubation Stage and Threshold Pressure

The first stage, $p<p_{thr}$, is an incubation stage. In this pressure range, no material removal takes place; although, the removal process is nonvisibly introduced in the material. The parameter p_{thr} is a threshold value that has to be overcome for a measurable material removal. For brittle materials, there exist some limited evidence to link this parameter to conventional material properties. For brittle materials, Evans et al. [170] develop a threshold concept for solid-particle erosion. In this model, the threshold velocity of the particles is

$$v_{Pthr} \propto K_{Ic}^2 \cdot c_M^{0.33}. \tag{7.9}$$

Momber et al. [252] carry out measurements on concrete and show that the concept of Evans et al. holds for the abrasive water-jet cutting of pre-cracked brittle materials.

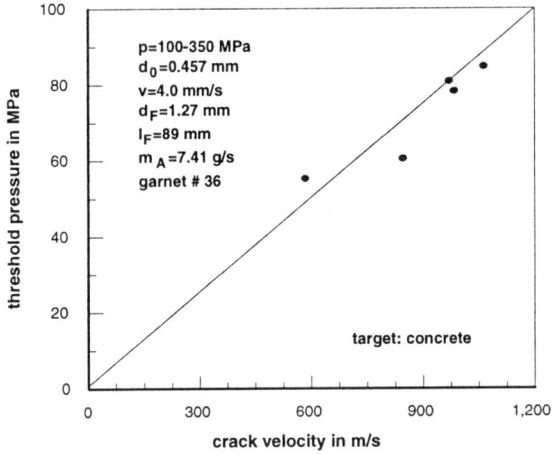

Figure 7.4 *Relation between the threshold pressure and crack velocity [252]*

Momber and Kovacevic find also a relation between the threshold pressure for abrasive water-jet cutting, the material crack-velocity [252] (Figure 7.4), and the energy absorbed during the compressive standard test [267], respectively. This type of material where the general failure process - micro-crack network generation and fracture - is similar in case of conventional testing as well as in abrasive water-jet cutting, offers some possibilities of linking the threshold pressure to general material properties. In contrast, for materials that are removed by micro-machining processes, such as chip formation, there is hardly any conventional standard test showing similar failure conditions.

Several authors show that the threshold pressure for a given material depends on the abrasive-mass flow rate [45, 282, 283], traverse rate [45, 69, 148, 151, 263, 254], water-orifice diameter [251, 264, 283, 284], and the focus diameter [282,

284]. Figure 7.5 presents some results. Nevertheless, the threshold pressure does not necessarily characterize the material resistance against abrasive water-jet cutting. It is shown that materials with identical threshold pressures react very differently as the material-removal process is introduced [189].

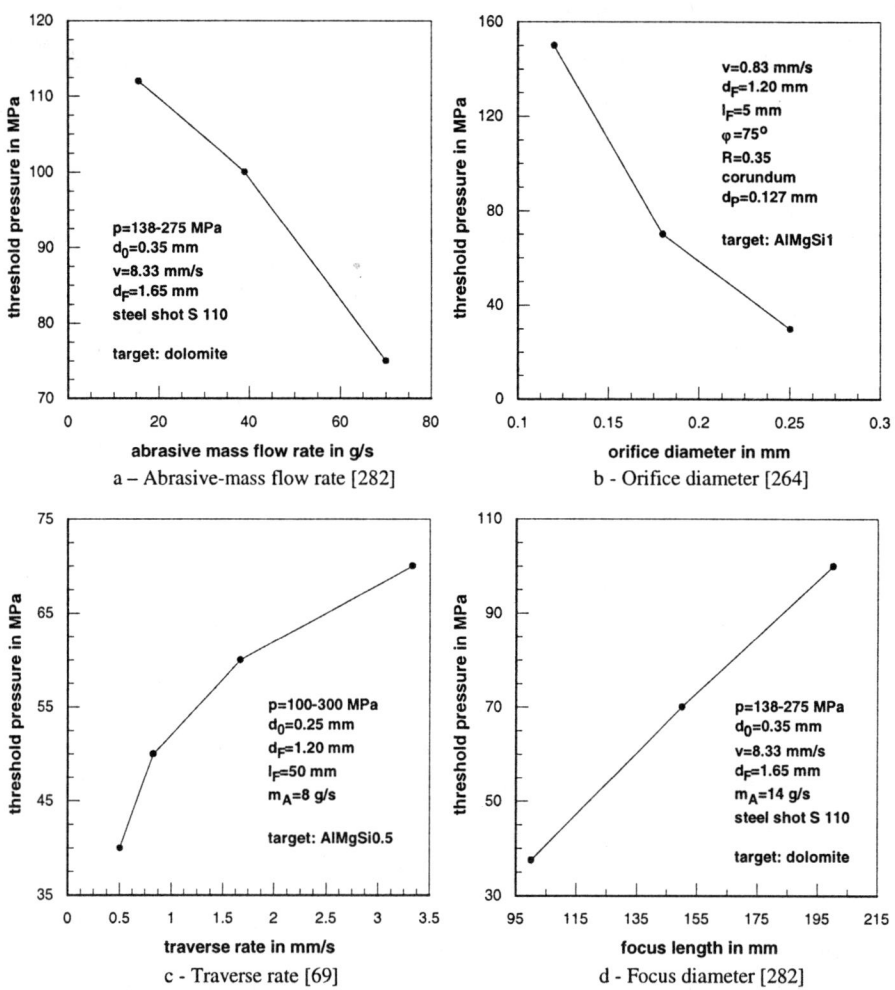

Figure 7.5 *Parameter influence on the threshold pressure*

7.2.1.3 Linear Stage and Decreasing Stage

In the second stage, $p_{thr}<p<p_{Cr}$, called the linear stage, the depth of cut increases linearly with an increase in the pump pressure. In this stage, $C_2=1$. Every increase in the pump pressure leads to a proportional increase in the depth of cut.

7.2 Influence of Hydraulic Process Parameters

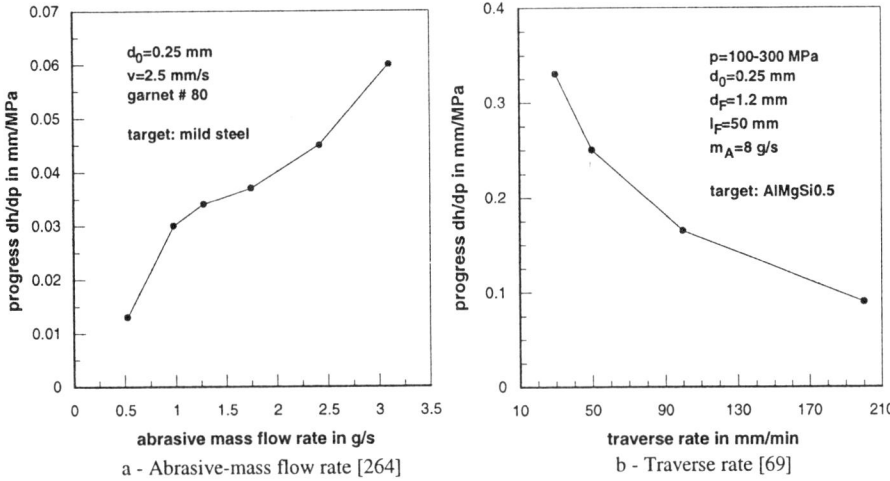

Figure 7.6 *Parameter influence on the progress parameter dh/dp*

The factor of proportionality is dh/dp>1 and depends on all the process conditions listed above as well as on the properties of the target material. Figure 7.6 presents some relations.

In the last stage, $p>p_{Cr}$, called the decreasing stage, the progress of the function drops. In this range, $0<C_2<1$. This behavior is significant at very high pressures [264] and for quasi-brittle materials, such as concrete [190, 191] rocks [253], and ceramics [188, 189]. The pressure $p=p_{Cr}$ is a limit for effective cutting. Beyond this pressure range, the cutting process becomes ineffective.

7.2.1.4 Optimization Aspects

With Eq. (7.6) and $E_A \propto p^{1.5}$ (section 3.1), the efficiency of the material removal can be estimated. For an effective process, the specific energy, $E_{Sp}=E_A/h$, must show a minimum

$$\frac{dE_A}{dh} = \frac{d(C_2 \cdot p^{1.5})}{d[C_1 \cdot (p - p_{thr})]} = 0. \tag{7.10}$$

The solution of Eq. (7.5) is

$$p(E_{SP\,min}) = 3 \cdot p_{thr}. \tag{7.11}$$

At a pressure range of about three times the threshold pressure, the efficiency of the material removal is maximum; whereas, the specific removal energy and specific cutting energy shows minimum values. For a non-linear relation as assumed for high pressures, the efficiency of the cutting process also decreases with an increase in the pump pressure. Figure 7.7 illustrates this situation. Momber et al. [228] alternatively prove Eq. (7.11) by sieve analysis of removed erosion debris. These authors found a decrease in the wear-particle size with an increase in the pump pressure (Figure 5.34). Experimental results from Gou et al. [254] suggest that Eq. (7.11) holds also for the specific cutting rate.

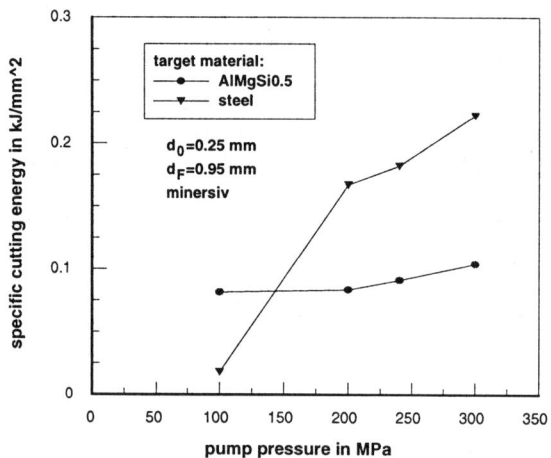

Figure 7.7 *Relation between the pump pressure and specific energy [254]*

Furthermore, if an optimum ratio between the water-mass flow rate and abrasive-mass flow rate is required, any change in the pump pressure requires a new calibration of the actual abrasive-mass flow rate.

Guo et al. [254] discover that the water exploitation almost linearly improves with an increase in the pump pressure.

7.2.2 Influence of Water-Orifice Diameter

7.2.2.1 General Trends

Figure 7.8a shows a typical relation between the depth of cut and the water-orifice diameter. The depth of cut increases with an increase in the diameter, but the progress of the function drops for large orifice diameters. A simple expression of this relation is

$$h(d_0) = C_4 \cdot d_0^{C_5}, \text{ with} \quad (7.12)$$
$$0 < C_5 < 1.$$

7.2 Influence of Hydraulic Process Parameters

In the range of small-sized and medium-sized orifice diameters, the cutting process is very sensitive to changes in this parameter.

The major contribution of the orifice diameter to the material-removal process is the determination of the water-flow rate as well as of the water-jet momentum. Therefore, the momentum-transfer capability of a water jet and the abrasive-particle velocity increases with larger orifice sizes. Chen and Geskin [46] prove this relation by laser-anemometer measurements on abrasive water jets.

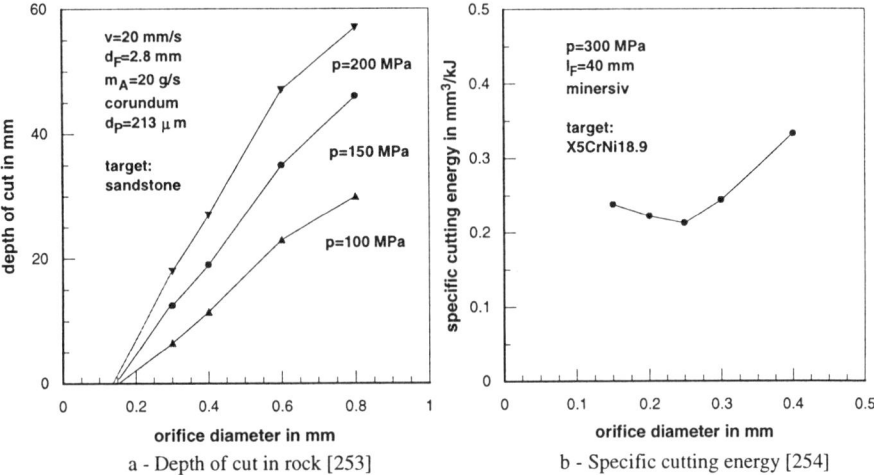

a - Depth of cut in rock [253]   b - Specific cutting energy [254]

Figure 7.8 *Relation between the orifice diameter and cutting results*

Simplified, the relation between the abrasive-particle velocity and orifice diameter is

$$v_P(d_0) \propto \frac{v_0}{1+\frac{Const}{d_0^2}} \quad . \tag{7.13}$$

Note that for high values of the orifice diameter, Eq. (7.13) approximates a saturation particle velocity. Therefore, an increase in the orifice diameter beyond a certain value does not significantly improve the abrasive-particle velocity. Blickwedel [69], Guo et al. [254], and Heβling [253] experimentally find a decrease in the progress of the depth-function at large orifices. During the cutting of ceramics by abrasive water jet, Zeng and Kim [158] even notice a decrease in the depth of cut at large orifice diameters

7.2.2.2 Threshold Orifice Diameter

Applying the concept of a threshold abrasive-particle velocity (section 7.2), Eq. (7.12) suggests the existence of a threshold orifice diameter, d_{thr}. Several authors observe this threshold diameter for metal cutting [157, 285] and for rock cutting [253]. Figure 7.8a shows some examples. Experimental results from Blickwedel [69], Guo et al. [254], and Laurinat et al. [286] evidence the presence of a threshold orifice diameter for cutting processes as well as for material-removal applications (section 9.2). Even if the concept of a threshold parameter is rejected, the experimental results clearly show that at least a range of very shallow depths of cut exists for very small orifices. In this range, the water-jet momentum is to low to effectively accelerate the abrasive particles. Therefore, at least a 'transition'-orifice diameter is proposed that has to overcome for an effective abrasive acceleration. This restriction is more pronounced at high abrasive-mass flow rates as Ansari [287] observes for the abrasive water-jet turning.

7.2.2.3 Optimization Aspects

For a given abrasive-mass flow rate, there is an optimum water-orifice diameter and, so, an optimum water-jet impulse for an optimum momentum transfer. Figure 7.8b shows this relation. In this figure, notice an optimum orifice diameter, d_{0opt}, at about $d_{0opt}=2.5 \cdot d_{0thr}$ that indicates optimum momentum-transfer conditions for a certain balance between the orifice diameter and pump pressure. For rock cutting, Heβling [253] finds similar results. Furthermore, if an optimum ratio between the water-mass flow rate and abrasive-mass flow rate is required, any change in the orifice diameter requires a new calibration of the actual abrasive-mass flow rate.

From the point of view of water consumption, the water exploitation is better for medium-sized orifice diameters. The cutting area produced per given water volume decreases for large orifice diameters [254].

7.3 Influence of Cutting Parameters

7.3.1 Influence of Traverse Rate

7.3.1.1 General Trends

Figure 7.9 shows typical relations between the traverse rate and depth of cut. This tendency is first discovered during the cutting of reinforced concrete by an abrasive water jet [288]. Walters and Saunders [136] show that this relation is the same for suspension-abrasive water jets. The dependence is very significant for low traverse rates. For high traverse velocities, the depth of cut approximates a saturation value or even crosses the abscissa. A simple mathematical expression of Figure 7.6 is

7.3 Influence of Cutting Parameters   205

$$h(v) = C_7 \cdot v^{C_8}, \quad C_8 < 1. \tag{7.14}$$

In this equation, C_8 is a negative number that lies usually between $-0.4 < C_8 < -1.0$ [148]. The constant C_7 considers the influence of other process parameters. Blickwedel [69] gives a more accurate approximation that considers a relation between C_8 and v (section 6.3.1),

$$h(v) = C_7 \cdot v^{-\left(0.86 + \frac{2.09}{v}\right)}. \tag{7.15}$$

The constant C_7 in Eqs. (7.14) and (7.15) depends on other process parameters, such as the pump pressure (Figure 7.9a), the abrasive-mass flow rate [251], the water-orifice diameter [69, 283], and the abrasive-grain diameter [184]. Figure 7.9a shows the relation between the traverse rate and depth of cut for different materials.

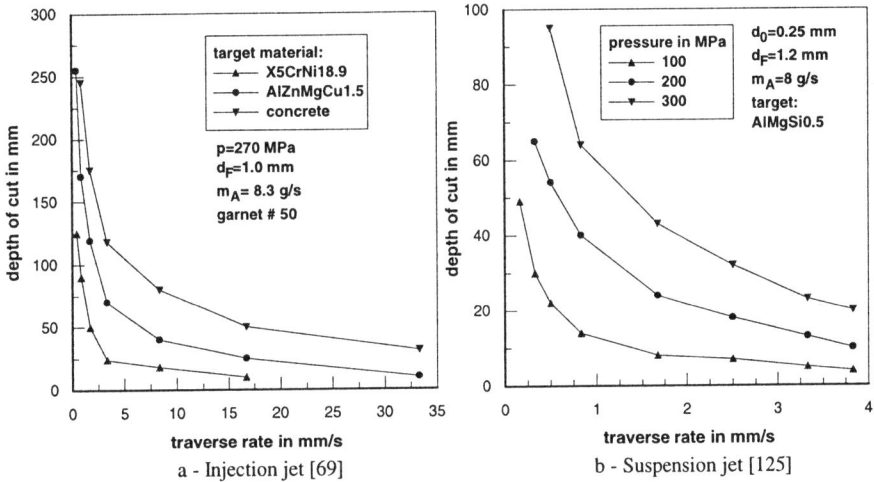

Figure 7.9 *Relation between the traverse rate and depth of cut*

7.3.1.2 Threshold Traverse Rate

Eq. (7.14) does not consider the critical case that the function crosses the abscissa at a critical traverse rate level. Heβling [253], for example, observes on rocks that the depth of cut is identical for water jet and abrasive water jet at high traverse rates. Momber and Kovacevic [194] consider this case and suppose

$$h(v) = h_{max} \cdot \left[-\left(\frac{\ln v}{\ln v_{cr}}\right) + 1\right]. \tag{7.16}$$

In the equation, v_{Cr} is a critical traverse rate for h=0. For brittle pre-cracked materials, this critical traverse rate is related to the crack velocity of the target materials [255].

The depth-of-cut-variation significantly decreases at higher traverse rates. The higher the traverse rate, the smoother the bottom of the generated kerf is [289].

7.3.1.3 Exposure Time

The major influence of the traverse rate on the machining process is the determination of the exposure time. Writing

$$v = \frac{dx}{dt}, \qquad (7.17)$$

yields

$$t = \frac{1}{v} \cdot \int_0^x dx = \frac{x}{v}. \qquad (7.18)$$

For cutting a workpiece the length $x=L_H$, the exposure time is $t=L_H/v$. The local exposure time that is the period of time over which the abrasive water jet cross-area acts on the material, is $t=d_F/v$. Therefore, the kinetic energy stored in the workpiece by an abrasive water jet is inversely proportional to the traverse rate.

7.3.1.4 Particle-Impact Frequency and Damping Effects

For a given abrasive-mass flow rate, an increase in the traverse velocity yields a reduction in the number of impacting abrasive particles as well as in the volume of the water that penetrates the erosion site (section 2.5). Nevertheless, the reduction in the abrasive-mass flow is not the only reason for the reduced cutting capability at high traverse rates. Figure 7.18 show typical relations between the depth of cut and abrasive-mass flow rate. In this figure, the depth of cut dramatically changes in the range of small abrasive-mass flow rates (low numbers of impacting abrasive particles). In contrast, high traverse rates do not significantly influence the depth of cut (low number of impacting abrasive particles).

A further impact of the traverse rate is its influence on damping effects at the bottom of the cut. This effect plays an important role for low traverse rates. The parameter h_{max} in Eq. (7.11) considers this case. Through acoustic-emission measurements, Momber et al. [190] find further experimental evidence for these damping effects (sections 10.7 and 10.8). Section 5.7 discusses general aspects of damping.

7.3.1.5 Influence on the Cutting Rate

Figure 7.10 plots typical functions for the cutting rate. From this figure, Eq. (7.14) is an oversimplification of the traverse problem because the equation yields $A_h(v) \propto v^C$, $C>0$. This result is valid only for low and medium traverse rates. Therefore, a more flexible function,

$$A_h(v) = C_9 \cdot v^{C_{10}} \cdot e^{C_{11} \cdot v}, \qquad (7.19)$$

better characterizes the relation between the traverse rate and cutting rate. This function is very sensitive to changes in the machining process. The equation also estimates the material-removal rate as a function of the traverse rate.

For the special case $C_{10}=C_{11}=0$, Eq. (7.19) yields A_h=constant. Matsui et al. [236] observe this case for cutting alumina ceramics by abrasive water jets. The case $0<C_{10}<1$ and $C_{11}=0$ characterizes the initial progress for low traverse rates. The maximum of the cutting rate comes from

$$\frac{dA_h}{dv} = 0 \rightarrow v_{Opt} = \frac{-C_{10}}{C_{11}}. \qquad (7.20)$$

Heßling [253] observes that under identical process conditions the values for v_{Opt} are much larger for water jets compared to abrasive water jets. This observation indicates that collision and damping effects due to abrasive-particle movements play an important role in the cut-formation process.

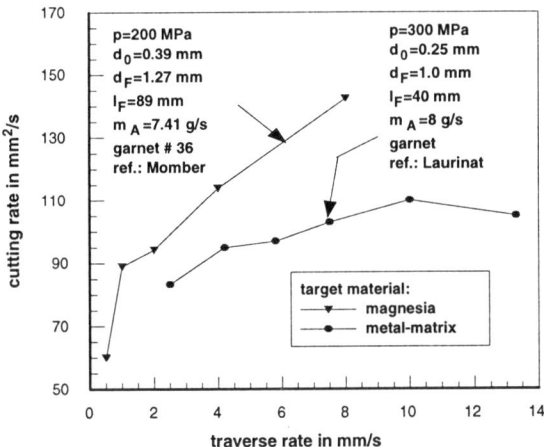

Figure 7.10 *Relation between the traverse rate and cutting rate [199, 290]*

7.3.2 Influence of Number of Passes

7.3.2.1 General Trends

The number of passes describes the effect of multipass cutting. The basis for this material-removal strategy is the relation between the depth of cut and exposure time as shown in Figure 7.11. In the figure, note that optimum cutting conditions exist at very low exposure times, t_{opt}. For larger time intervals, the efficiency drops due to damping and friction effects. The damping can be avoided, and the efficiency can be increased by cutting a workpiece several times with comparatively high traverse rates. As a result, a typical maximum in the cutting rate appears (Figure 7.10).

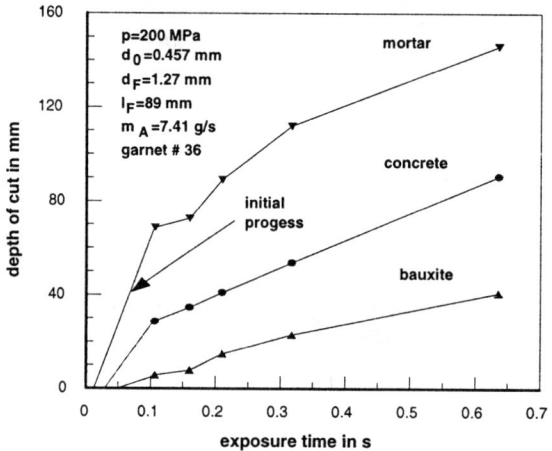

Figure 7.11 *Relation between the exposure time and depth of cut [191]*

From Eq. (7.11), the optimum traverse rate is

$$v_{opt} = \frac{L_h}{t_{opt}}. \tag{7.21}$$

The optimum number of passes, n_{Popt}, is

$$n_{Popt} = f(v) = \frac{t}{t_{opt}} = \frac{v_{opt}}{v}, \tag{7.22}$$

with $n_P \in (G)$.

7.3.2.2 Multipass Cutting

Figure 7.12a plots a typical relation between the number of passes and the depth of cut. Notice that the relation is almost linear in the beginning, but the progress drops at a certain critical number of passes. The same trend is known for water-jet cutting applications [291]. Hashish [283] shows that the effect of the multi-pass cutting is determined by the abrasive-particle diameter.

Adding lines of equal energies in a number-of-passes diagram (Figure 7.12b), an optimum combination between the number of passes and the traverse rate exists that yields a maximum depth of cut. This result is due to a balance between the impact damping and wall friction at a certain depth of cut [292]. The difference between the measured depth of cut and the depth of cut obtained from an idealized linear relation between the depth of cut and the number of passes, Δh_P, is a result of friction effects. At a given number of passes, n_{cr}, the friction losses become significant and cover the positive effect of reduced damping [292].

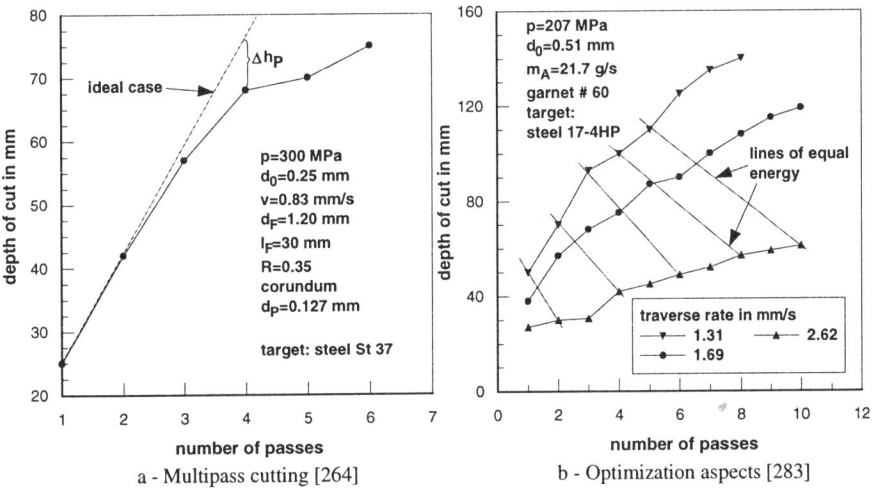

Figure 7.12 *Relation between the number of passes and depth of cut*

The cutting rate almost linearly decreases with an increase in the number of passes [283]. Therefore, the multi-pass strategy is efficient for deep-cut generation. The strategy is inefficient for obtaining high cutting rates. On the other side, Hu et al. [293] show that the material-removal rate increases with multi-pass cutting.

7.3.3 Influence of Standoff Distance

7.3.3.1 General Trends

Barton [294] first investigates the influence of the standoff distance on the depth of cut in abrasive water-jet cutting. He finds an almost linear decrease in the depth of cut with an increase in the standoff distance. Blickwedel [69], Guo et al. [254], Chung et al. [270], Kovacevic [271], and Kovacevic et al. [272] confirm this result for cutting different materials by injection-abrasive water jets. Figure 7.13a gives typical examples. Mathematically, a negative function with a low power-coefficient fits the experimental results,

$$h(x) = C_{12} \cdot x^{C_{13}}, \quad C_{13}<0. \tag{7.23}$$

For suspension-abrasive water jets, Brandt et al. [125] and Liu et al. [134] observe the same relations (Figure 7.13b).

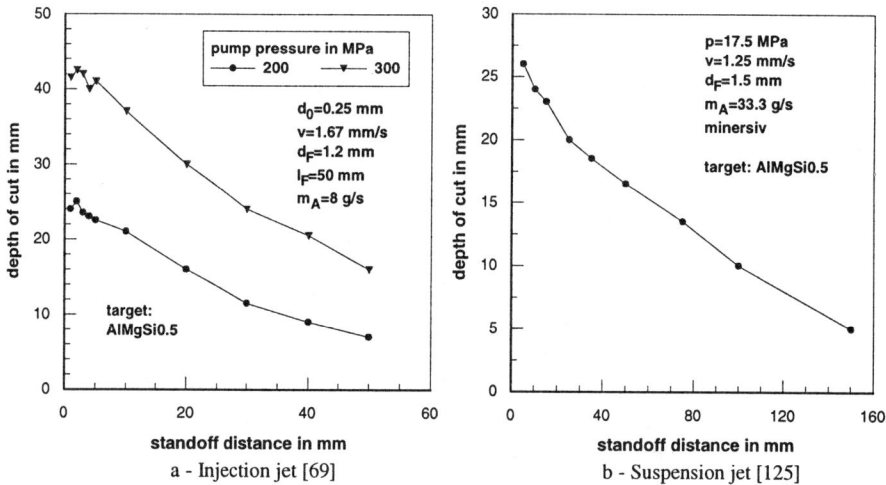

Figure 7.13 *Relation between the standoff distance and depth of cut*

Changes in the standoff distance do not significantly influence the velocity of the abrasive particles [46, 47, 58, 295]. Therefore, the optimum standoff distance that is noticed for material-removal processes, is the result of the radial expansion of the abrasive water jet that yields a larger exposed area. The general increase in the width of cut with an increase in the standoff distance expresses this effect.

7.3.3.2 Special Observations

Under certain process conditions, Oweinah [264] and Guo et al. [203] notice peaks in the depth-of-cut functions. Oweinah explains this observation due to back-flow effects in the focus. If the distance between the nozzle outlet and the specimen is to small, the abrasive-water flow is damped or decelerated by the target surface that generates shallower depths of cut. To verify this assumption, Oweinah [264] carries out cutting experiments under different impact angles. He notices the 'peak'-behavior at impact angles of $\varphi=90°$. If the impact angle changes to $\varphi=75°$, the 'peak'-effects disappears. Gou et al. [254] suggest an optimum standoff distance for cutting at about $x_{opt}=2.0$ mm.

7.3.4 Influence of Impact Angle

7.3.4.1 Influence on Ductile-Behaving Materials

Barton [294] first investigates the influence of the impact angle on the abrasive water-jet cutting. For mild steel, he notices a maximum depth of cut at angles of $\varphi=90°$. Hashish [296] and Oweinah [264] make similar observations. For cutting stainless steel and aluminum by abrasive water jets, these references detect maximum depths of cut at impacting angles between $\varphi=75°$ and $\varphi=80°$. Figure 7.14a gives an example. The material in this figure generally shows a ductile behavior. Assuming a micro-cutting material-removal mechanism, as Hutchings [164] suggests, an inclination of the jet improves the chip generation-process. Therefore, in linear cutting of this group of materials, a jet forward-angle of about 10° increases the depth of cut.

7.3.4.2 Influence on Brittle-Behaving Materials

It is known from solid-particle erosion experiments that brittle-behaving materials often show maximum erosion rates at impact angles of $\varphi=90°$. Figure 7.14b proves this tendency for abrasive water-jet cutting of several brittle composite materials. For cutting ceramics with abrasive water jets, Wada and Kumon [297] find that the depth of cut has a maximum value at perpendicular impact angles. Ramulu et al. [126] observe during the machining of metal-matrix composites that the material-removal mechanism changes from a micro-cutting of the matrix at shallow impact angles to a cracking process at large angles. For impact angles of $\varphi=20°$, significantly more surface cracks appear in the material compared to conditions of very shallow ($\varphi=5°$) impacting abrasives.

Wada and Kumon [298] observe that for 'soft' abrasive materials, the material-removal performance in ceramics is optimum at $\varphi=90°$. For harder abrasive materials, especially silica carbide, the optimum material removal takes place at $\varphi=120°$. Wada and Kumon explain this fact by a secondary erosion effect by the

212   7. Process-Parameter Optimization

abrasive-water mixture back-flow at very high impact angles. Because the weaker abrasive material breaks during the primary impact erosion, it can not contribute to the secondary erosion.

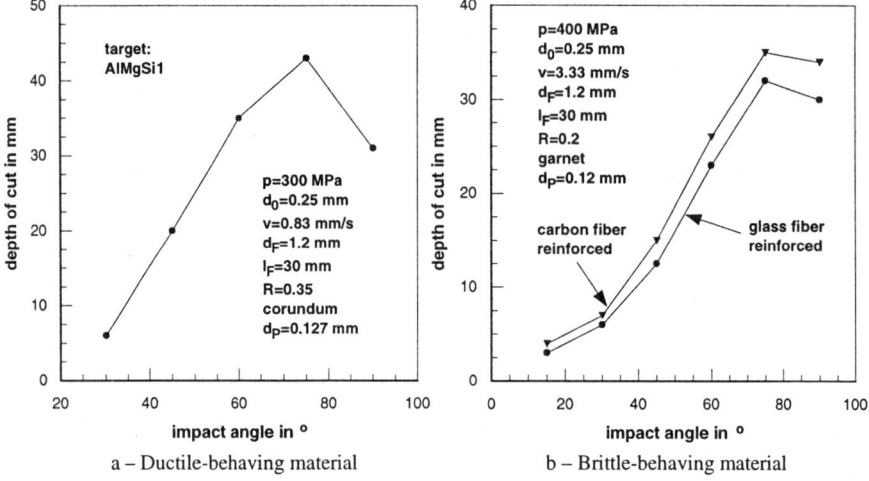

a – Ductile-behaving material   b – Brittle-behaving material

Figure 7.14 *Relation between the jet impact-angle and depth of cut [264]*

7.4 Influence of Mixing Parameters

7.4.1 Influence of Focus Diameter

7.4.1.1 General Trends

Figure 7.15 gives general tendencies between the focus diameter and the depth of cut. The depth of cut decreases with an increase in the focus diameter. For ceramics, Zeng et al. [251] obtain similar results. In contrast, Oweinah [264] observes an increase in the depth of cut at large focus diameter that indicates a more complex relation between the both parameters.

Laser-anemometer measurements on abrasive water jets show that the abrasive-particle velocity decreases with an increase in the focus diameter [46, 68]. In an analysis of the abrasive water-jet mixing process in the focus, Blickwedel [69] finds that the final abrasive-particle velocity depends on the density of the abrasive-water-air mixture: the denser the mixture, the higher the abrasive exit-velocity. This condition exists for small-diameter focuses. Figure 3.13 shows that more air is delivered at large focus diameters. This air results in a lower mixture density and in a reduced abrasive velocity. Himmelreich [47] proves that large focus diameters are characterized by comparatively high degrees of turbulence (Figure 4.10). These arguments cause the drop in the depth of cut beyond a certain focus diameter.

On the other hand, a small diameter focus supports particle collision, friction, and abrasive fragmentation that contribute to an ineffective mixing-and-acceleration process. Himmelreich [47], who shows that the average jet velocity decreases with a decrease in the focus diameter for practical abrasive-mass flow rates, verifies this assumption.

7.4.1.2 Optimum Focus Diameter

From these results, an optimum focus diameter exists. Blickwedel [69], Heßling [253], and Kovacevic [271] observe this optimum focus diameter (Figure 7.15b). This optimum value depends on the machined material as well as on the other process conditions.

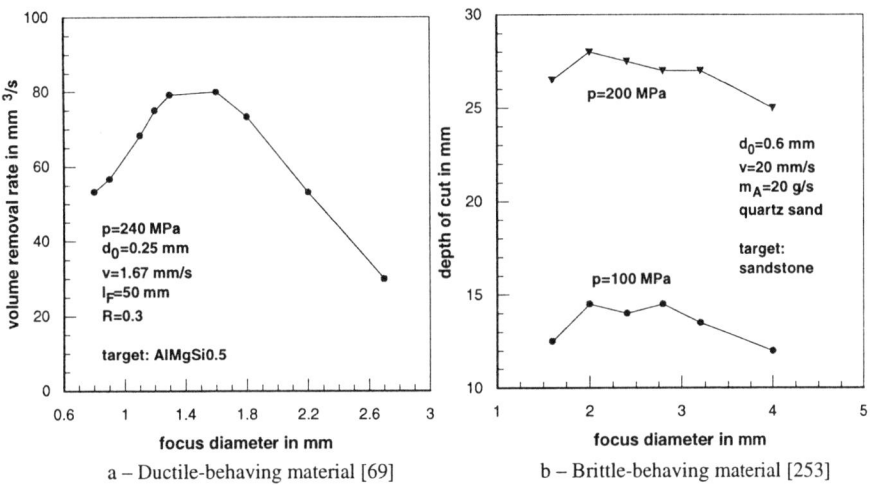

a – Ductile-behaving material [69]   b – Brittle-behaving material [253]

Figure 7.15 *Relation between the focus diameter and depth of cut*

Blickwedel [69] supposes an optimum ratio between the focus diameter and water-jet orifice diameter,

$$d_{Fopt} = (3...4) \cdot d_0 . \tag{7.24}$$

Mazurkiewicz et al. [67] suggest an optimum ratio between the abrasive-particle diameter and focus diameter,

$$d_{Fopt} = 3 \cdot d_P . \tag{7.25}$$

During the abrasive water-jet cutting of rocks, Heßling [253] observes that optimum cutting occurs at this ratio.

214   7. Process-Parameter Optimization

Blickwedel [69] investigates the relation between the focus diameter and abrasive-mass flow rate. The author notices an optimum abrasive exploitation for a small-diameter focus. The ratio between the depth of cut and abrasive-mass flow rate shows maximum values at small focus diameters.

7.4.2 Influence of Focus Length

7.4.2.1 General Trends

Figure 7.16 shows typical relations between the focus length and depth of cut. Initially, the depth of cut linearly increases with an increase in the focus length. This is due to the fact that a certain acceleration distance is necessary to accelerate the injected abrasive particles (Figure 3.19). Beyond this critical acceleration distance, no further increase in the abrasive velocity happens, but now friction due to the spreading water jet starts to act. This friction reduces the abrasive velocity that leads to a drop in the depth of cut.

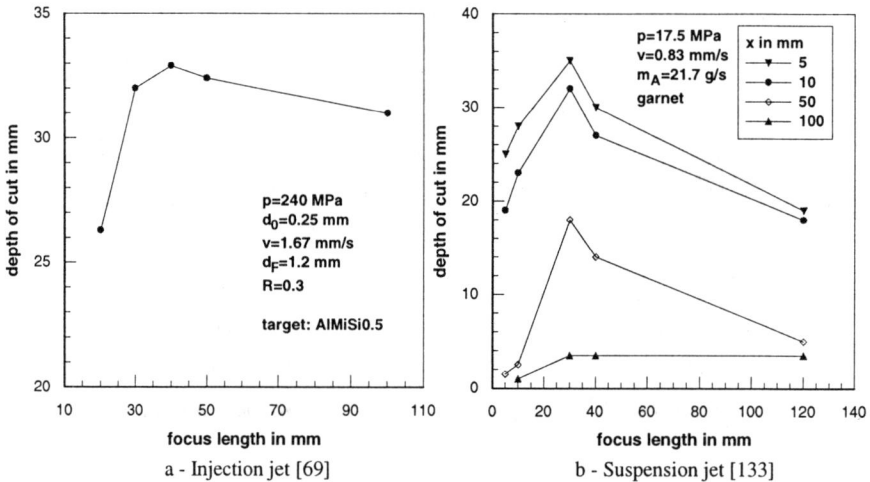

Figure 7.16 *Relation between the focus length and depth of cut*

The particular structure of the focus-length-function as well as the location of the optimum focus length depends on the pump pressure [282], the standoff distance [133], and the abrasive-mass flow rate. Generally, the optimum length increases with an increase in the pump pressure and standoff distance, and with a decrease in the abrasive-mass flow rate. The influence of the pump pressure explains by the fact that a higher-velocity water jet has the capability to accelerate the abrasives to comparatively high velocities. This process requires longer focusing nozzles. The reduced optimum focus length for high abrasive-mass flow rates is a result of the increased interaction between the focus wall and abrasive particles. Figure 7.16b illustrates that these relations are also valid for suspension-abrasive water jets.

7.4.2.2 Optimum Focus Length

Heβling [253] shows that the optimum acceleration distance is strongly influenced by the abrasive-material density. Heavy abrasives, such as steel shot, require longer acceleration distances. Galecki and Summers [282] observe that larger abrasive particles require longer focus lengths, which is substantially the same trend as that between the abrasive-material density and acceleration distance. The heavier the particle, the longer the required acceleration distance (Figure 7.17). Figure 7.17 also shows that the advantage of a long focus significantly reduces if broken abrasive material is used compared to round particles. This result is due to the increased friction force acting on the non-regular particles. Nevertheless, the influence of the abrasive-material density on the optimum focus length is much more pronounced.

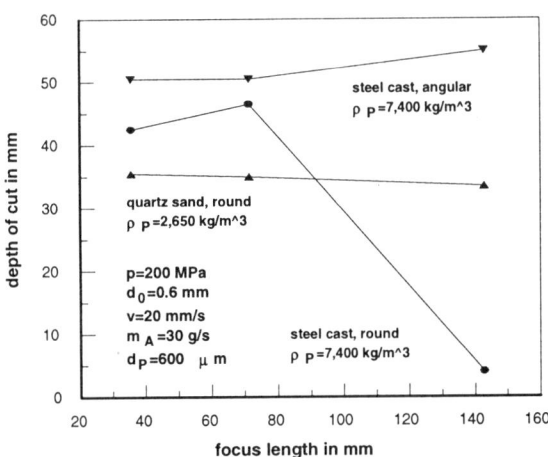

Figure 7.17 *Optimizatiion of the focus length [253]*

From the point of view of abrasive consumption, longer focus nozzles are recommended [69].

Results of fluid-dynamic experiments [47] and of cutting tests with abrasive water jets [69] recommend a relation between the focus length and the focus diameter in order to obtain optimum cutting results,

$$l_{Fopt} = (25...50) \cdot d_F \qquad (7.26)$$

7.5 Influence of Abrasive Parameters

7.5.1 Influence of Abrasive-Mass Flow Rate

7.5.1.1 General Trends

Figure 7.18 shows typical relations between the abrasive-mass flow rate and depth of cut. Momber [148] investigates the structure of this function in detail. Based on a reaction kinetics model, Momber derives

$$h(\dot{m}_A) = k \cdot \dot{m}_A^m . \tag{7.27}$$

In the equation, the power exponent, m, is a function of the abrasive-mass flow rate (section 6.5). For small abrasive-mass flow rates, m=1. The value of m decreases up to m=0 for optimum abrasive-mass flow rates and becomes m<0 for very high abrasive-mass flow rates. From the point of view of abrasive consumption, the optimum range is at low abrasive-mass flow rates and m=1. In this range, every increase in the abrasive-mass flow rate leads to a proportional increase in the depth of cut and in the material-removal rate [293].

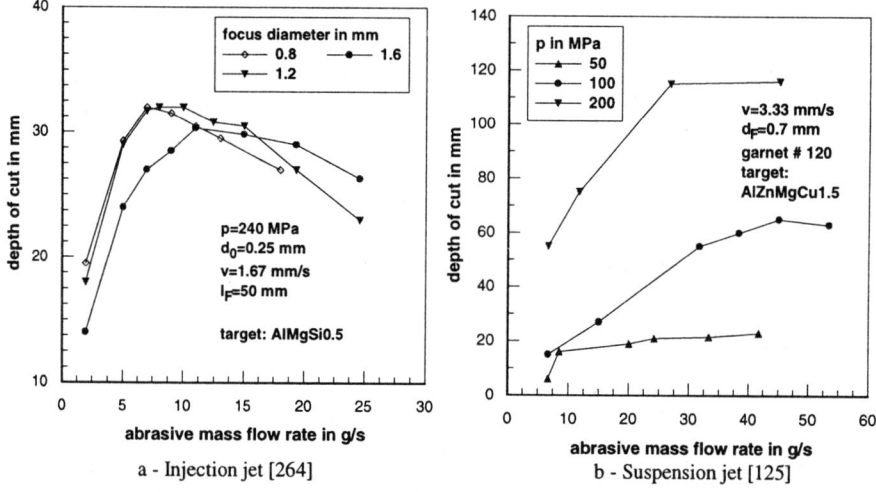

a - Injection jet [264]   b - Suspension jet [125]

Figure 7.18 *Relation between the abrasive-mass flow rate and depth of cut*

The typical abrasive-mass-flow-rate function is a result of several effects. First of all, the abrasive-mass flow rate determines the number of impacting abrasive particles as well as their kinetic energies. The higher the abrasive-mass flow rate, the higher the number of particles involved in the mixing-and-cutting processes. Assuming no contact between the single abrasive grains in the course of mixing and cutting, every increase in the abrasive-mass flow rate leads to a proportional increase in the depth of cut. Figure 7.18 shows that this holds for relatively low abrasive-mass flow rates. Hu et al. [293] find a linear increase in the depth of cut as the number of impacting abrasive particles increases. For higher abrasive-mass flow rates, some damping mechanisms, such as particle collision and the generation of water-solid films, occur in the mixing chamber, in the acceleration focus, and in the cut. Also, the limited kinetic energy of the water jet distributes over a very high number of particles that leads to a decrease in the kinetic energy of the single particles. This effect cancels the positive effect of the higher impact frequency. Figure 2.9 schematically illustrates these relations.

Chen and Geskin [46], Miller and Archibald [78] and Himmelreich and Rieß [147] measure a drop of the abrasive-particle velocity with an increase in the abrasive-mass flow rate. In contrast, Neusen et al. [48] does not find any influence of the abrasive-mass flow rate on the abrasive velocity. Measurements from Kovacevic et al. [272] show both tendencies. The second tendency is valid for relatively low abrasive-mass flow rates; whereas, the first relation holds for comparatively high abrasive-mass flow rates. From Eq. (3.25), the relation between the abrasive velocity and abrasive-mass flow rate is

$$v_P(\dot{m}_A) = \eta_T(\dot{m}_A) \cdot \frac{v_0}{1 + \frac{\dot{m}_A}{\dot{m}_w}}. \tag{7.28}$$

There is some experimental evidence that the efficiency parameter, η_T, decreases with an increase in the abrasive-mass flow rate (Figure 3.17b). Also, the turbulence in an abrasive water jet increases with an increase in the abrasive-mass flow rate (Figure 4.10). Nevertheless, mathematically, if the abrasive-mass flow rate is comparatively high, any additional increase does not contribute to a notable modification of the abrasive-particle velocity.

7.5.1.2 Optimization Aspects

Figure 7.18 shows that the function for the abrasive-mass flow rate generally exhibits a maximum at an optimum abrasive-mass flow rate. This optimum depends on the material as well as on the process parameters, including the pump pressure [125, 251, 254, 282, 284), the orifice diameter [254, 283, 284], the traverse rate [157], the focus diameter [125, 251, 284], the focus length [264], the abrasive-particle diameter [207, 264, 283], the abrasive-particle shape [264], the abrasive type [207, 264, 299] and the target material [207, 254, 264, 285]. Figure 7.19 plots some relations.

According to Momber et al. [189], the location of the optimum abrasive-mass flow rate depends on the deformation behavior of the target materials. Whereas, materials with the ability of plastic deformation reach the optimum at comparatively high abrasive-mass flow rates, very brittle materials reach the optimum region at low abrasive-mass flow rates. This difference is due to the higher sensitivity of brittle materials to the impact energy of the abrasive particles. In contrast, a material that reacts with plastic deformation is more sensitive to the number of the impacting particles. Therefore, an increase in the abrasive-mass flow rate is beneficial for these materials to overcome the plastic-deformation capability.

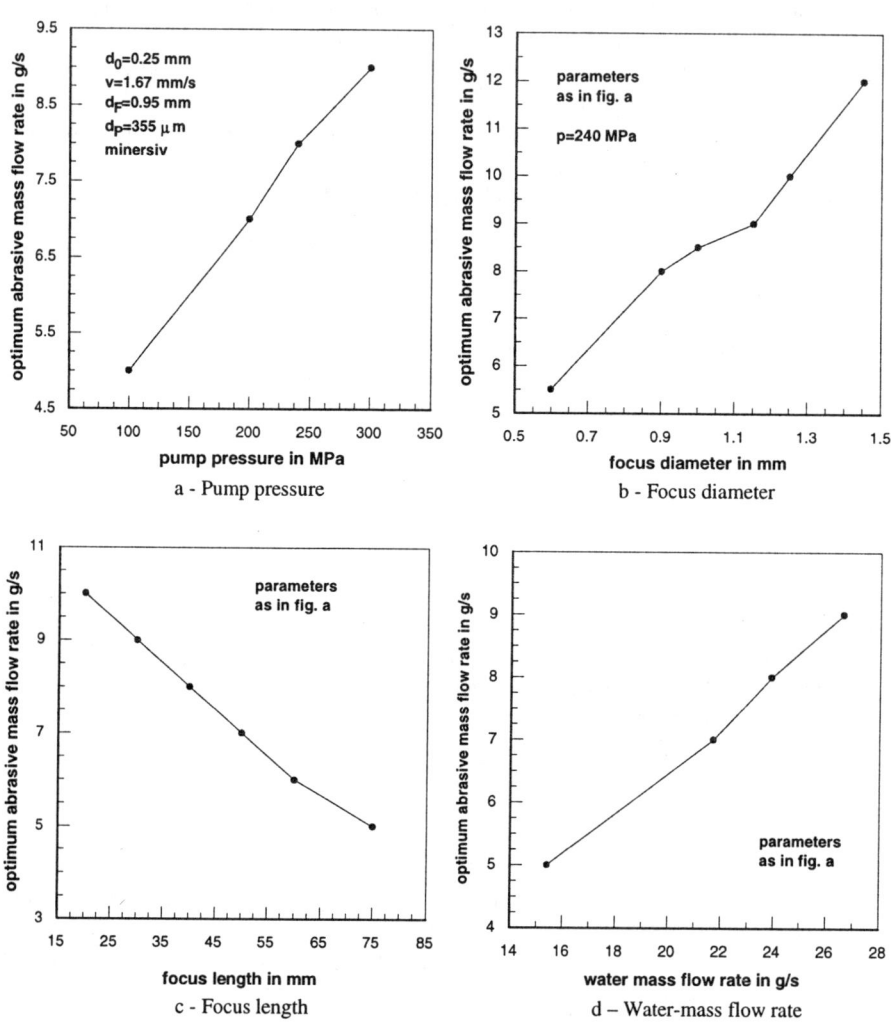

Figure 7.19 *Parameter influence on the optimum abrasive-mass flow rate [180]*

7.5.1.3 Influence on the Cutting Rate

Figure 7.20a shows the relation between the abrasive-mass flow rate and cutting rate. This trend is identical to that of the depth of cut: a typical maximum appears at an optimum abrasive-mass flow rate. From the point of view of abrasive exploitation, small abrasive-mass flow rates are recommended (Figure 7.20b).

The specific energy significantly reduces by using higher abrasive-mass flow rates [189].

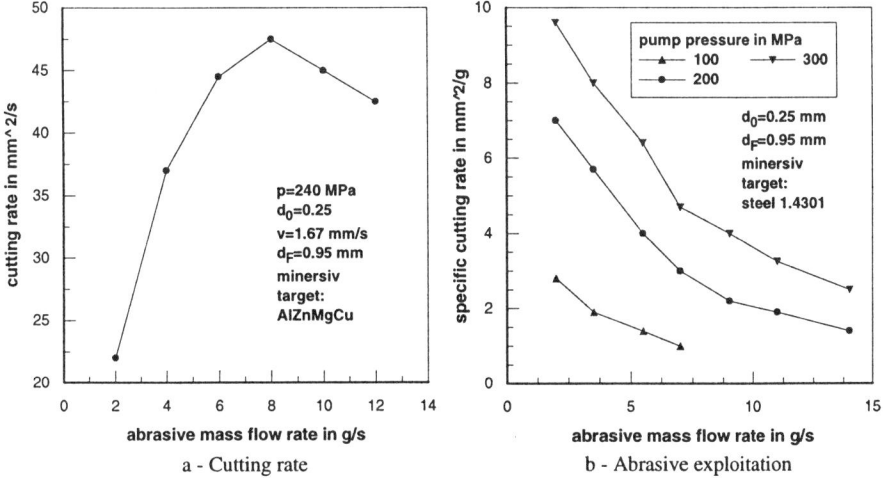

Figure 7.20 *Optimization of the abrasive-mass flow rate [254]*

7.5.2 Influence of Abrasive-Particle Diameter

7.5.2.1 General Trends

Figure 7.21 illustrates typical relations between the abrasive-particle diameter and depth of cut. Mathematically, the trend is

$$h(d_P) = C_{14} \cdot d_P^{C_{15}} \cdot \exp(C_{16} \cdot d_P). \tag{7.29}$$

In the equation, four qualitatively ranges are distinguished. For small particle diameters, $C_{15}=1$ and $C_{16}=0$. This conditions yields a linear relation between the abrasive-particle diameter and depth of cut. The reason for this relation is the higher kinetic energy of the larger particles that is expressed by $E_P \propto d_P^3$. On the other hand, the number of particles reduces with an increase in the particle diameter (section 2.4).

These effects become more important in the second range. In this range, $0<C_{15}<1$ and $C_{16}=0$. In the range, the progress of the function drops because of the reduced impact frequency. The optimum balance between the kinetic energy of a single abrasive grain and the number of impacting particles is exceeded.

In the third range, optimum conditions exist. In this range, $C_{15}=C_{16}=0$. In the range, the maximum depth of cut, $h_{max}=C_{14}$, is reached. The location and the width of this range depend on the properties of the target material as well as on the process parameters. The range is comparatively large for brittle-behaving materials (glass in Figure 7.21a) This wide range indicates that this material group is less sensitive to the impact frequency. This range generally offers the possibility to influence the cut quality without a significant influence in the depth of cut.

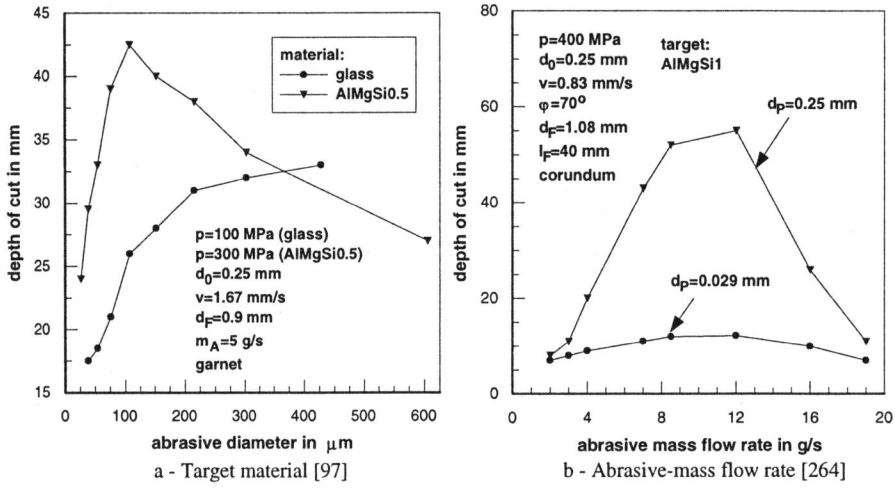

Figure 7.21 *Relation between the abrasive-particle diameter and depth of cut*

In the last range, the depth of cut drops with an increase in the abrasive-particle diameter. Guo et al. [44], Momber et al. [42], and Nakamura et al. [300] observe this effect. This result is due to the reduced impact frequency and due to the reduced abrasive-particle velocities (Figure 3.19a).

7.5.2.2 Optimization Aspects

The structures of the curves plotted in Figure 7.21 are sensitive to the traverse rate. In rock cutting experiments, Foldyna and Fialova [301] find that the positive effect of large abrasive diameters gets lost for high traverse rates. For this case, the number of abrasive particles becomes too low to be effective.

Figure 7.21b shows that the influence of the abrasive-particle diameter on the cutting process depends on the abrasive-mass flow rate. Small abrasive grains are not sensitive against changes in the abrasive-mass flow rate; whereas, larger abrasive grains lead to a significant optimum range for the cutting process. Some other aspects that determine the influence of the abrasive-particle diameter on the cutting process are the higher impact fracture-probability of larger grains, and the relation between the grain size and grain shape (Figure 2.4).

7.5.3 Influence of Abrasive-Particle Size Distribution

The abrasives entrained in the water jet are characterized by a large range of different diameters [section 2.3]. Generally, the average or mean size of the grains is only one aspect of a particle distribution. More often, such distributions are characterized by two or more parameters.

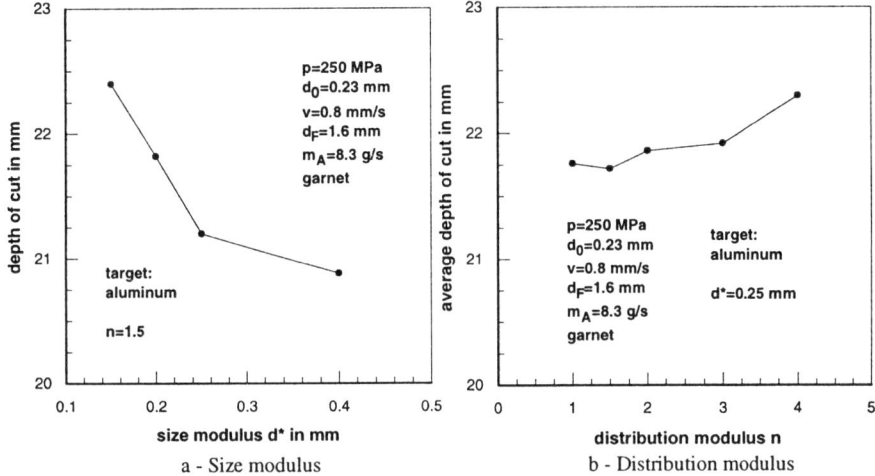

Figure 7.22 *Influence of the abrasive-particle size distribution on the depth of cut [42]*

Momber et al. [42] carry out an investigation into the influence of the parameters of abrasive-particle size distributions on the abrasive water-jet cutting process. In their study, they choose a Rosin-Rammler-Sperling (RRSB) distribution according to Eq. (2.9b). Momber et al. find that the influence of the size modulus is more significant than that of the distribution modulus. Figure 7.22 shows that changes in the distribution modulus, n, generates deviations in the depth of cut less than 5 %. In contrast, the smoothness of the kerf bottom considerably improves by varying the parameters of the abrasive-particle size distribution.

7.5.4 Influence of Abrasive-Particle Shape

7.5.4.1 General Trends

Section 2.3 defines the shape parameters of an abrasive grain. More general, the particle shape distinguishes between spherical and broken abrasive particles.

Systematic studies in the field of solid-particle erosion [35, 302] show that the abrasive-grain shape has an important influence on the material-removal regime. Bahadur and Badruddin [35] relate the particle-shape influence to the different removal mechanisms, such as micro-cutting for angular particles, and micro-ploughing for spherical particles. Cousins and Hutchings [302] show that the usually used terms 'ductile' and 'brittle' behavior are determined by the abrasive-grain shape.

7.5.4.2 Influence on Ductile-Behaving Materials

The few results reported in abrasive water-jet cutting do not show any general tendencies between the abrasive shape and depth of cut. For cutting aluminum alloys, Oweinah [264] compares glass beads and broken glass particles with approximately identical grain sizes. Figure 7.23 illustrates the results.

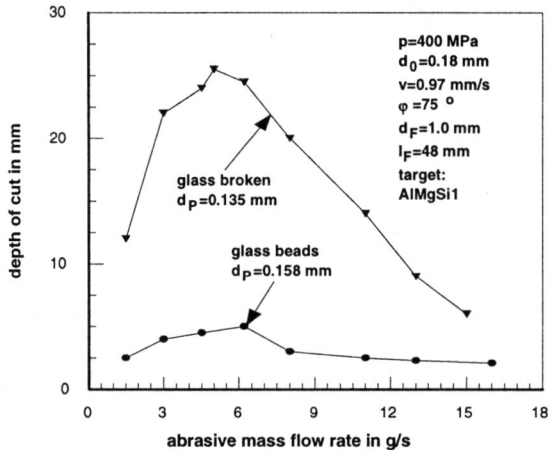

Figure 7.23 *Relation between the abrasive-grain shape and depth of cut [264]*

The broken glass particles significantly improve the cutting process. The depth of cut increases by a factor of 5. Nevertheless, the location of the maximum depth of cut with respect to the abrasive-mass flow rate does not depend on the abrasive-particle shape. In both cases, the maximum depth of cut is at the same level of the abrasive-mass flow rate. Nevertheless, the broken glass is more sensitive to changes in the abrasive-mass flow rate. These results suggest that materials that are generally removed by a micro-cutting mechanism are effectively cut by sharp-edges abrasive grains rather than by spherical particles.

7.5.4.3 Influence on Brittle-Behaving Materials

For other material combinations, this tendency can be the contrary. For rocks, minerals, or ceramics, a micro-cutting action contributes but does not play a major role in the material-removal process because of the brittle behavior of these materials. Therefore, the action of stress waves that are induced during the particle impact that generate fracture in the material is more important. Heßling [253] observes that the depth of cut in rocks increases if round quartzite is used instead of broken quartzite. Heßling also detects a relation between the grain-shape influence and focus length. Whereas, for short focuses broken grains are more effective, for round and broken abrasive particles the depth of cut is identical with the use of a long focus (Figure 7.17). The explanation gives Eq. (3.40) that shows that the particle-acceleration efficiency increases as the friction coefficient increases. The friction coefficient is higher for non-regular particles. Therefore, for a short focus and identical particle diameter, these particles accelerate more efficiently than spherical particles.

Broken abrasive grains also contribute to the abrasive and erosive wear of several parts of the abrasive water-jet equipment, such as focus nozzles and hoses.

7.5.5 Influence of Abrasive-Material Hardness

7.5.5.1 General Trends

Table 2.5 lists typical hardness values of abrasive materials. From investigations of abrasion and solid-particle erosion, a 'transition stage' exists at the point of comparable hardness of abrasive material and target material [303],

$$\frac{H_M}{H_P} \to 1. \tag{7.30}$$

In this region, erosive material-removal processes are very sensitive to changes in the hardness relation.

7.5.5.2 Observations in Abrasive Water-Jet Cutting

Figure 7.24 shows similar relationships for the abrasive water-jet cutting. In the figure, the progress of the function drops with an increase in the abrasive-material hardness. The application of corundum or silica carbide does not increase the depth of cut significantly, but leads to an accelerated focus wear. Oweinah [264] shows that aluminum and glass-fiber-reinforced plastics are cut at comparatively depths with broken glass particles owning a low hardness. Switching from broken glass to corundum as a very hard abrasive material, the depth of cut in aluminum is 40% higher than in the plastic material. This result indicates a higher sensitivity of ductile-behaving materials to changes in the abrasive-material hardness.

Figure 7.24b shows a result for brittle-behaving material. Notice a significant improvement of the material-removal process in the range of $1.0 < H_M/H_P < 1.1$. Beyond this range, the progress of the function decreases. Therefore, a further increase in the abrasive hardness does not substantially improve the material-removal performance.

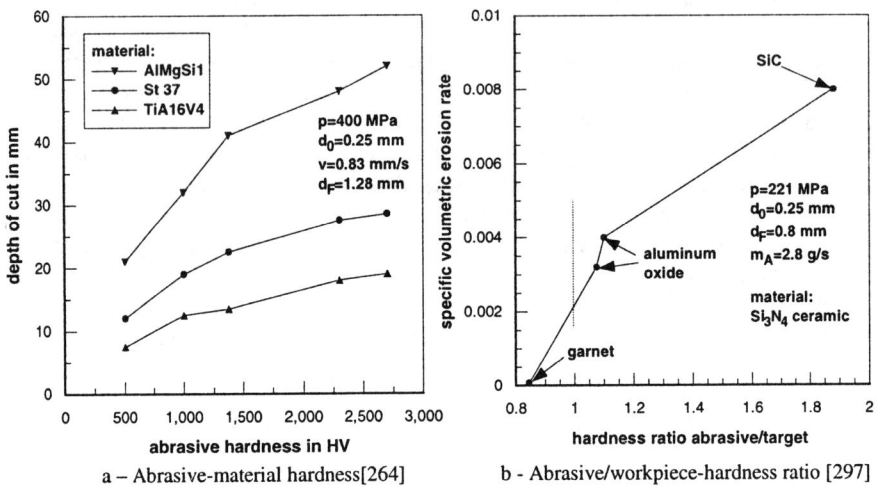

a – Abrasive-material hardness[264]   b - Abrasive/workpiece-hardness ratio [297]

Figure 7.24 *Influence of the abrasive-material hardness on the machining results*

The abrasive-material hardness influences the fracture behavior of the abrasive particles. The harder the material, the higher is the probability of particle fracture [93]. Therefore, hard abrasive grains fracture during the mixing-and-acceleration process into sharp-edged particles that support the micro-cutting process in ductile-behaving materials.

7.5.6 Recycling Capacity of Abrasives

7.5.6.1 Early Observations

In the field of the demolition of concrete structures by abrasive water jets, Nakaya et al. [109] find that viscous abrasive materials (steel grit) exhibit a good recycling capability compared to brittle abrasive materials (aluminum oxide) because they maintain their size and shape after the cutting process. Echert et al. [304] verify these results. These authors find very good cutting results for recycled steel shot; whereas, recycled garnet abrasives reduce the efficiency up to 50% compared to the original material. Konno et al. [305] show that the efficiency of steel grit remains more or less constant after three recycling cycles. Matsumoto et al. [306] analyze the change in the shap of the abrasives during concrete cutting as well as during steel cutting. These authors suggest that the final shape of the abrasive particles depends on the abrasive material as well as on the target material. Kokaji et al. [307] show that even garnet can successfully be recycled if the small-size fraction (in their investigation mesh # 100) is removed. Under these condition, they obtain a recovery rate of 90% for the cutting of concrete. For manufacturing applications, Hunt et al. [247] carry out the first investigation. For piercing aluminum, this reference finds that more than 60% of all abrasives are destroyed.

7.5.6.2 Parameter Influence on Disintegration

Figure 7.25a illuminates the influence of the pump pressure on the disintegration number. As the pressure increases, the disintegration increases. Figure 7.25b shows the relation between the abrasive diameter and particle fragmentation. The larger particle classes (d_P=300 μm to d_P=500 μm) fracture more intensely than the smaller fractions (d_P=150 μm to d_P=300 μm). This result is in agreement with observations from Simpson [90]. Labus et al. [308] observe an intense particle fragmentation at high traverse rate. These authors also show that an increase in the abrasive-mass flow rate slightly reduces the abrasive disintegration.

The amount of target material that is removed during a given cutting time influences the disintegration process because the single abrasive grain performs more destructive work for the removal of a larger quantity of target material. Thus,

$$\phi_D \propto \dot{V}_M, A_h. \qquad (7.31)$$

Figure 7.25c that shows the relationships between the specific cutting rate, target material, and disintegration number, experimentally verifies Eq. (7.31). As the cutting rate increases, the fragmentation increases. Also, particle fragmentation is at a maximum for steel followed by titanium and aluminum which is the same line as for the material's machinability number (Table 5.8) and their specific cutting-efficienciy (Table 5.9). Thus,

$$\phi_D \propto N_m. \qquad (7.32)$$

Labus et al. [308] notice an increased abrasive fragmentation for high target-material hardness.

Table 7.1 *Average abrasive-particle diameter and disintegration numbers for garnet after different stages of the cutting process [44]*

Stage	$\bar{d}_{Pin}$ =355-500 μm		$\bar{d}_{Pin}$ =500-710 μm	
	$\bar{d}_{Pout}$ [μm]	ϕ_D	$\bar{d}_{Pout}$ [μm]	ϕ_D
After focus	277	0.35	302	0.50
Quality cut	245	0.43	323	0.47
Rough cut	172	0.60	225	0.63

The recycling capability strongly depends on the cutting task, such as kerfing, rough cutting, and quality cutting (Table 7.1). Figure 7.25d shows a linear relationship between the workpiece thickness on the disintegration number. The disintegration number is almost untaught for high-quality cuts, suggesting that just a low number of abrasive particles is directly involved in the material-removal process. Labus et al. [308] and Foldyna and Martinec [38] notice the same qualitative trend for garnet, staurolite, ilmenite, and rutile.

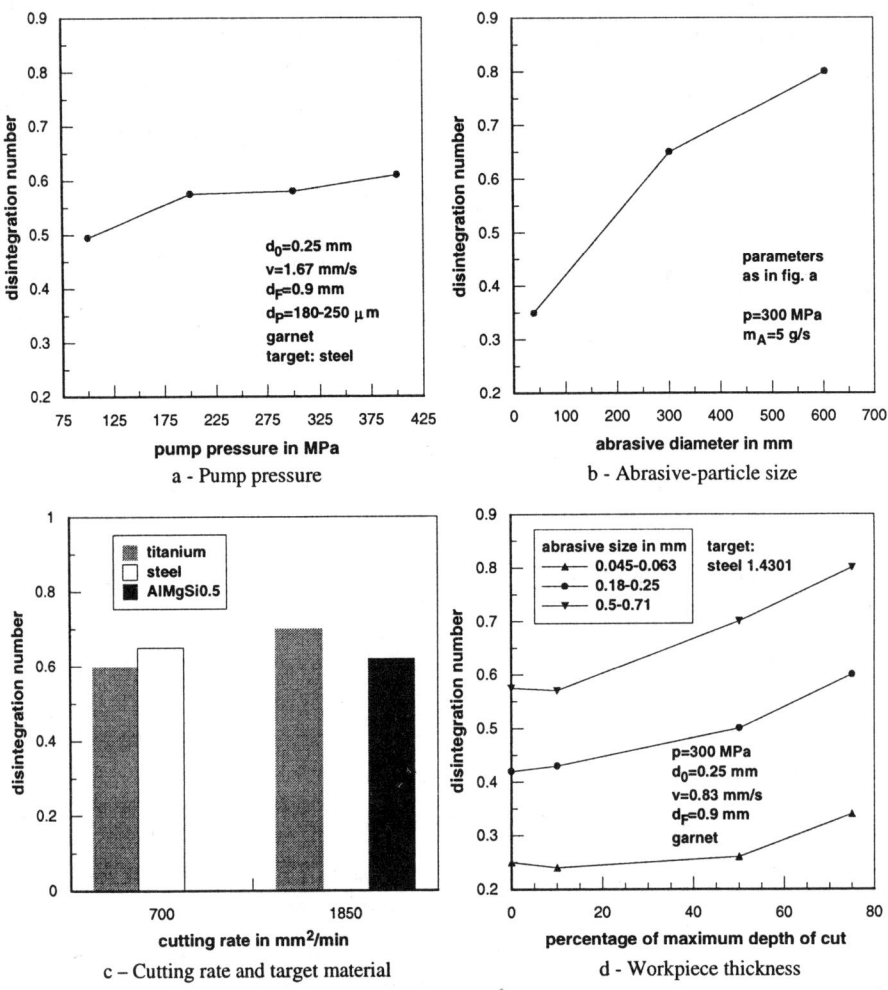

Figure 7.25 *Parameter influence on the abrasive-particle disintegration [97]*

7.5.6.3 Particle-Shape Modification

Table 7.2 illuminates aspects of the abrasive-particle shape modification during the cutting process. The shape factor of the particles generally reduces after cutting. Thus, sharp corners and edges are removed. An exception is grossulare Z where the abrasive grains become more non-regular due to the cutting process.

Table 7.2 *Abrasive-particle shape modifications after cutting steel by an injection-abrasive water jet [98]*

Size fraction	Shape factor F_{shape}	
[μm]	Before cutting	After cutting
	Almandite B	
> 200	0.73	0.74
100 - 125	0.68	0.73
< 63	0.67	0.68
	Almandite K	
> 200	0.75	0.77
100 - 125	0.70	0.70
< 63	0.70	0.70
	Grossulare Z	
> 200	0.75	0.75
100 - 125	0.70	0.67
< 63	0.70	0.65
	Andradite VC	
> 200	0.75	0.77
100 - 125	0.73	0.74
< 63	0.71	0.69

7.5.6.4 Suspension-Abrasive Water Jets

Walters and Saunders [136] observe an intense particle fragmentation during steel cutting by suspension-abrasive water jets (Figure 7.26). Yazici [273] carries out investigations into the recycling capability of abrasives in suspension-abrasive water jetting of rocks. This author finds that the disintegration number increases as the abrasive particle diameter increases. Table 7.3 illustrates the recycling capability of garnet used for rock cutting with a suspension-abrasive water jet. The recycling capability reduces as the abrasive particle size increases.

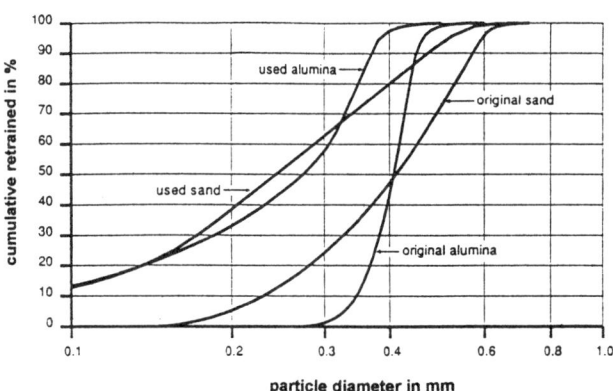

Figure 7.26 *Abrasive-particle fragmentation during cutting by a suspension-abrasive water jet [136]*

Table 7.3 *Recycling capability of garnet in rock cutting with suspension-abrasive water jets [273]*

Average particle diameter [μm]	Percent of the first cut
296	90
586	79
867	71
1,250	64

7.5.6.5 Modeling of Recycling Processes

In order to evaluate the recycling capability of an abrasive mixture under given process conditions, a modeling of the disintegration process is necessary. Table 3.3 presents some models for the impact comminution of single mineral grains. Ohlsen [97] develop a recycling model. This model is partly published by Guo et al. [309]. The mathematical model is constructed as a matrix multiplication,

$$B_{i+1} = A_i \cdot Z. \tag{7.33}$$

In the equation, A_i characterizes the abrasive-particle distribution before cutting, Z is a disintegration matrix, B_{i+1} characterizes the abrasive-particle distribution after cutting, and i is the number of cycles. The disintegration matrix Z includes the particle-size distribution of each particle-size range after cutting with a given parameter combination. Each row of this matrix represents the particle-size distribution of a separate particle range after cutting. The main input-parameter range is the value on the main diagonal. To simulate the exchange of particles, the amount of abrasive material that is replaced is calculated by the scalar product of B_{i+1} and an exchange matrix E. This exchanged matrix characterizes the percentage of every particle range that is separated. This range has the same row number as Z. If, for example, E=[1 0.5 0 0 0], 100% of the first particle-size range and 50% of the second particle-size range are separated. The particle-size distribution for the next cutting cycle, A_{i+1}, is

$$A_{i+1} = B_{i+1} - \underbrace{B_{i+1} \cdot E}_{\text{seperated abrasives}} + \underbrace{SP_i \cdot NM}_{\text{new abrasives}}. \tag{7.34}$$

In the equation, SP_i is the scalar product of B_{i+1} and E, and NM is the particle-size distribution of the newly added abrasives. Fig. 7.27 presents an experimental verification of this model. Table 7.4 shows how to optimize process parameters for maximum recycling capacity.

7.5 Influence of Abrasive Parameters   229

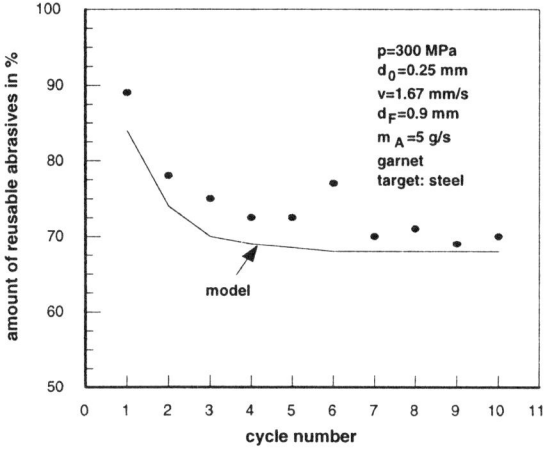

Figure 7.27 *Comparison between recycling tests and calculations from the model for abrasive-particle recycling [97]*

Table 7.4 *Parameter optimization for abrasive-particle recycling [97]*

Parameter	Parameter range	Optimum
Pump pressure p	100 MPa – 390 MPa	Maximum
Orifice diameter d_0	2.25 mm	No influence if R=const. and d_F/d_0=const.
Traverse rate v	0.17 mm/s – 6.67 mm/s	A function of the other parameters
Focus diameter d_F	0.55 mm – 1.2 mm	$d_F = 4 \cdot d_0$
Focus length l_F	20 mm – 100 mm	No influence
Abrasive-mass flow rate m_A	1 g/s – 12 g/s	Minimum
Abrasive diameter d_P	0 – 710 µm	d_P=355 µm-500 µm

8 Geometry, Topography and Integrity of Abrasive Water-Jet Machined Parts

8.1 Cut Geometry and Structure

8.1.1 Definition of Cut Geometry Parameters

Figure 8.1 defines the parameters that characterize the geometry of a cut generated by abrasive water-jet cutting. These parameters are [180, 185]:

- top width of the cut (b_T)
- bottom width of the cut (b_B)
- taper of the cut (T_R)
- flank angle (φ_F)
- initial-damage width (b_{IDZ})
- initial-damage depth (h_{IDZ})

The taper of the cut is usually defined as a non-dimensional ratio between the top cut-width and the bottom cut-width,

$$T_R = \frac{b_T}{b_B}. \tag{8.1}$$

Oweinah [264] and Kitamura et al. [310] give an alternative definition:

$$T_R = \frac{b_T - b_B}{2}. \tag{8.2}$$

Chung et al. [270] define

$$T_R = \arctan\left(\frac{b_T - b_B}{16}\right). \tag{8.3}$$

The taper is divergent ($b_T < b_B$), convergent ($b_T > b_B$) or even divergent-convergent.

The flank angle is [180]

$$\varphi_F = \arctan\left(\frac{b_T - b_{0.4 \cdot h}}{2 \cdot (0.4 \cdot h - 0.5)}\right). \tag{8.4}$$

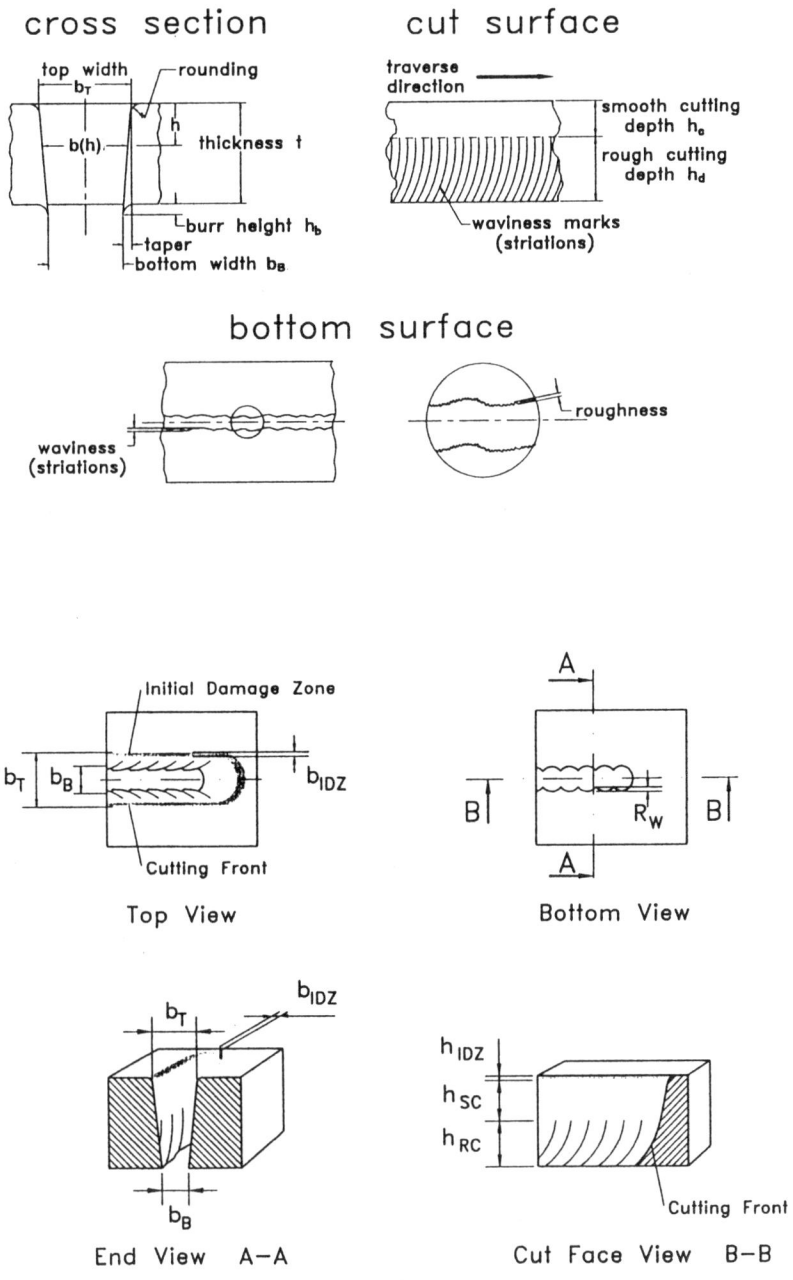

Figure 8.1a *Geometry of a cut generated by abrasive water-jet cutting [180, 185]*

232   8. Geometry, Topography and Integrity of Abrasive Water-Jet Machined Parts

Surface Topography (left) and cut geometry (right)

Top area of a generated cut

Initial-damage zone

Figure 8.1b *Geometry features of cuts generated by abrasive water-jet cutting [180]*

8.1.2 Width on Top of the Cut

8.1.2.1 Ductile-Behaving Materials

Figure 8.2 summarizes the parameter influence on the top cut-width for ductile-behaving materials.

As Figure 8.2a illustrates, an increase in the pump pressure almost linearly increases the top width of the cut [180]. Figure 8.2b shows that the traverse rate significantly influences the top width as a decrease in the top width occurs from an incerase in the traverse rate [180, 264, 270, 311, 312]. By regression analysis, Chung et al. [270] find a linear relation (Table 8.1). From Figure 8.2c, the top width also strongly depends on the standoff distance. This realation is experimentally estimated by Chung et al. [270] and Guo [180] for injection-abrasive water jets, and Laurinat et al. [133] for suspension-abrasive water jets. This effect is simply due to the increasing diameter of the abrasive water jet with an increase in the standoff distance. Figure 8.2d that shows the influence of the focus diameter on the top cut-width supports this explanation. In the figure, the width almost linearly increases with the focus diameter [69, 180, 264, 270]. From Figure 8.2e, the top width of the cut reduces if a long focus is used [180, 264]. Chung et al. [270] and Guo [180] notice a slight increase in the top cut-width with an increase in the abrasive-mass flow rate (Figure 8.2f). In both cases, the abrasive-mass flow rate influence is not very significant [313]. The abrasive-particle diameter does not show any significant influence on the top cut-width.

Table 8.2 summarizes the parameter influence on the top cut-width in ductile-behaving materials.

Sano et al. [312] show for abrasive water-jet cutting of thin metal sheets (h=28 mm-40 mm) that the top cut-width depends on the number of layers simultaneously being cut. The width increases with the layer number up to 50 layers. Beyond this number, the width becomes larger if the material is cut by very small abrasive particles. The same tendency appears by varying the layer number and traverse rate. For certain traverse rates, the top width increases if a critical number of layers is exceeded.

8.1.2.2 Brittle-Behaving Composite Materials

Arola and Ramulu [185] and Ramulu and Arola [187] perform a systematic study into the top cut-width in brittle-behaving graphite-epoxy composites. Figure 8.3 shows the effects of several process parameters on the cut width at different depths of cut. In the figure, the standoff distance accounts for the largest degree in variation of the entrance cut-width at shallow depths, and the pump pressure, traverse rate, and abrasive-particle diameter assume secondary roles. This result agrees with Table 8.2. With an increase in the depth of cut, the influence of the traverse rate increases.

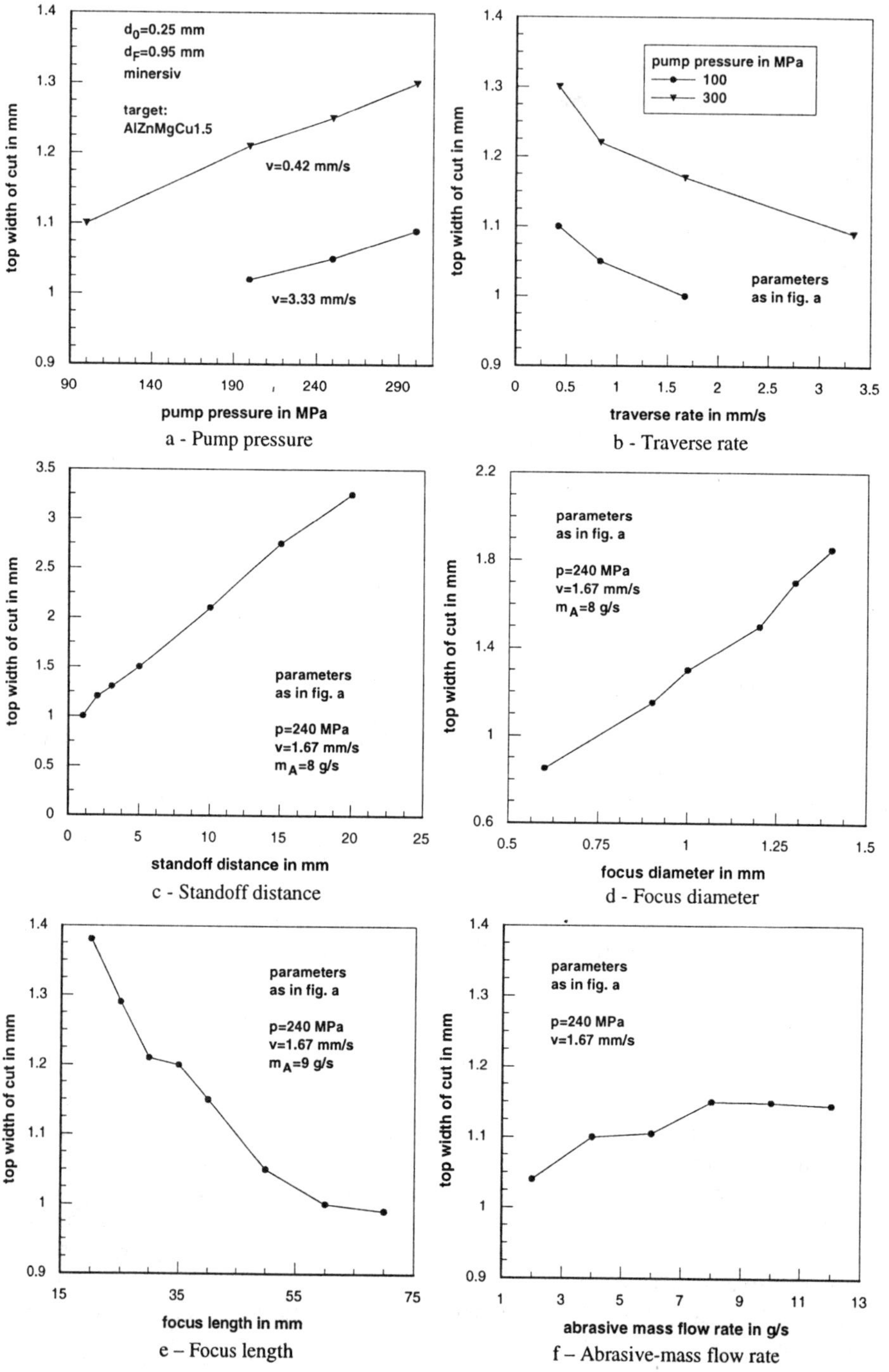

234   8. Geometry, Topography and Integrity of Abrasive Water-Jet Machined Parts

Figure 8.2 *Parameter influence on the top cut-width in a ductile-behaving material* [180]

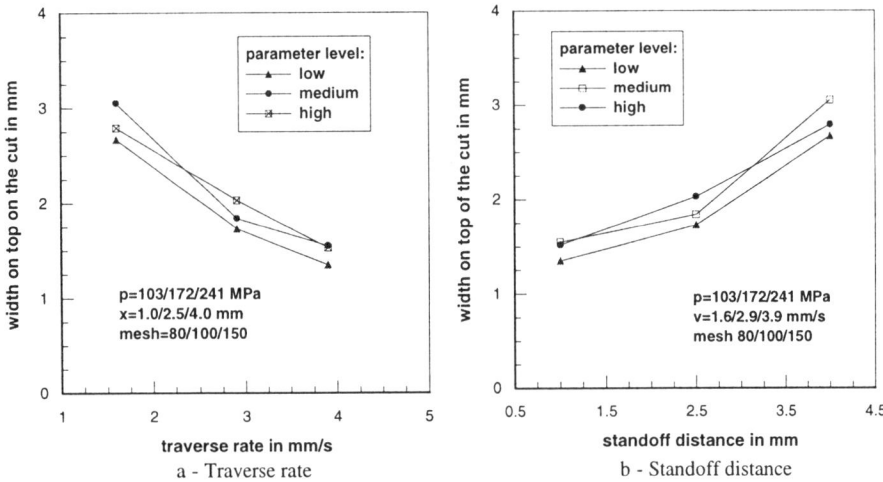

Figure 8.3 *Levels of the parameter-influence on the top cut-width in brittle composites [185]*

At cutting depths greater than h=8 mm, the pump pressure, traverse rate, and abrasive-particle size almost equally influence the top cut-width. However, the influence of the standoff distance is insignificant. Arola and Ramulu [185] also find that the top width of the cut in composite materials increases with an increase in the standoff distance. This effect is very significant for high pump pressures (Figure 8.ba). Figure 8.5a shows that the traverse rate does not have a major effect on the top width. Also, the direction of the influence of this parameter depends on the level of the other process parameters. For high traverse rates (such as, v>4 mm/s), the effect of the other parameters becomes insignificant.

8.1.2.3 Ceramics, Glass and Metal-Matrix Compounds

For ceramics, Hochheng and Cheng [314] and Momber et al. [189] investigate the top width of the cut. Figure 8.4 shows some results.

From Figure 8.4a, no general relation exists between the pump pressure and the top width of the cut in ceramics. For a material with a high machinability number, the width reduces at very high pressures. In contrast, for a material with a lower resistance, the width slightly increases [189]. Figure 8.4b illustrates that the top cut-width linearly decreases as the traverse rate increases. Geskin et al. [315] make the same observation for cutting glass with an abrasive water jet. The cut geometry of thicker ceramic plates is more sensitive to changes in the traverse rate [314]. Several authors [189, 314, 315] observe a notable influence of the abrasive-mass flow rate on the top cut width in ceramics and glass. From Figure 8.4c, the larger the abrasive-mass flow rate, the larger the top cut-width. This effect is more pronounced for low traverse rates.

For metal-matrix composites, Hashish [316] shows that the volumetric percentage of reinforcing particles does not influence the top cut-width.

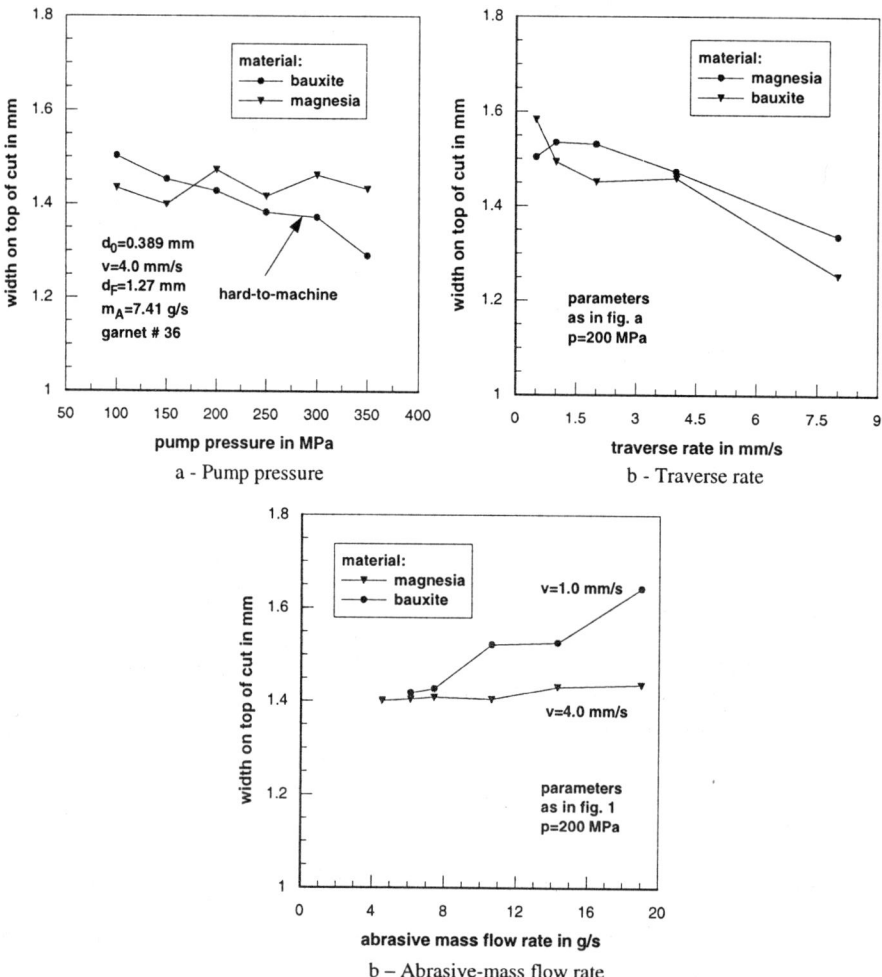

Figure 8.4 *Parameter influence on the top cut width in brittle materials [189]*

8.1.2.4 Models for Top Cut-Width Estimation

Laurinat et al. [286] develop a regression model for the estimation of the width on top of the cut as a function of the standoff distance,

$$b_T = a_1 \cdot x + a_2. \tag{8.5}$$

8.1 Cut Geometry and Structure

Section 9.1 discusses this model in more detail. Table 9.1 gives the regression parameters for the model. Chung et al. [270] who also develop regressions for the influence of the focus diameter and abrasive-mass flow rate, find the same relation. Table 8.1 lists these models.

8.1.3 Width on the Bottom of the Cut

8.1.3.1 Ductile-Behaving Materials

Chung et al. [270] carry out a systematic study for ductile-behaving materials. Figure 8.5 and Table 8.2 summarize the results of these authors. The bottom cut-width reduces with an increase in the traverse rate (Figure 8.5b). All the other process parameters do not significantly influence this geometry parameter (Figures 8.5a,c-e). These results agree with measurements from Oweinah [264] who also finds that the bottom width of the cut is not sensitive to changes in the focus length.

8.1.3.2 Brittle-Behaving Composite Materials

Arola and Ramulu [185] investigate the parameter influence for the cutting of brittle-behaving composite materials. The bottom width significantly decreases with an increase in the traverse rate. The influence of the traverse rate on the exit width is comparatively high. The influence of the traverse rate increases with an increase in the level of the other process parameters. For the influence of the abrasive-mass flow rate, the authors notice a range with a maximum exit cut-width. In contrast to the top width of the cut, the influence of the standoff distance on the exit cut width is insignificant.

8.1.3.3 Ceramics and Glass

Figure 8.6 illustrates results for ceramics [189, 314] and glass [317].

Figure 8.6a shows a general increase in the bottom width of refractory ceramics at high pump pressure. Figure 8.6b shows that the bottom cut-width decreases with higher traverse rates. This tendency is more pronounced for thick specimens and for low-resistant ceramics. From Figure 8.8d, the same relationship holds for glass. The bottom cut-width in ceramics slightly augments with an increase in the abrasive-mass flow rate [189, 314]. Figure 8.6c shows an example. Again, this relation seems to be more pronounced for low-resistant ceramics.

In glass, the bottom cut-width increases with a decrease in the traverse rate and an increase in the abrasive-particle diameter (Figure 8.8d).

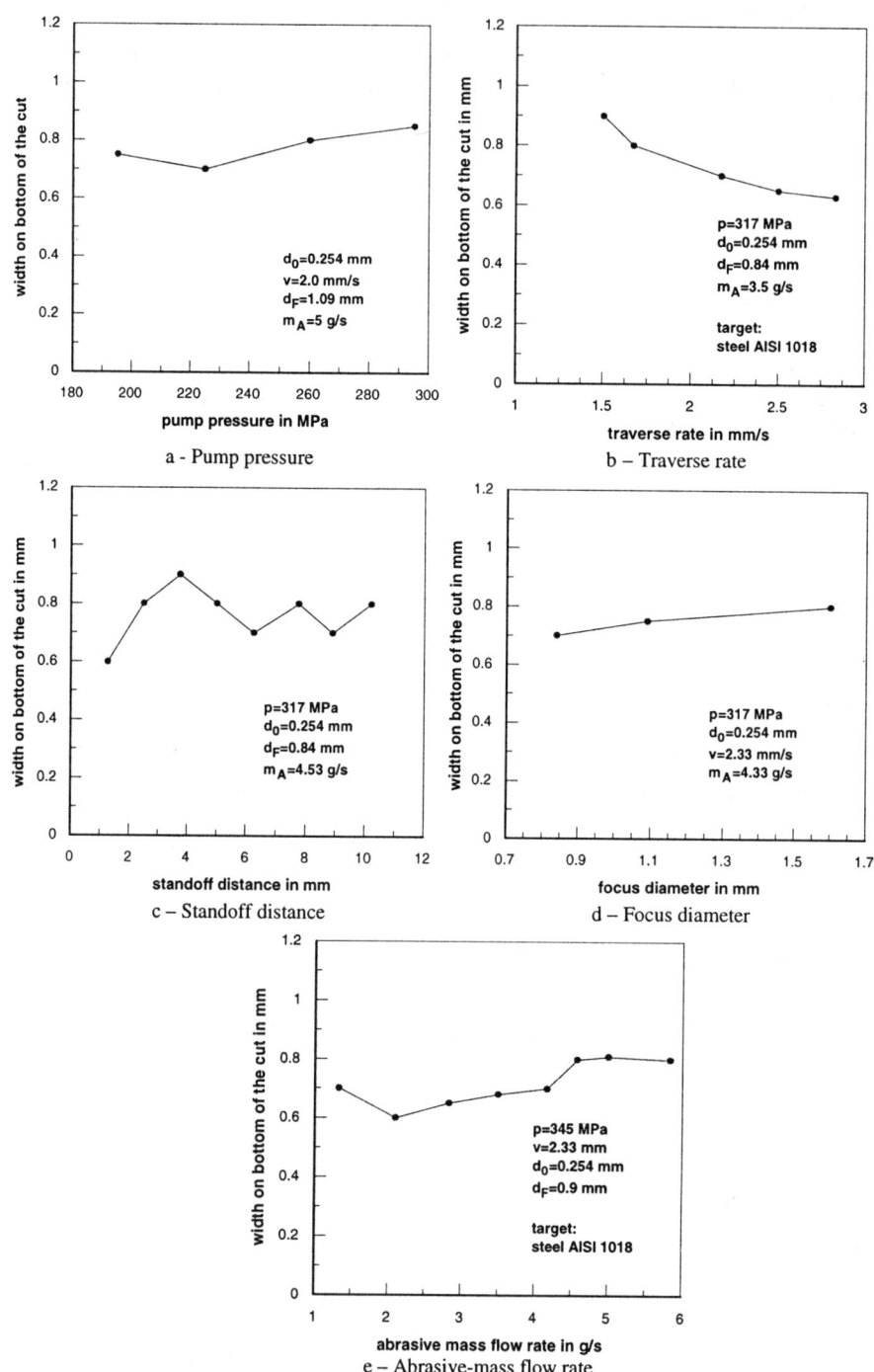

Figure 8.5 *Parameter influence on the bottom cut-width in aluminum [270]*

8.1 Cut Geometry and Structure   239

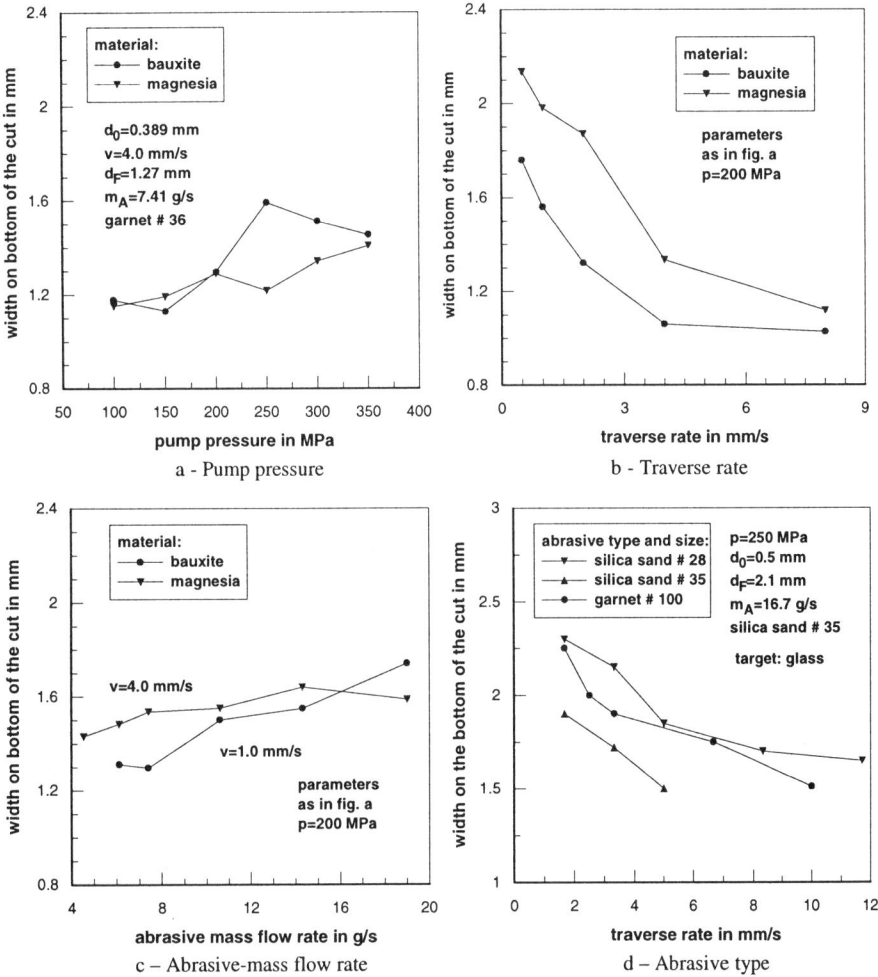

Figure 8.6 *Parameter influence on the bottom cut-width in brittle materials [189, 317]*

8.1.4 Taper of the Cut and Flank Angle

8.1.4.1 Ductile-Behaving Materials

Chung et al. [270] who use the taper definition according to Eq. (8.3) perform a general investigation in the parameter influence for ductile-behaving materials. Figure 8.7 summarize the results. The authors find that the taper increases with an increase in the traverse rate, the focus diameter, and the standoff distance.

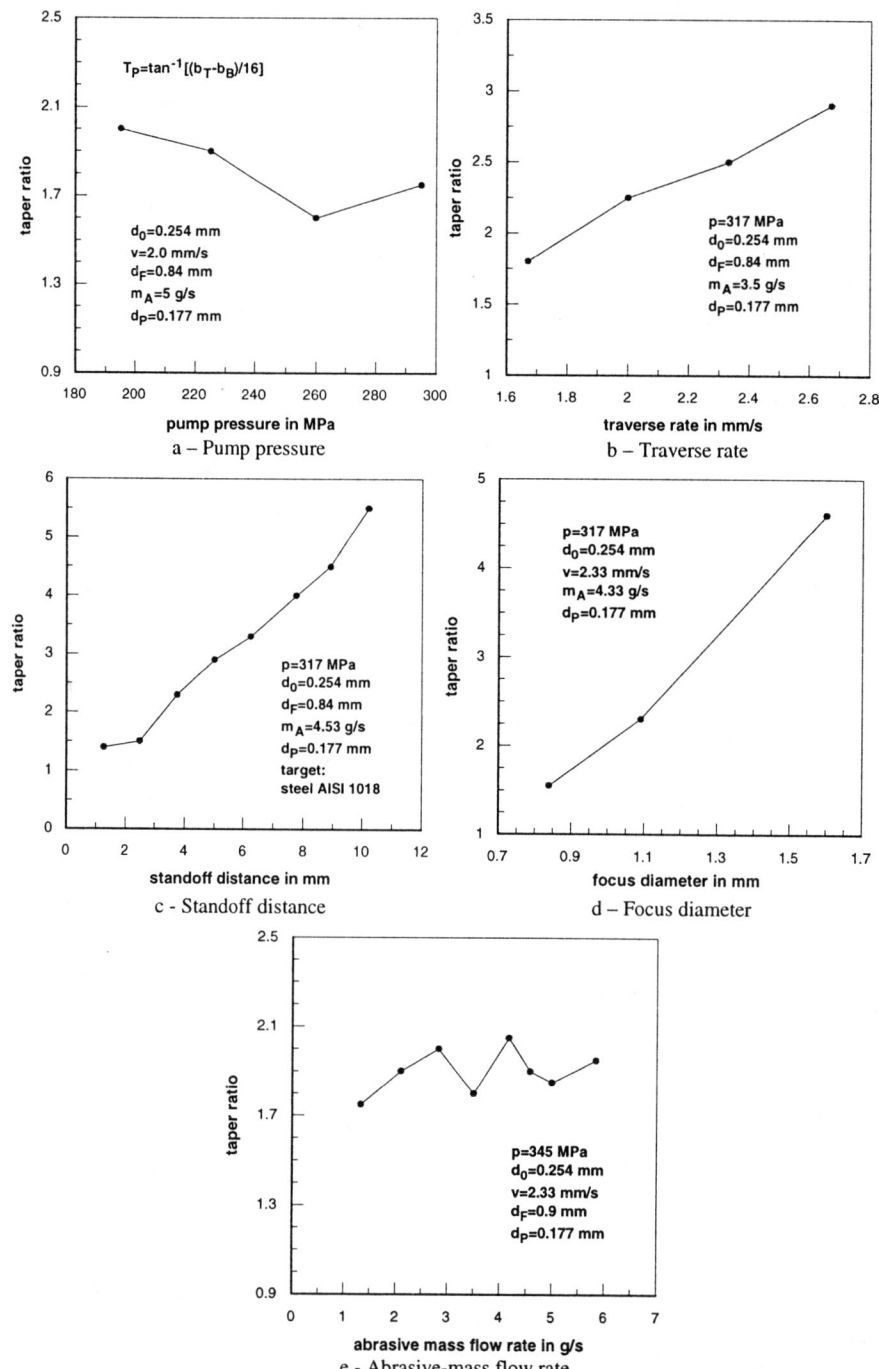

Figure 8.7 *Parameter influence on the cut taper in ductile-behaving material [270]*

The taper decreases with an increase in the pump pressure. The abrasive-mass flow rate does not show a clear relation to the taper of the cut (Figure 8.7e). These results suggest that the taper reduces with an increase in the abrasive water-jet kinetic energy.

Chung et al. [270] develop a regression equations for the parameter influence on the cut taper. Table 8.1 lists the results of this study. These regressions are valid only for the given process conditions. In Figure 8.7b, the influence of the traverse rate is evaluated by only four experimental points.

Table 8.1 *Correlation between the abrasive water-jet parameters and cut geometry in ductile-behaving materials [270]*

Geometry parameter	Process parameter	Range	Regression
Top cut-width	Focus diameter	0.838 - 1.6 mm	$b_T = a_1 \cdot d_F + b_1$
	Abrasive-mass flow rate	0.7 - 6.0 g/s	$b_T = a_2 \cdot \dot{m}_A + b_2$
	Standoff distance	1.0 - 12.5 mm	$b_T = a_3 \cdot x + b_3$
Cut taper	Traverse rate	0.16 - 10 mm/s	$T_R = a_4 \cdot v + b_4$
	Standoff distance	-	$T_R = a_5 \cdot x + b_5$
	Focus diameter	-	$T_R = a_6 \cdot d_F + b_6$

Kitamura et al. [310] and Matsui et al. [311] show that the taper non-linearly increases with an increase in the traverse rate. For high traverse rates, the progress of the function drops (Figure 8.8). Also, from Figure 8.8, the taper increases with an increase in the specimen thickness. Figure 8.8 also shows a more complex relation between the taper, based on Eq. (8.4), and the traverse rate. In the figure, the taper changes from convergent for high traverse rates to divergent for low traverse rates. A 'transition' traverse rate exists at v=0.2 mm/s. At this traverse rate, the taper is about $T_R=0$.

Table 8.2 *Parameter influence on the cut geometry in ductile-behaving material [270]*

Geometry parameter	Process parameter						
	p	d_0	d_F	d_P	$\dot{m}_A$	x	v
b_T	-	-	●	-	+	●	-
b_B	●	+	+	+	●	●	●
T_R	●	+	+	+	●	●	●

● strong effect, + light effect, - no effect

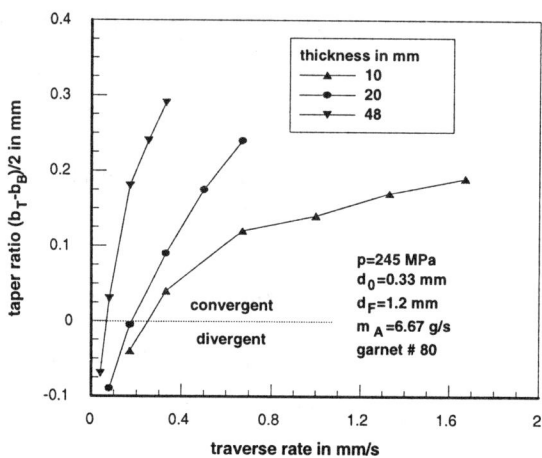

Figure 8.8 *Relation between the traverse rate, specimen thickness, and cut taper for steel [311]*

Guo [180] investigates the formation of the flank angle in ductile-behaving materials. Figure 8.9 shows some results. In this figure, an angle $\varphi_F > 0$ characterizes a divergent taper, and $\varphi_F < 0$ stands for convergent taper. $\varphi_F = 0$ is a straight cut.

Figure 8.11a plots the influence of the pump pressure and traverse rate. High pressure and low traverse rate, or, high-energy jets, always produce convergent tapers. For a certain parameter combination, the taper disappears. Figure 8.9b shows the influence of the standoff distance. The figure illustrates the overwhelming importance of this parameter. The taper reduces if the standoff distance is minimized. Figure 8.9c displays the same significant influence of the focus diameter. As for the standoff distance, taper minimizes if the focus diameter reduces. The focus length that is plotted versus the flank angle in Figure 8.9d does not notably influence this geometry parameter. Figures 8.11e,f illustrates the influence of the abrasive parameters, namely abrasive-mass flow rate and abrasive diameter on the flank angle. High abrasive-mass flow rate and large abrasive diameter reduce divergent taper effects. However, if a certain abrasive diameter is exceeded, the tapering just switches from divergent to convergent.

8.1.4.2 Brittle-Behaving Composite Materials

Arola and Ramulu [185] and Ramulu and Arola [187] investigate the taper formation in graphite/epoxy-composites. They define the taper according to Eq. (8.3). Figure 8.10 summarizes the results of these authors.

8.1 Cut Geometry and Structure 243

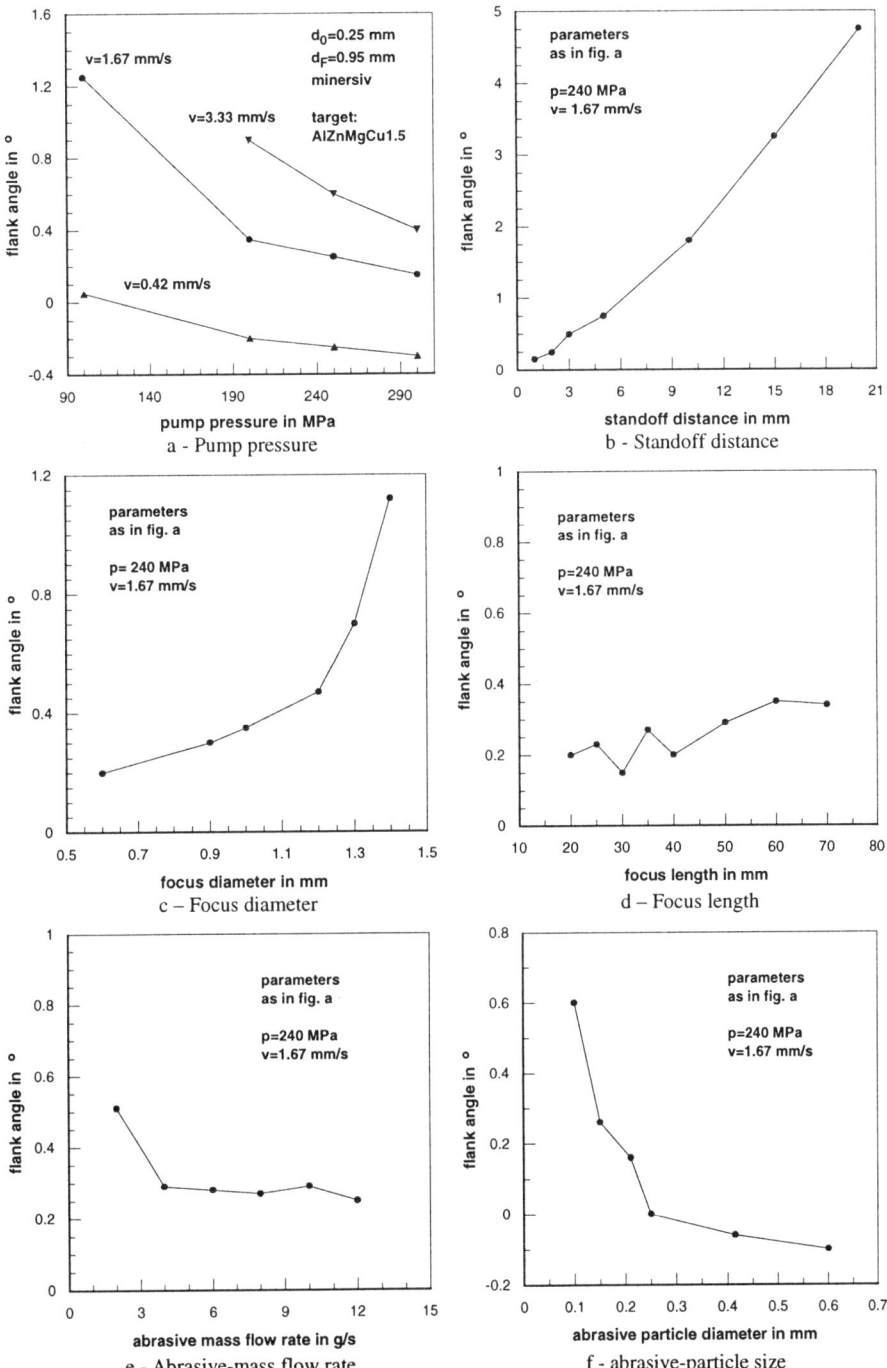

Figure 8.9 *Parameter influence on the flank angle in ductile-behaving materials [180]*

Figure 8.10a shows that the taper significantly decreases with an increase in the pump pressure. The influence of the traverse rate on the taper is also complex. For comparatively low pump pressures, the taper increases with high traverse rates (Figure 8.10b). For high pressure ranges, the trend is opposite. In that case, the taper increases with an increase in the traverse rate. Ramulu and Arola [187] explain these discrepancies by an increased probability of jet deflection at lower pump pressures. Similar relations count for the influence of the standoff distance. Whereas, for comparatively low pump pressures the taper shows a typical maximum at a standoff distance of x=2.5 mm [187], the situation changes with an increase in the pump pressure. As Figure 8.10c shows, the taper has minimum values at this standoff distance.

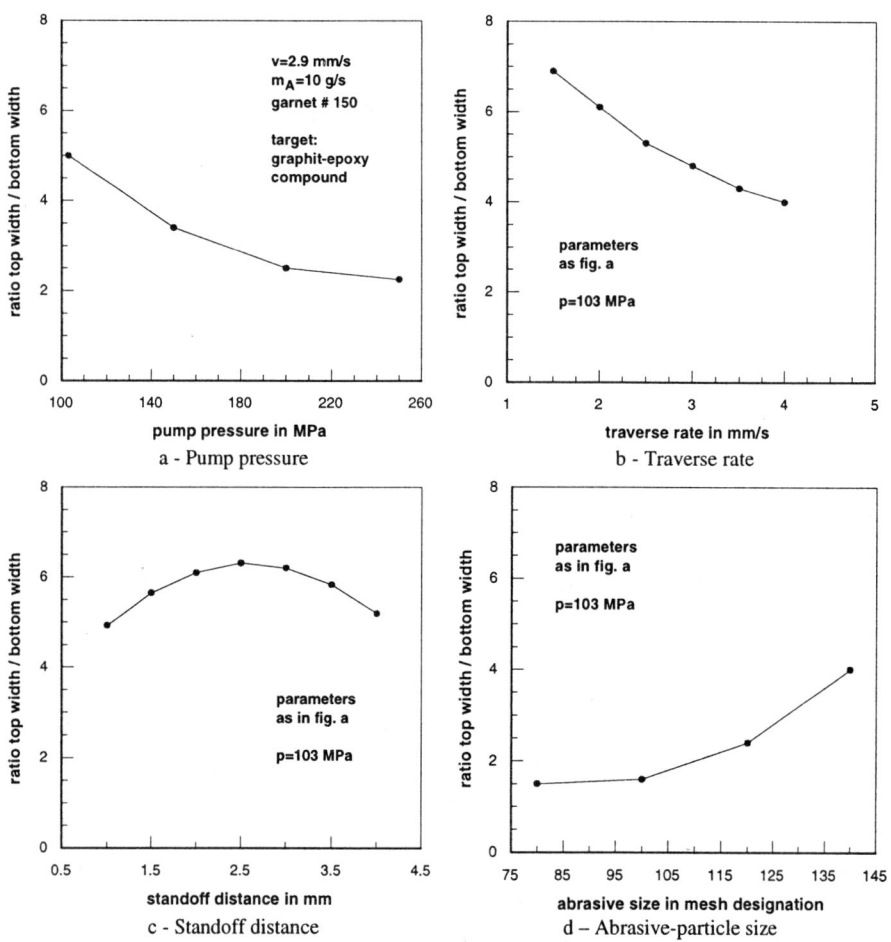

Figure 8.10 *Parameter influence on the cut taper in graphite/epoxy-composite [185, 187]*

It is noticed that larger abrasive diameter reduces taper formation. Agus et al.[318] observe this effect for granite cutting. Figure 8.10d shows that the influence of the abrasive-particle diameter decreases with an increase in the pump pressure. For low pump pressures (p=103 MPa), a typical minimum in the taper appears at a certain abrasive-particle diameter [185].

All the trends discussed above are very sensitive to the thickness of the materials being cut. Arola and Ramulu [185] statistically investigate this aspect.

When cutting materials less than h=5 mm thick, the standoff distance primarily controls the cut taper. At any thickness greater than h=5 mm, the variation in taper is more affected by a combination of the levels of the standoff distance and abrasive-particle size. The effects of the pump pressure and traverse rate on taper do not become significant until the depths of cut increases h=10 mm. For cutting small specimens, sufficient cutting energy is available for a stable cutting process so that changes in the pump pressure and the traverse rate do not dramatically influence the cutting process. At cutting depths greater than h=15 mm, no single parameter clearly dominates the taper formation process.

8.1.4.3 Ceramics, Glass and Metal-Matrix Compounds

Hochheng and Cheng [314] and Momber et al. [189] investigate the taper formation in ceramics. Figure 8.11presents some results. The taper linearly increases with an increase in the traverse rate (Figure 8.11a). For lower pump pressures and reduced abrasive-mass flow rates, the cut taper generally increases.

Figure 8.11b shows the taper formation in a metal-matrix composite. In this figure, the taper decreases with larger abrasive grains. This result agrees with Figure 8.10d and with experimental results from rock cutting [318]. An optimum traverse rate exists for given abrasive-particle sizes and abrasive-mass flow rates for this material group. At this optimum, the cut is not tapered ($T_R=1$). At slow traverse rates, the taper switches from divergent to convergent which is explained by the relative long exposure time at low traverse rates [184]. For metal-matrix composites, Hashish [316] finds that the volumetric percentage of reinforcement particles significantly influences the cut taper (Figure 8.11c). When the particle content increases, the cut becomes more tapered. This result indicates a greater resistance to abrasive water-jet penetration. This tendency is independent on the traverse rate.

Tapered cuts are also observed in glass samples cut by an abrasive water jet [315]. In this case, the taper also almost linearly increases with an increas in the traverse rate and decreases with an increase in the abrasive-mass flow rate.

Figure 8.12 shows results of flank-angle measurements in ceramics. The flank angle decreases as the pump pressure increases. At very high pump pressures, the flank angle tends against zero (Figure 8.12a).

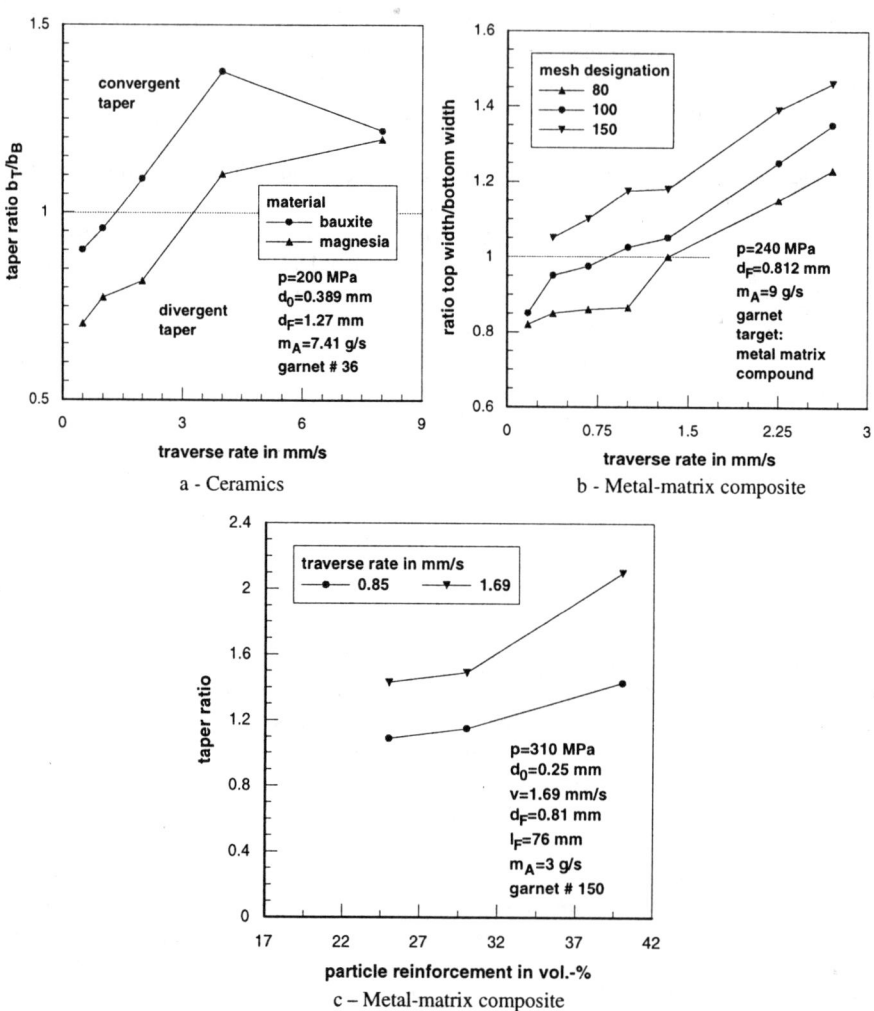

Figure 8.11 *Cut taper-formation in difficult-to-machine materials [184, 189, 316]*

The influence of the traverse rate is very pronounced. As in case of metal-matrix compounds, the taper switches from divergent to convergent for low traverse rates (Figure 8.12b). Also, a high traverse rate generates almost parallel kerf walls in all investigated materials. The high-resistant ceramic predominantly shows divergent tapers.

8.1 Cut Geometry and Structure

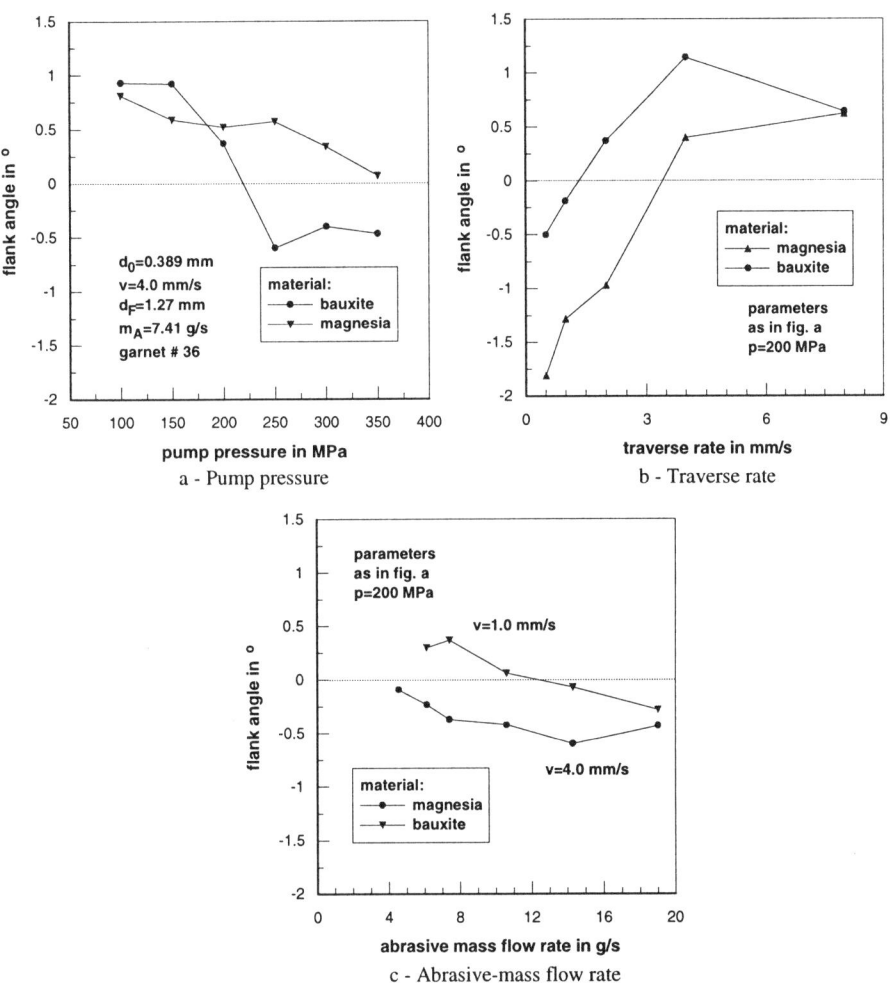

Figure 8.12 *Flank-angle formation in ceramics [189]*

Low abrasive-mass flow rates generate almost untapered cuts except for the high-resistant material. The taper of the high-resistant material is much lower at high abrasive-mass flow rates compared to the other materials (Figure 8.12c).

8.1.4.4 Models for Taper Estimation

By using a second-order polynomial regression, Ramulu and Arola [187] develop a mathematical model for the estimation of the taper in composite materials,

$$T_R = c_0 + \sum_{i=1}^{4} C_i \cdot x_i + \sum_{i=1}^{4} C_{ii} \cdot x_i^2 + \sum_{i=1}^{3} \sum_{j=i+1}^{4} c_{ij} \cdot x_i \cdot x_j . \tag{8.6}$$

In the model, x_1 is the pump pressure, x_2 is the abrasive-grain diameter, x_3 is the traverse rate, x_4 is the standoff distance, and x_6 is the depth of cut (here h=16 mm).

8.1.5 General Cut Profile

8.1.5.1 Experimental Results

Figure 8.13a shows examples for cut structures that are generated with different traverse rates. This figure illustrates very well that neither top and bottom width nor cut taper describes the entire cut profile. For a given material, several parameters influence the profile of the cut. As the figure shows, the low traverse rate generates a bumped geometry with convergent taper (called S_1 here); whereas, the high traverse rate generates a linear cut profile that is tapered divergently (called S_2 here). Figure 8.13b illustrates the influence of the target material on the cut profile. In the figure, the low-resistant material tends to the convergent profile-type (S_1). In contrast, a linear divergent cut-profile is generated in the high-resistant ceramic (S_2).

Guo (1994) experimentally shows that the profile-type S_2 usually appears at a low pump pressure, high traverse rate, low abrasive mass flow rate, small abrasive diameter, and high standoff distance. Thus, the lower the available jet energy, the higher is the probability of type S_2. From Figure 8.13b, this statement can be extended: as the ratio between the available jet energy and material resistance decreases, the cut profile tends to type S_2.

8.1.5.2 General Cut-Geometry Model

In order to develop a complex model for the cut profile in terms of process parameters and depth of cut, Arola and Ramulu [185] analyze cut-width measurements. This statistical model is

$$b(h) = f(v, d_P, x, p, h) . \tag{8.7}$$

8.1 Cut Geometry and Structure   249

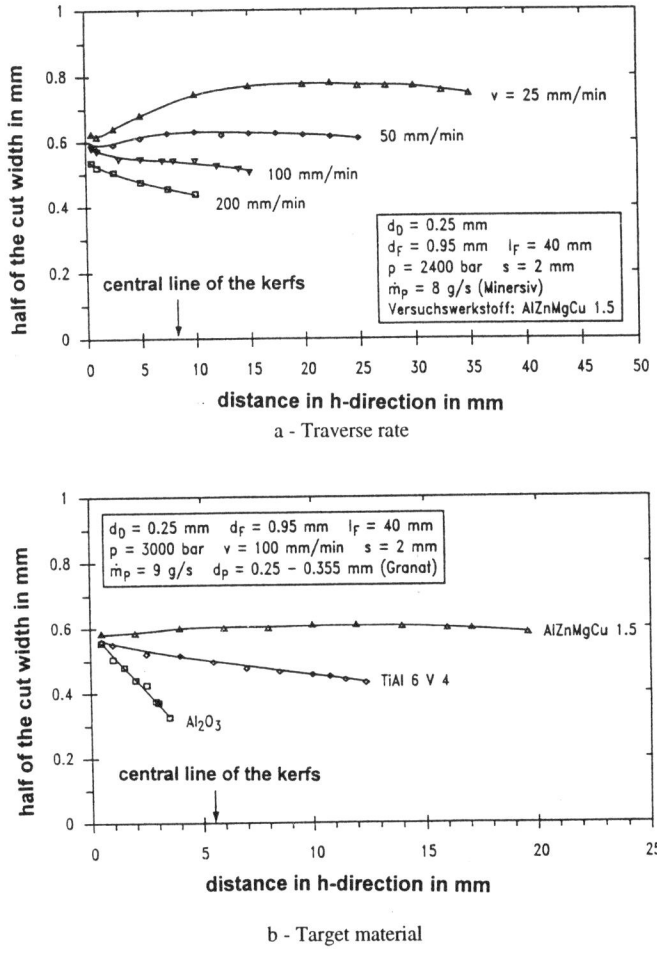

Figure 8.13 *Parameter influence on the cut structure [180]*

Figure 8.14 provides a comparison of experimental measurements of the cut width and the corresponding cut profiles that are generated from the model for three different parameter cases. A comparison of the three parameter conditions in Figure 8.14 illustrates the relative effects of the standoff distance, pump pressure, and abrasive-particle diameter on the profile of the cut. In Figure 8.14c, the profile remains nearly constant along the depth of cut due to the high pump pressure, the larger abrasive-particle size, and the low standoff distance. In Figures 8.14b and 8.14c, the cut width provided by the model at the entry slightly deviates from the experimental results due to the influence of an 'initial damage zone'. Section 8.1.6 discusses this zone.

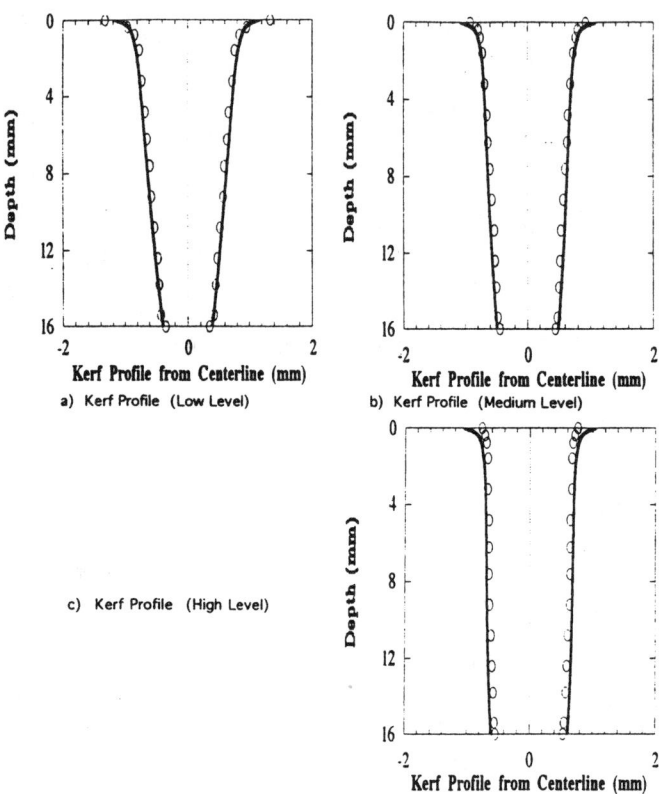

Figure 8.14 *Verification of Arola-Ramulu's model [185]*

8.1.6 Initial-Damage Zone Geometry

8.1.6.1 General Relations

Figure 8.1 illustrates the geometry of the initial-damage zone. In this zone, abrasive particles impact the surface at normal angles promoting a large degree of constituent disruption. Arola and Ramulu [185] observe fiber pullout and fiber-matrix-debounding in this zone. Also, small penetration craters and shallow abrasive-wear tracks appear. In ductile-behaving materials, extensive plastic deformation characterizes the initial-damage zone [181].

8.1.6.2 Ductile-Behaving Materials

Figure 8.15 illustrates general relations between the width of the initial-damage zone in a ductile-behaving material. From Figure 8.15a, the standoff distance has a dramatic impact on the zone width. Arola and Ramulu [181] also notice this effect. The same relation counts for the focus diameter influence (Figure 8.15b). From these figures, the larger the diameter of the abrasive water jet, the larger the width of the initial-damage zone. Guo [180] notices that the focus length has a certain influence on the width. All other process parameters do not significantly contribute to the width of the damage zone. For a standoff distance of about x=2 mm, the width of the initial-damage zone is between b_{IDZ}=150 µm to b_{IDZ} =350 µm for a wide range of cutting parameters [180].

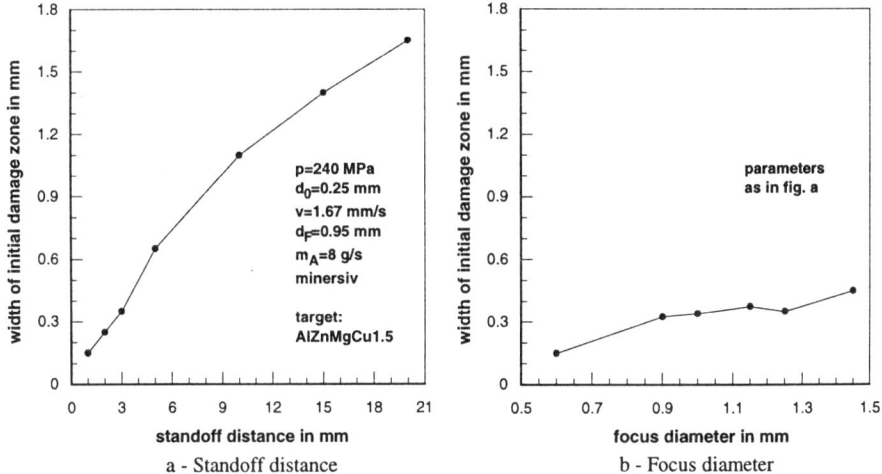

Figure 8.15 *Parameter influence on the initial-damage zone width [180]*

8.1.6.3 Brittle-Behaving Composite Materials

Arola and Ramulu [185] study the geometry of the initial-damage zone in composite materials. As for ductile materials, the standoff distance is the most influencing parameter. Table 8.3 illustrates this result. The initial-damage depth and width both increase with an increase in the standoff distance. If large standoff distances are required, the resulting damage can be disminished by the selection of smaller abrasive particles.

8.1.6.4 Model for Initial-Damage Zone Geometry

Arola and Ramulu [185] develop a regression model for the estimation of the initial-damage zone geometry. For the depth of the initial-damage zone, they find

$$h_{IDZ} = a_1 \cdot p^{a_2} \cdot x^{a_3} \cdot v^{a_4} \cdot d_P^{a_5}. \tag{8.8}$$

For the width of the initial-damage zone, they derive

$$b_{IDZ} = b_1 \cdot p^{b_2} \cdot x^{b_3} \cdot v^{b_4} \cdot d_P^{b_5}. \tag{8.9}$$

Table 8.3 lists the influence of the process parameters on the initial-damage zone geometry. Figure 8.16 presents a comparison between the model and experimental data.

Table 8.3 *Process-parameter influence on the initial-damage zone-geometry [185]*

Parameter	b_{IDZ} (% effect)			h_{IDZ} (% effect)		
	low	medium	high	low	medium	high
Pump pressure	5	3	8	7	2	8
Standoff distance	65	92	88	90	88	90
Traverse rate	10	1	1	1	5	2
Abrasive diameter	20	4	3	2	5	0

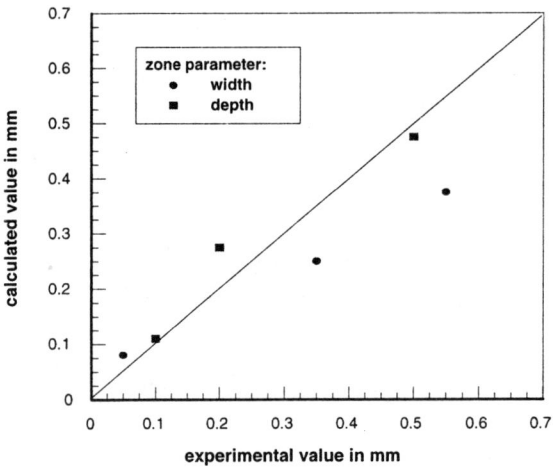

Figure 8.16 *Verification of Arola-Ramulu's model [185]*

8.2 Topography of Abrasive Water-Jet Generated Surfaces

8.2.1 General Characterization

8.2.1.1 Introductory Aspects

Generally, solid surfaces divide into homogeneous and non-homogeneous surfaces. Abrasive water jet generated surfaces divide into a zone of small, randomly distributed surface heights (smooth cutting zone), and a more regular, wavy zone (rough cutting zone). A transition zone separates both ranges (Figure 8.1).

Homogeneous surface separate into deterministic and random surfaces. Based on the distribution of the surface heights, random surfaces further divides into Gaussian or non-Gaussian surfaces. Gaussian surfaces are isotropic and non-isotropic. Zhou et al. [202] use auto-correlation functions, surface height distributions, and isotropic criteria and conclude that the surface topography of the smooth cutting zone generated by abrasive water jet is random, Gaussian, and weakly isotropic.

8.2.1.2 Static Characterization

The analysis of the surface profiles generated under different abrasive water jet cutting conditions is conducted qualitatively in terms of *static characteristics*, including the roughness average (R_a), the peak-to-valley height (R_y), and the root mean square roughness (R_q). Figure 8.17 plots typical roughness averages R_a of an abrasive water-jet generated surface profile, that are measured at three levels.

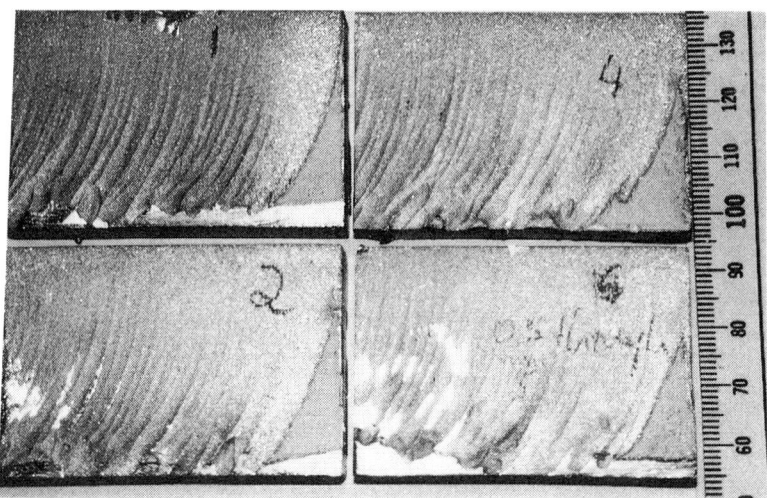

Figure 8.17 *Typical surface profiles generated by an abrasive water jet in aluminum*

Figure 8.18 displays a diagram of skewness versus kurtosis for various manufacturing processes. This diagram compares the profiles generated by abrasive water jets with that of other manufacturing processes in terms of their shape parameters, skewness (R_{Sk}), and kurtosis (R_{ku}). Note a considerable overlap between the profiles generated by an abrasive water jet and that of grinding as well as electro-discharge machining (EDM). The results indicate that abrasive water jet is capable of generating surfaces of quality comparable to that of those methods. A value of zero for skewness (R_{Sk}) and three for kurtosis (R_{ku}) is typical for a random, Gaussian profile. The plot of R_{Sk} against R_{ku} of abrasive water jet generated profile is centered around the random Gaussian profile. Zhou et al. [202] obtain the same quantitative results.

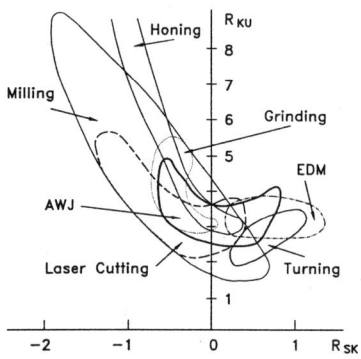

Figure 8.18 *Skewness-Kurtosis diagram for different manufacturing processes [206]*

Table 8.4 lists comparative roughness measurements for different manufacturing methods. From this table, abrasive water jet generates the highest roughness values. Nevertheless, considered that abrasive water-jet machining is used simultaneously for workpiece sectioning and finishing.

Table 8.4 *Surface roughness of titanium alloys machined by different methods [177]*

Method	Cutoff [mm]	R_a [μm]	R_y [μm]	R_q [μm]	Skewness
AWJ*	0.8	3.0	18.8	3.9	-0.5
	3.5	4.6	33.9	5.8	
Slab mill	0.8	0.4	2.3	0.6	-0.6
	3.5	0.7	4.5	0.9	
Ground Grinding	0.8	0.7	4.9	1.0	-0.3
	3.5	1.1	9.5	1.5	

* Abrasive water-jet cutting, p=207 MPa, x=2.5 mm, v=0.7 mm/s
$\dot{m}_A$ =10 g/s, garnet # 50

8.2.1.3 Dynamic Characterization

To understand more about the underlying principle of surface-profile generation in abrasive water jet cutting, the dynamic characteristics of the generated surface is discussed. These dynamic characteristics are derived using time-series analysis [206]. Hence, the surface profile data are modeled using an auto-regressive modeling technique. For further analysis, the Green's function, the auto co-variance function and the power-spectrum density of the models are computed [319]. Figure 8.19 displays typical plots of the Green's function, the auto co-variance function and the power-spectrum density for auto-regressive models for an abrasive water jet cut profile. Table 8.5 lists some peak values of the Green's function, the auto co-variance function and the power-spectrum density for different cutting conditions at three characteristic surface levels. Mean lag (M.Lag) indicates the lag for which the respective functions reach the mean position.

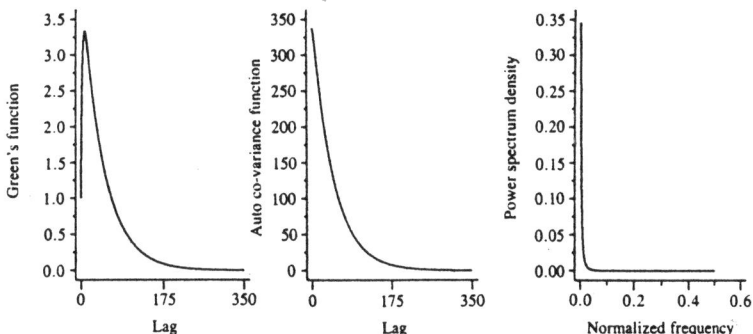

Figure 8.19 *Typical plots of dynamic surface-characterization parameters [204]*

The peak values of Green's function, the auto co-variance function and the power-spectrum density exhibit the same trend as that of surface roughness average with a change in cutting parameters. As the pump pressure increases, the roughness average reduces; so do the peak values of Green's function, the auto co-variance function and the power-spectrum density. With an increase in the standoff distance, the surface becomes smoother until a standoff distance of about x=7 mm. For larger standoff distances, the surface roughens. Similarly, the peak values of Green's function, the auto co-variance function and the power-spectrum density initially reduces until a standoff distance of x=7 mm is reached, and then the peak values increase. As the traverse rate increases, the surface roughness average as well as the peak values of Green's function, the auto co-variance function and the power-spectrum density increase. With an increase in the abrasive-mass flow rate, the peak values of Green's function, the auto co-variance function and the power-spectrum density have the same trend as that of change in the roughness average. Thus, Green's function, auto co-variance function and power-spectrum density peaks give a well defined and quantified surface roughness value for qualitative analysis of surface profiles.

256   8. Geometry, Topography and Integrity of Abrasive Water-Jet Machined Parts

Table 8.5 *Dynamic parameters of abrasive water-jet generated surface profiles [206]*

Parameter	Green's function		Auto CoV. function		Spectrum-density peak
	Peak	Mean Lag	Peak	Mean Lag	
p=248 MPa	1.44@2.6	42	16.9	42	1.36E0.3
p=331 MPa	1.40@2.5	55	17.2	55	1.70E03
x=5.08 mm	1.38@2.2	42	13.5	42	9.27E02
x=12.7 mm	1.53@3.0	52	23.0	52	2.31E03
v=0.42 mm/s	1.42@2.3	58	15.6	58	1.41E03
v=1.69 mm/s	1.51@2.7	55	20.6	55	2.01E03
m_A=2.27 g/s	1.59@2.8	56	21.8	56	2.16 E03
m_A=5.54 g/s	1.38@2.3	60	16.2	60	1.39E03

8.2.1.4 Wavelength Decomposition

In order to derive more information about the generated surface profile and to relate the surface to the cutting process in a more tangible way, wavelength decomposition of the roots of the surface structure signal models are performed (section 5.4). The characteristic frequencies of the discrete roots give the corresponding wavelengths. The relative power of the roots are obtained from the variance decomposition. Tables 5.3 and 8.6 list characteristic roots of auto-regressive models obtained from very different abrasive water jet cutting process conditions as well as their wavelength decompositio. Table 8.4 also gives the break frequencies of the real roots and the damped natural frequencies of the complex roots. The relative power of the discrete roots is also given.

Table 8.6 *Wavelength-decomposition of abrasive water-jet generated Surface profiles in the smooth-cutting zone [206]*

Parameter	Root	Discrete roots		Frequ.	Power	Wavelength (mm)
		Real	Imagin.			
p=248 MPa	1	0.8718	0.0000	2.1830	0.7660	0.2290
	2	0.6222	0.0000	6.5600	0.2340	0.0762
p=331 MPa	1	0.8943	0.0000	1.7885	0.8285	0.2811
	2	0.6207	0.0000	7.5894	0.1715	0.0659
x=5.08 mm	1	0.8542	0.0000	2.5089	0.7602	0.1993
	2	0.6338	0.0000	7.2568	0.2398	0.0689
x=12.7 mm	1	0.9073	0.0000	1.5488	0.8354	0.3228
	2	0.6387	0.0000	7.1354	0.1646	0.0701
v=0.43 mm/s	1	0.9042	0.0000	1.6028	0.9118	0.3120
	2	0.3729	0.2388	9.0653	0.0441	0.0552
	3	0.3729	-0.2388	9.0653	0.0441	0.0552
x=1.69 mm/s	1	0.9045	0.0000	1.5973	0.8532	0.3130
	2	0.5895	0.0000	8.4113	0.1468	0.0594
m_A=2.27 g/s	1	0.8979	0.0000	1.7146	0.8287	0.2916
	2	0.6181	0.0000	7.6563	0.1713	0.0653
m_A=4.54 g/s	1	0.8975	0.0000	1.7206	0.8541	0.2906
	2	0.5765	0.0000	8.7668	0.1459	0.0570

8.2 Topography of Abrasive Water-Jet Generated Surfaces

8.2.2 Surface Roughness

8.2.2.1 General Relations

The surface roughness is coupled with the texture of the smooth cutting zone.

Table 8.6 lists results of wavelength decomposition of a surface profile from the smooth-cutting zone. For all cutting conditions, the primary wavelength (section 5.4) of the model representing the smooth-cutting zone is of the order of $1 \cdot d_P$ to $1.5 \cdot d_P$. Hence, the surface roughness in this region is primarily caused by the abrasive particles. The secondary wavelength at the smooth-cutting zone is of the order of $0.33 \cdot d_P$ to $0.5 \cdot d_P$. Thus, in the smooth cutting region, the waviness (or striations) of the jet stream does not have any significant effect on the surface profile. The power of the primary wavelength is between 75% and 85% with one exception of very low traverse rate when it is 91%. All the primary as well as secondary wavelengths at this level have real roots except at very low traverse rate.

8.2.2.2 Influence of Hydraulic Parameters

Several authors [174, 180, 202, 320] perform systematic studies into the influence of process parameters on the surface roughness. Figure 8.20a shows the effect of pump pressure on the surface roughness. An increase in the pressure, in general, improves the surface quality. This is partly consistent with results reported by Kovacevic et al. [174] on metals, and Hocheng and Chen [314] and Momber et al. [189] on ceramics. This results are due to the increased fragmentation probability of the abrasive particles with an increase in their velocity. This fragmentation reduces the size of the impacting abrasive particles. Also, an increase in the pump pressure increases the abrasive water jet kinetic energy. This process allows part of the excess energy to smoothen the surface. However, Kovacevic et al. [174] show that the influence of the pump pressure on the roughness also depends on the traverse rate. Only for high traverse rates, an increase in the pump pressure decreases the roughness. From Figure 8.20a, the pump pressure does not have a significant influence on the roughness in the upper range of the cut. Singh et al. [320] and Guo [180] who find that the influence of the pump pressure strongly depends on the depth of cut, make a similar observation. The deeper the cut, the more pronounced is the pump pressure influence.

Figure 8.20b illustrates the influence of the orifice diameter on the surface finish. The differences in the roughness are not large, despite about a 100 % difference in the orifice diameter. Zhou et al. [202] obtains identical result. These observations support results from Kovacevic et al. [206] who find that the orifice diameter influences the waviness of the surfaces but not their roughness.

258   8. Geometry, Topography and Integrity of Abrasive Water-Jet Machined Parts

8.2.2.3 Influence of Cutting Parameters

Figure 8.20b also plots typical relations between the traverse rate and the surface roughness. The roughness significantly increases as the traverse rate increases. This observation agrees with results on aluminum [180, 263, 320], ceramics [189, 321], metal-matrix composites [182, 184], glass [315], and fiber-reinforced composites [322]. Singh et al. [315] and Blickwedel et al. [263] observe that the relation between the depth of cut and surface roughness changes dramatically at high traverse rates (ca. v=200 mm/min). At comparatively low traverse rates (ca. v=10 mm/min), the roughness shows almost stable values even at deep cuts. The increase in the number of impacting particles at lower traverse rates contributes to the improved surface finish. The additional particles serve to smoothen the surface that forgoing particles generated.

Table 8.7 *Quality levels for surfaces generated by abrasive water jet [262]*

Quality level	Description
$q_M=1$	Criteria for separation cut. Usually, $q_M>1.2$ should be used.
$q_M=2$	Rough surface finish with striation marks at the lower half surface.
$q_M=3$	Smooth/rough transition criteria. Slight striation marks may appear.
$q_M=4$	Striation free for most engineering materials.
$q_M=5$	Very smooth surface finish.

In order to rigorously analyze the influence of the traverse rate on the surface quality, Zeng and Kim [262] introduce several surface-quality levels. Based on their 'machinability number' concept (section 5.9.2), the authors derive

$$v = \left[\frac{N_m \cdot p^{1.25} \cdot \dot{m}_W^{0.687} \cdot \dot{m}_A^{0.343}}{c_0 \cdot q_M \cdot h \cdot d_F^{0.618}} \right]^{1.15} . \qquad (8.10)$$

In this equation, q_M is a quality-level parameter between $q_M=1$ and $q_M=5$, depending on the desired quality level. A quality level of $q_M=1$ is referred to as a separation cut in which the abrasive water jet is just capable of separating the workpiece. Examinations of cut surfaces of separation cuts revealed that striation marks appear at about one third of the total workpiece thickness (Table 8.7).

8.2.2.4 Influence of Mixing Parameters

Figure 8.20d illustrates the influence of the focus diameter on the surface roughness. As a certain focus diameter is exceeded, the surface roughness increases with an increase in the focus diameter. The deeper the cut, the more pronounced becomes the influence of the focus diameter.

8.2 Topography of Abrasive Water-Jet Generated Surfaces

Kovacevic et al. [174] observe that for very low and very high traverse rate, the roughness is independent on the focus diameter. Singh et al. [320] find almost identical roughness values for large focus nozzles; whereas, the roughness of the small diameter focus is significantly lower.

From Figure 8.20c, the focus length does not show any significant influence on the surface roughness especially in the range of shallow cuts.

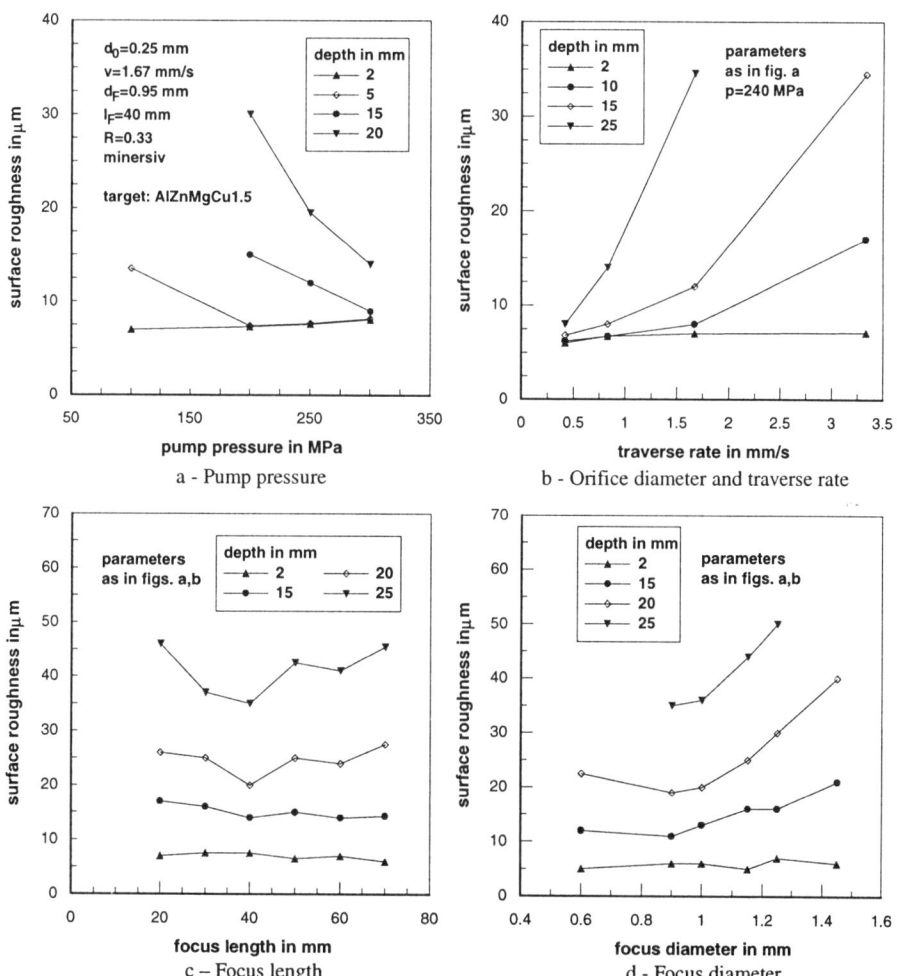

Figure 8.20 *Parameter influence on the surface roughness [180]*

8.2.2.5 Influence of Abrasive Parameters

Figure 8.21a illustrates the influence of the abrasive-mass flow rate on the surface roughness. In the figure, the roughness decreases with an increase in the abrasive-mass flow rate. Several authors obtain identical results for aluminum [180], glass [315], and refractory ceramics [189]. A high number of abrasive particles involved in mixing increases the probability of particle collision that decreases the average diameter of the impacting particles. Nevertheless, Kovacevic et al. [174] observe under certain conditions, such as comparatively low pump pressure, that the surface roughness increases as far as a critical abrasive-mass flow rate (for stainless steel about 11.3 g/s) is exceeded. This result indicates that a certain ratio between the abrasive-particle number and abrasive-particle velocity is a necessary condition for effective surface finishing. From Figure 8.21a, this argument holds for cut depths larger than h=15 mm.

Figure 8.20b shows the relation between the surface roughness and abrasive-particle diameter. In the range of small diameters, the roughness almost linearly decreases as the particle diameter increases, except for very shallow cuts. Ohlsen [97] also notices this trend. This effect is very pronounced in deeper cuts. The individual particle-impact characteristics is overcome by the general material-removal characteristics as a certain depth of cut is obtained. Other effects, such as a a decrease in the jet instability with a decrease in the abrasive diameter [47] contributes to the surface structure. Nevertheless, as a certain abrasive diameter is exceeded (90 μm for steel, 200 μm for aluminum), the roughness increases. Guo et al. [203] and Kovacevic et al. [206] show through wavelength decomposition that the wavelength of the roughness profile in the upper cutting zone significantly relates to the average abrasive-particle diameter. Zhuo et al. [202] assume that the surface generated by an abrasive water jet in the smooth-cutting zone is formed by the superposition of micro-dimples that are created by individual abrasive grains. The size of these dimples directly relates to the abrasive-particle diameter

Figure 8.21c shows the influence of the abrasive type on the surface roughness. The use of the harder aluminum-oxide substantially reduces the surface roughness of the ceramic material being cut. The relation between the traverse rate and surface roughness depends on the abrasive type. For garnet, the roughness is much more sensitive to changes in the traverse rate compared to aluminum oxide. The cutting performance of the hard aluminum oxide is that high that it partly compensates the worse surface quality from the high traverse rate. Singh et al. [323] detect a relation between the abrasive type, target material, and surface roughness. An abrasive material that generates a low roughness in aluminum, creates a comparatively high roughness level in ceramics.

Momber et al. [42] find that the surface roughness substantially improves by selecting an appropriate abrasive-particle size distribution (Figure 8.21d). For comparable cut depths, the average surface roughness improves up to 60 % if an appropriate abrasive-particle size distribution is applied.

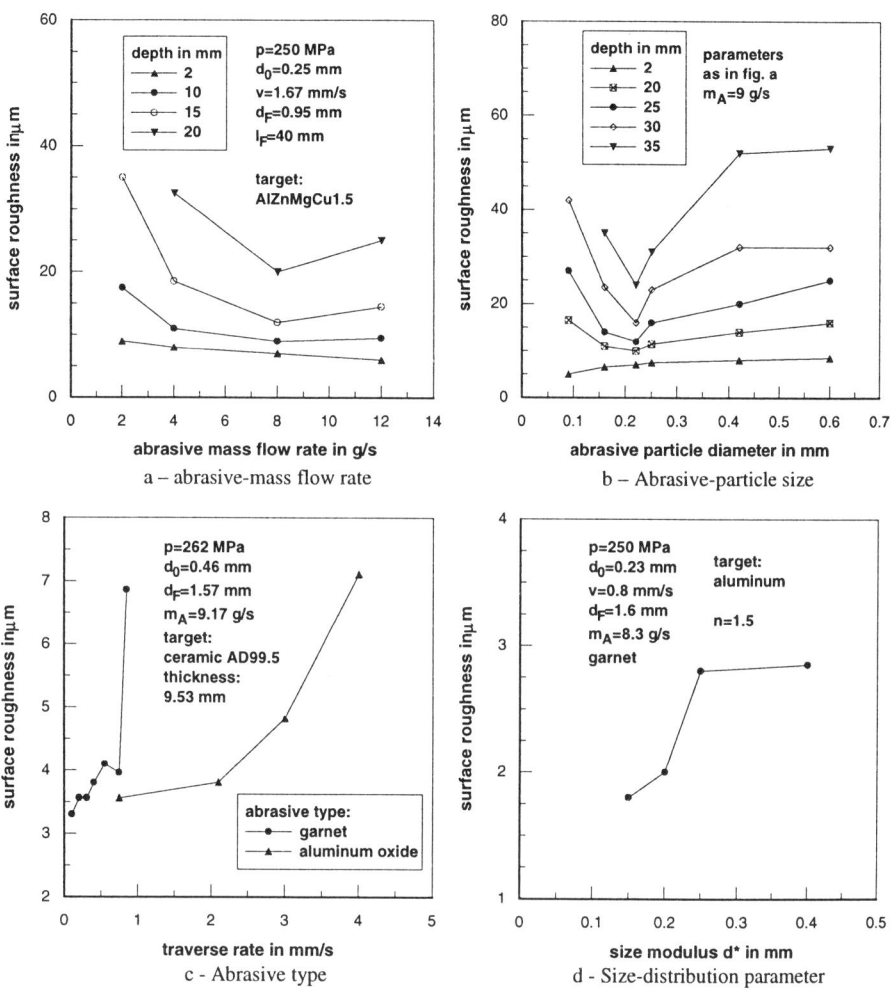

Figure 8.21 *Abrasive-parameter influence on the surface roughness [42, 180, 263]*

8.2.2.6 Influence of Target-Material Structure

Figure 8.22 illustrates the influence of some target-material structural parameters, such as fiber orientation in reinforced plastics, lamina number in ceramics, and silicon-carbide content in metal-matrix compounds. From experiments on refractory ceramics, the roughness is low as the resistance of the material against abrasive water-jet erosion is high [189].

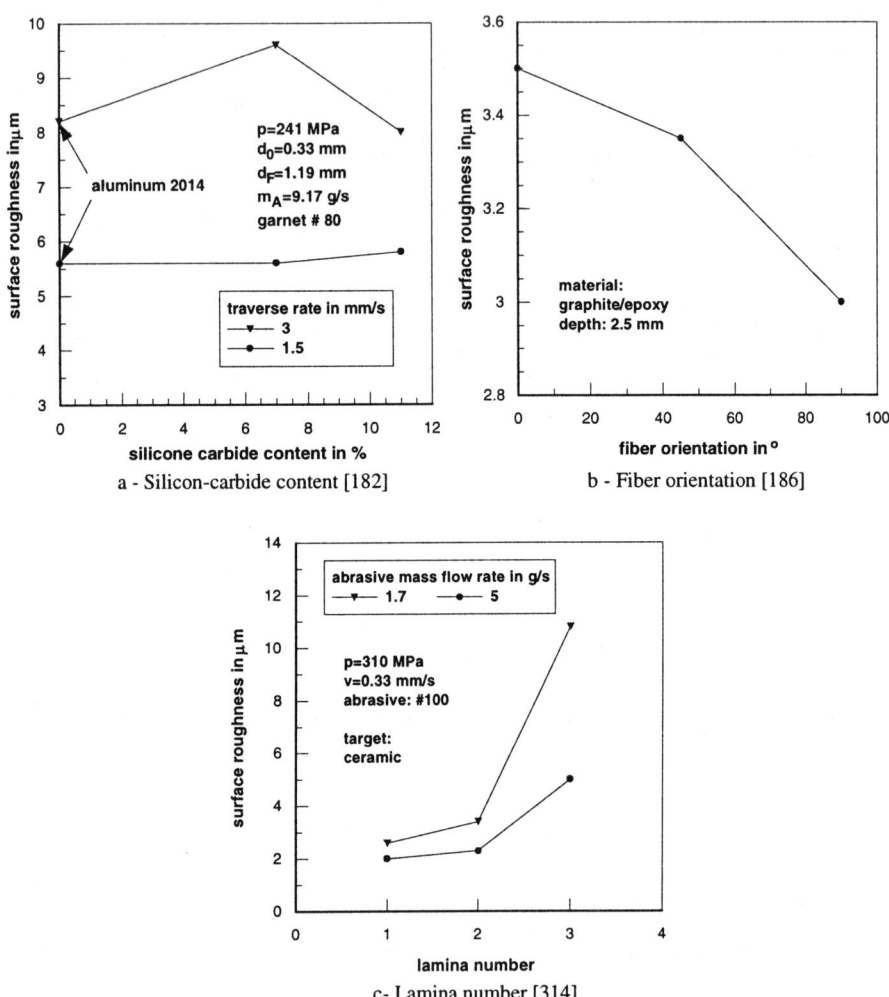

Figure 8.22 *Influence of the specimen structure on the surface roughness*

6.2.2.7 Models for Roughness Estimation

Curham et al. [324] establish a simple regression analysis

$$R_a = a_1 \cdot h^{a_2} \cdot v^{a_3}. \tag{8.11}$$

Burnham and Kim [321] find

$$R_a = b_1 + b_2 \cdot h + b_3 \cdot h \cdot v. \qquad (8.12)$$

Webb and Rajukar [175] include several further process parameters in their regression model,

$$R_a = c_1 \cdot (v \cdot d_P)^{c_2} \cdot (\dot{m}_A \cdot p)^{c_3}. \qquad (8.13)$$

Kovacevic [176] and Ramulu and Arola [187] who exploit second-order models develop a more complex relation,

$$R_a = C_0 + \sum_{i=1}^{n} C_i \cdot X_i + \sum_{i=1}^{n} C_{ii} \cdot X_{ii}^2 + \sum_{i=1}^{n} C_{ij} \cdot X_i \cdot X_j. \qquad (8.14)$$

In this model, C_0 is a constant, and C_i, C_{ii}, and C_{ij} are the first order, second order, and interaction coefficients, respectively. X_i are the investigated process parameters. In order to solve Eq. (8.14), both references use a commercial SAS-software.

Blickwedel et al. [263] who approximate the roughness by a continuous curvature, introduce an empirical model

$$R_a = R_{a0} + A_1 \cdot h^{A_2}. \qquad (8.15)$$

The constants R_{a0}, A_1, and A_2 are estimated by regression. This analysis replaces the depth of cut by Eq. (6.33). Thus,

$$R_a = R_{a0} + \left[A_3 \cdot \frac{p - p_{thr}}{v^{0.303}} \right] \cdot \left[\frac{h}{h_{max}} \right]^{A_2}. \qquad (8.16)$$

Figure 8.23 gives a comparison between Eq. (8.15) and experimentally estimated roughness values.

Based on waviness height measurements, Arola and Ramulu [185] develop a regression model for the estimation of the depth of the smooth cutting zone. These authors assume that the beginning of the rough cutting zone corresponds to a waviness height of about 40 µm. The model is

$$h_{SC} = [-3{,}677.7 - 2.04 \cdot p + 2{,}886 \cdot x - 2{,}658 \cdot v + 92.8 \cdot d_P - 165.8 \cdot h + 1.11 \cdot p^2 \\ - 30.8 \cdot x^2 + 688.6 \cdot v^2 + 28.6 \cdot h^2 - 1.17 \cdot p \cdot x - 1.11 \cdot p \cdot d_P - 1.14 \cdot p \cdot h - \\ 13.67 \cdot x \cdot d_P - 14.08 \cdot v \cdot d_P + 53.1 \cdot v \cdot h + 1.94 \cdot d_P \cdot h]^{0.33}. \qquad (8.17)$$

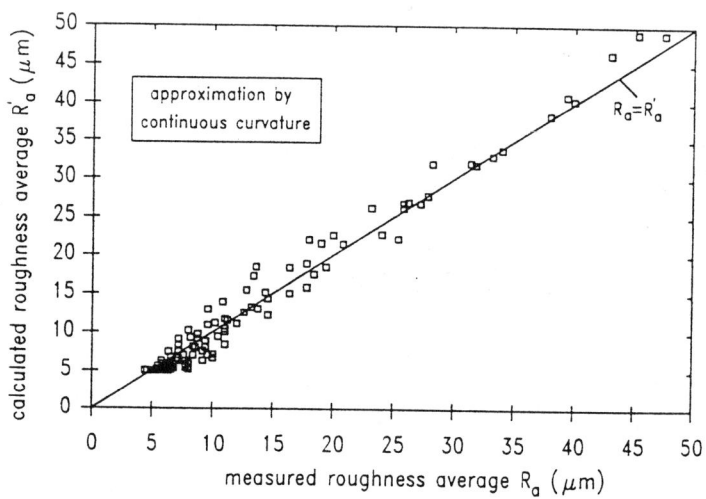

Figure 8.23 *Verification of Blickwedel's model [263]*

8.2.3 Surface Waviness

8.2.3.1 General Relations

The waviness, or striation marks, corresponds to the rough cutting zone of an abrasive water jet generated surface. This zone is characterized by a comparatively regular surface profile.

Kovacevic et al. [206] show that the power of the primary wavelength in the rough cutting zone is above 90% for all cutting conditions and most of them are considerably above 95%. This indicates that the effect of the jet diameter becomes more and more predominant downwards along the kerf wall. Also, the primary wavelength of all profiles have real roots.

Figure 8.24 shows the influence of several parameters on the primary wavelength. At lower pump pressures, the primary wavelength of the roots is about $3 \cdot d_{jet}$. However, as the pump pressure increases, the primary wavelength decreases. The influence of standoff distance is considerable on the primary wavelength at the rough cutting zone. At lower standoff distances, the wavelength is approximately $5 \cdot d_{jet}$. At a standoff distance of about x=7.6mm, the wavewlength decreases. Nevertheless, with further increase in the standoff distance, the primary wavelength at the rough-cutting zone increases.

At very low traverse rates, the primary wavelength in the rough cutting zone is about $0.33 \cdot d_{jet}$. At a traverse rate of v=0.85 mm/s, the primary wavelength is about $1.5 \cdot d_{jet}$. The primary wavelength increases as the traverse rate increases. At lower abrasive-mass flow rates, the primary wavelength is about $6 \cdot d_{jet}$. With an increase in the abrasive-mass flow rate, the wavelength reduces initially; again it increases and finally reduces. The secondary wavelength at the rough-cutting zone is about $0.5 \cdot d_P$ to $0.17 \cdot d_P$. This result attributes to the particle fragmentation rather than improper penetration. The power of the secondary wavelength is only about 1% at this level. Hence, the effect of secondary wavelength on the surface profile can be ignored. For all cutting parameters, if the primary wavelength of the profile at the rough cutting zone is smaller, the surface is smoother.

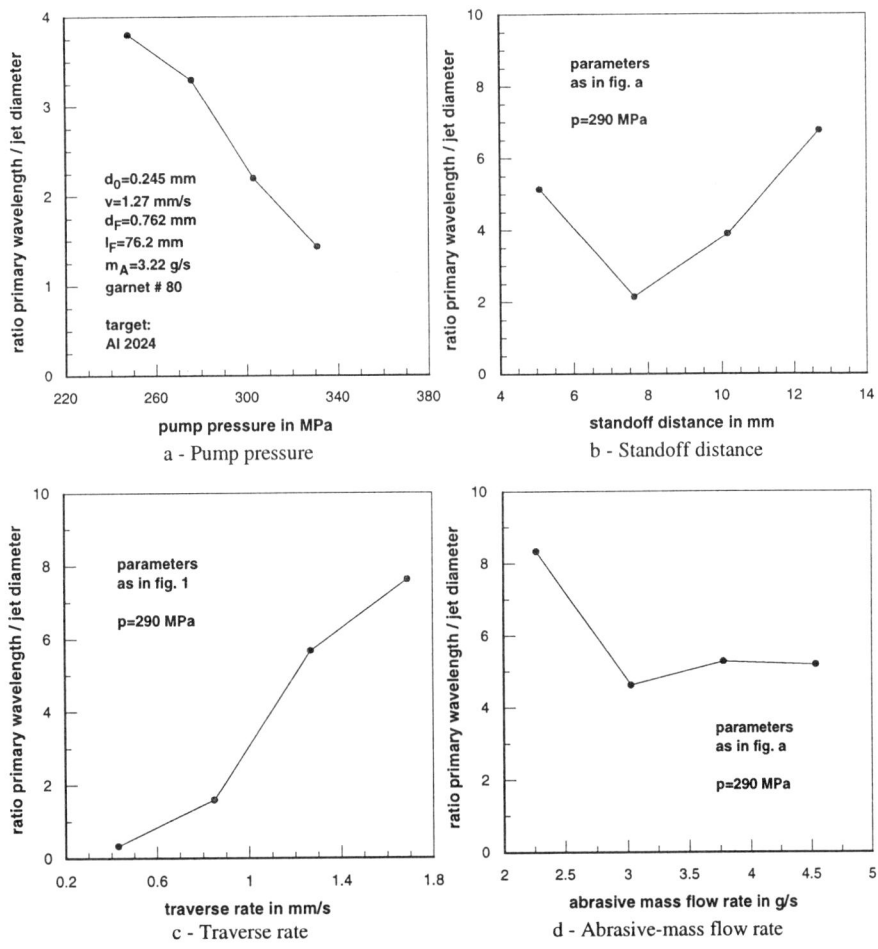

Figure 8.24 *Parameter influence on the primary wavelength [206]*

Figure 8.25 displays a plot of the amplitude of dominant striation marks against the depth of cut. This relation fits into

$$R_W(h) = a_1 \cdot h^2 + a_2 \cdot h + a_3. \qquad (8.18)$$

In this equation, the values of the constants a_1 to a_3 depend on the traverse rate. The higher the traverse rate, the higher the amplitude of the striation marks. Tan [325] obtains the same relation. The structure of Eq. (8.18), a second-order polynomial, actually supports the energy-dissipation models in sections 5.4 and 5.5. As the amount of abrasive water-jet energy that is absorbed during the cutting increases, striation formation increases.

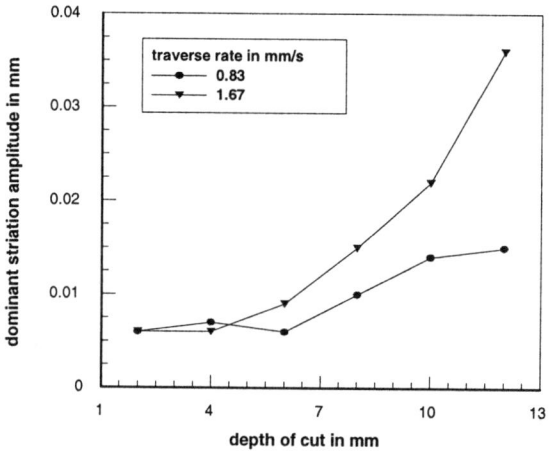

Figure 8.25 *Relation between the depth of cut, traverse rate, and amplitude of dominant striation marks [213]*

8.2.3.2 Influence of Process Parameters

Figure 8.26 shows the influence of the traverse rate and abrasive-mass flow rate on the waviness. With an increase in the abrasive-mass flow rate, the waviness substantially reduces. For high abrasive-mass flow rates, the waviness is less sensitive to changes in the traverse rate. Additionally, for very low traverse rates, the influence of the abrasive-mass flow rate on the waviness is almost eliminated. Therefore, an abrasive water jet with a high cutting capability generates a regular striation structure since external effects, such as vibrations and fluctuations, are terminated. Also, the waviness is proportional to the jet diameter. Thus, reductions in focus diameter and standoff distance reduces the striation height.

8.2 Topography of Abrasive Water-Jet Generated Surfaces

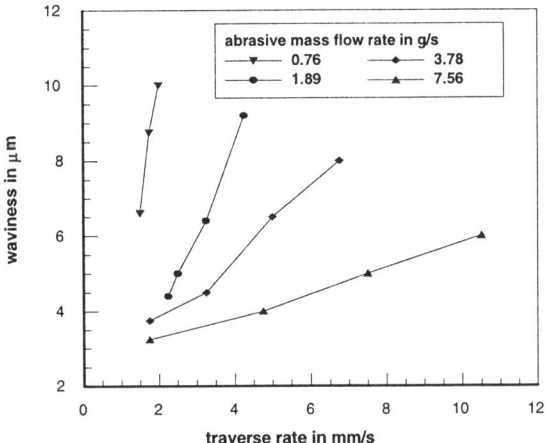

Figure 8.26 *Parameter influence on the waviness [214]*

8.2.3.3 Models for Waviness Estimation

Tan [325] develops the first geometrical waviness model. Based on a cyclically changing jet diameter over each single cutting cycle (Figure 8.27a), Tan derives a system of equations. For the depth of jet penetration, he derives

$$h = h_{max} \cdot \left[1 - \exp\left(-a_1 \cdot \frac{x_0}{\lambda_S - x_0}\right)\right]. \qquad (8.19a)$$

In this equation, h_{max} is the maximum depth of cut as the traverse rate approximates $v=0$, λ_S is the wave length of the striation marks, x_0 is the delay of the point where the jet diameter is $d_{jet}=0$ within a cycle. The change in the jet diameter is

$$d_{jet}(x) = d_F \cdot \left[1 - \exp\left(-a_2 \cdot \frac{(h - h_{max})^2}{h^2} \cdot \frac{\Delta x - x_0}{\lambda_S - x_0}\right)\right]. \qquad (8.19b)$$

In the equation, d_F is actually the jet diameter for $h=0$ (it is assumed here as the focus diameter), Δx is the displacement within each cycle. For $\Delta x = \lambda_S$, the diameter of the jet at the end of each cycle is

$$d_{jet}(x = cycle) = d_F \cdot \left[1 - \exp\left(-a_2 \cdot \frac{(h - h_{max})^2}{h^2}\right)\right]. \qquad (8.19c)$$

In this equation, $a_2=0.005 \cdot p \cdot d_P / v^2$. Comparisons between experiments and model show a good correspondence (Figure 8.28). The model is also agrees with Figure 8.25 and Eq. (8.17). The geometrical basic of the model explains that the values of estimated primary wavelengths of abrasive water jet generated surfaces exceed the jet diameter. Due to the cyclical changes in the jet diameter, the secondary wavelength is identical to the length of the envelope shown in Figure 8.27a. In Figure 8.29, three typical cases are distinguished:

$$x_0 = 0, \qquad (8.20a)$$

$$0 < x_0 < \frac{d_{jet}(x = cycle)}{2}, \qquad (8.20b)$$

$$x_0 > \frac{d_{jet}(x = cycle)}{2}. \qquad (8.20c)$$

If the delay is zero, a waviness profile is created. The larger the delay, the higher the waviness values. Figure 8.29b characterizes the critical case for which the envelopes are tangent to each other. In this case, the maximum possible waviness is reached. In Figure 8.29c, the jet only partially penetrates the workpiece.

Hashish [214] develops a simplified physical waviness model. Figure 8.27b shows the idealized (circular) geometry for this model. From this figure, the waviness height is

$$R_W = \frac{d_F}{2} - \sqrt{\frac{d_F^2}{4} - \frac{x^2}{4}} = \frac{d_F}{2} \cdot \left[1 - \sqrt{1 - \left(\frac{x}{d_F}\right)^2}\right]. \qquad (8.21)$$

If the jet traverse distance, x, is related to the abrasive water-jet penetration rate,

$$R_W = \frac{d_F}{2} \cdot \left[1 - \sqrt{1 - \left(\frac{\pi}{4}\right)^2 \cdot \left[\frac{2 \cdot d_F \cdot (h - h_C) \cdot v \cdot \varepsilon_{Mdef}}{\dot{m}_A \cdot v_P^2}\right]^2}\right]. \qquad (8.22)$$

With $d_F \cdot (h - h_C) \cdot v = \dot{V}_M$ and $\frac{\dot{m}_A \cdot v_P^2}{2 \cdot \varepsilon_{Mdef}} = \dot{V}_{ch}$, three characteristic cases distinguish [214]. This result agrees with Tan's [325] model. The cases are

$$\dot{V}_{ch} > \frac{\pi}{4} \cdot \dot{V}_M, \qquad (8.23a)$$

$$\dot{V}_{ch} = \frac{\pi}{4} \cdot \dot{V}_M, \qquad (8.23b)$$

$$\dot{V}_{ch} < \frac{\pi}{4} \cdot \dot{V}_M. \tag{8.23c}$$

Eq. (8.23a) characterizes the most common case. This case occurs when the adjacent locations of the abrasive water jet show some overlap that results in a continuous cutting through over the entire thickness of the workpiece (Figure 8.29a).

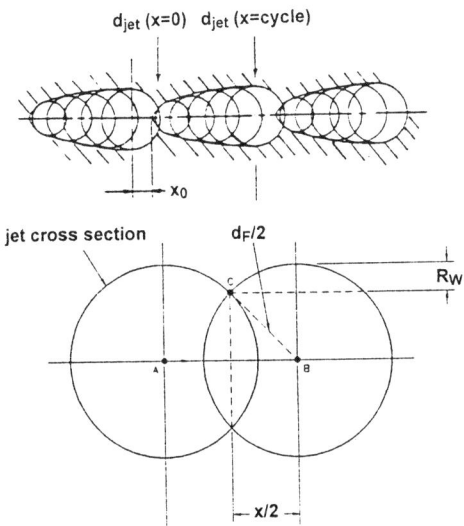

Figure 8.27 *Basics for abrasive water-jet waviness models*

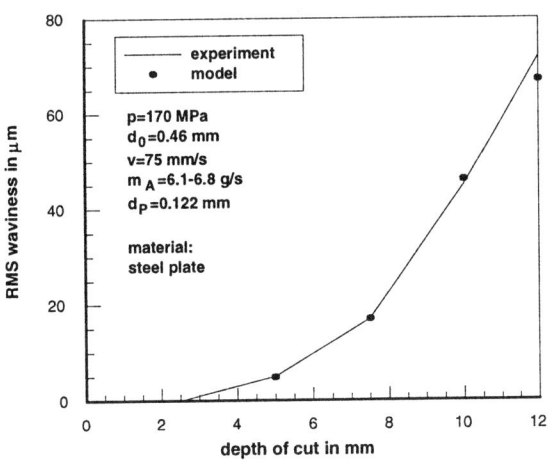

Figure 8.28 *Verification of Tan's model [325]*

The less the overlap, the higher the waviness. Eq. (8.23b) is a limiting case in which the adjacent locations of the jet at the bottom of the cut are just tangent to each other (Figure 8.29b). Finally, case (8.23c) occurs when there is no cutting through and the adjacent locations of the jet in the rough cutting zone are some distance apart (Figure 8.29c). Eq. (8.22) just qualitatively agrees with experimentally estimated waviness values; the values calculated from the model are up to 20 times too high. Therefore, the model is refined [214] by incorporating cut taper effects (section 8.1.3),

$$R_{W\,mod} = \frac{1}{2} \cdot (d_F - b_B) + R_W .\qquad(8.24)$$

Figure 8.30 that is a comparison between Eq. (8.23) and experimental data, shows reasonable correlation, but there are still deviations up to 300 %.

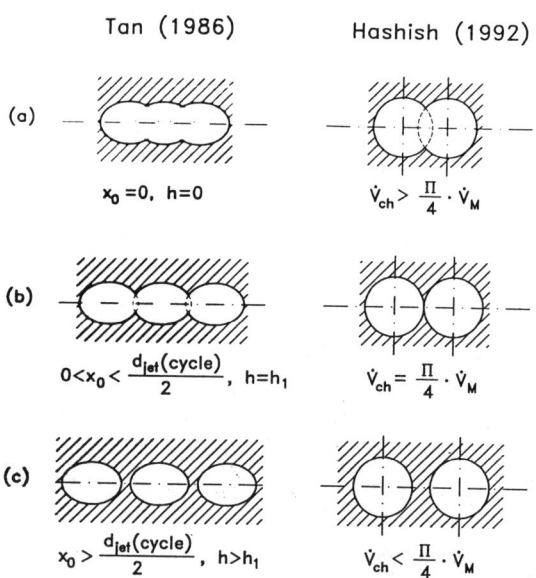

Figure 8.29 *Critical waviness-model conditions*

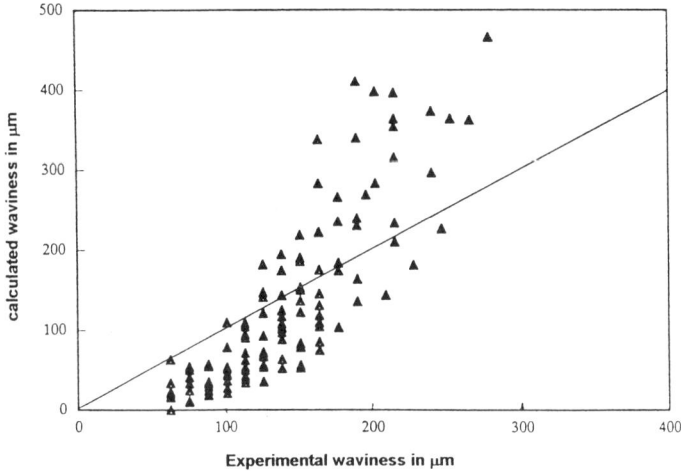

Figure 8.30 *Verification of Hashish's waviness model [241]*

8.3 Integrity of Abrasive Water-Jet Generated Surfaces

8.3.1 Fatigue Life

The machining processes demand that the machining method not negatively influences the fatigue life of the material.

Holland [120] compares the fatigue life of samples of stainless steel, inconel, and titanium cut by shearing and abrasive water jet cutting. He shows that the fatigue life of abrasive water jet cut materials is sensitive to the abrasive-particle size. Table 8.8 shows that the fatigue life for titanium is doubled by using smaller abrasive particles. Singh and Jain [326] perform high-cycle fatigue tests on titanium alloys as well as on steel samples cut by an abrasive water jet. The authors find fatigue cracks. Obviously, these cracks result from abrasive grains embedded in the target material during the material-removal process (section 8.3.4).

Investigations on titanium and aluminum samples cut by an abrasive water jet [327] show that the cutting process not negatively influences the fatigue behavior of these materials. Figure 8.31 shows results of titanium tests. In the figure, the Wöhler-curves obtained after abrasive water-jet cutting under different conditions (a,b,c) locates inside a small band below the reference Wöhler-curve (d) of the plain material sample. Arola and Ramulu [328] find that the deformation resulting from abrasive water-jet cutting, remains localized at the free surface. Therefore, the materials's ductility is unaltered during the cutting and serves to retard the propagation of surface flaws under cyclic loading. Furthermore, stress fields resulting from abrasive water-jet cutting are compressive [328]. Compressive residual stress fields superposed with tensile service loads reduce the effective stress at the machined surface. This effect provides a longer fatigue life through deterring crack propagation and stress-corrosion cracking.

Table 8.8 *Fatigue life of different materials cut by abrasive water jets (AWJ) [120]*

Material	Pump pressure p [MPa]	Edge preparation	Life cycle	
			Low	High
CRES 321	70	AWJ (# 60)[1]	100	122
		AWJ (# 150)	157	188
		Shear	107	132
		Polish	200	319
Inconel 625	90	AWJ (# 60)	52	83
		AWJ (# 150)	75	112
		Shear	120	137
		Polish	177	286
Titanium	110	AWJ (# 60)	20	30
		AWJ (# 150)	38	42
		Shear	10	20
		Polish	73	108

[1] referring to the abrasive-particle size in Mesh destination

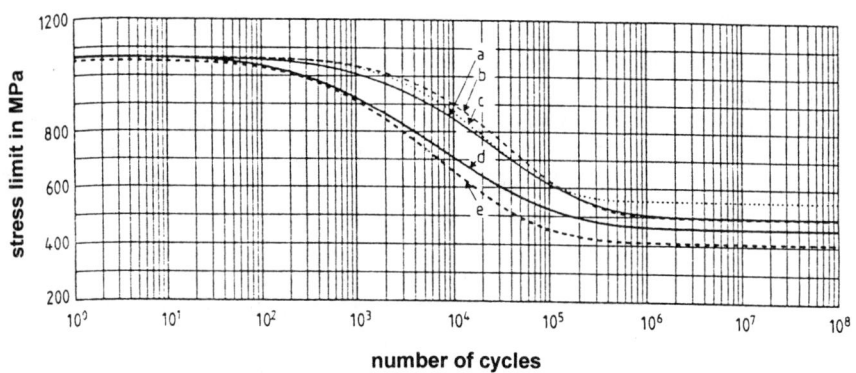

Figure 8.31 *Fatigue Wöhler-curves for titanium [327]*
a,c - AWJ, v=3.17 mm/s, no additional treatment
b - AWJ, v=2.17 mm/s, edge deburring
d,e - Reference lines (Ti6Al4V)

8.3.2 Surface Hardening

8.3.2.1 Hardness Measurements

The impact of spherical particles and even of plain water jets [22] is used to alter the surface hardness in the process of peening. Therefore, some degree of surface hardening could occur in abrasive water jet-machining.

Hashish [313] and Savrun and Taya [183] show through micro-hardness surveys on several materials that no appreciable hardness alteration occurs during the abrasive water jet machining-process (Table 8.9).

Lavender and Smith [329] investigate metal-matrix composites and find similar microhardness distributions on samples machined by diamond tools and by abrasive water jets (Figure 8.32a). Kitamura et al. [330] compare the behavior of Cr-Mo steel cut by EDM and abrasive water jet. From Figure 8.32b, there is no increase in the surface hardness of the samples treated by an abrasive water jet. Generally, the hardness alteration is even lower than that for the EDM-machined sample.

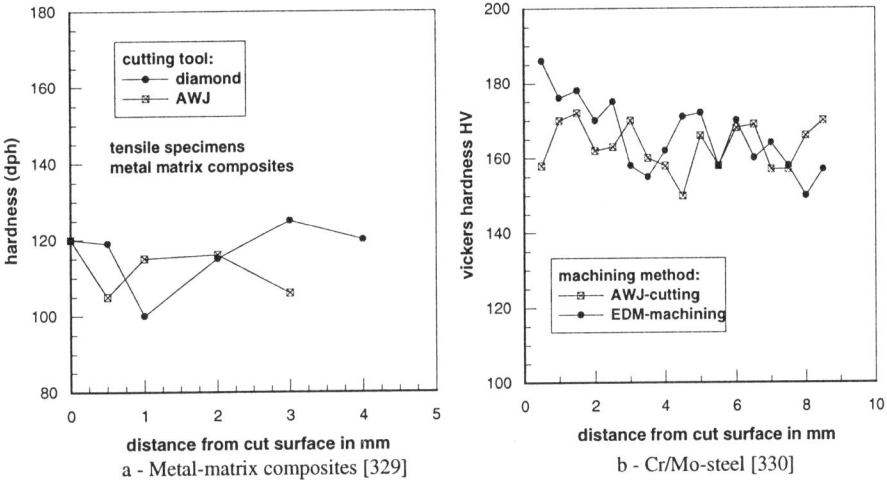

Figure 8.32 *Microhardness measurements on machined surfaces*

8.3.2.2 Stress Measurements

Arola and Ramulu [177, 328] who carry out residual stress measurements on several metals after abrasive water jet cutting, find that the stresses generated by the abrasive particles are compressive, similar to those induced by slab milling. Guo [180] obtains the same result. The value of the compressive stress increases as the diameter of the abrasive particles increases (Table 8.10).

Table 8.9 *Hardness of abrasive water-jet (AWJ) machined surfaces, according to ASTM Standard E-384 [313]*

Material	Hardness
Steel 4340 cut surface AWJ machined	R_C 23.2/24.5/21.7/24.5/20.8 20.3/19.7/22.2/22.3/22.9
Carbon steel A 572 cut surface AWJ machined	R_B 82.0/83.8/83.2/83.0/82.3 84.0/84.8/85.0/84.8/84.3
Tool steel A2 cut surface AWJ machined	R_B 92.5/93.0/90.6/89.2/92.2 92.2/92.7/93.0/91.5/92.2
Titanium cut surface AWJ machined	R_C 33.0/33.8/35.5/32.8/35.7 33.5/33.0/34.5/35.4/33.8
Aluminum 6061-T6 cut surface AWJ machined	R_F 49.2/54.3/56.7/57.5/56.3 56.0/59.0/60.5/59.0/59.0
Magnesium cut surface AWJ machined	R_F 51.5/56.2/54.3/51.0/57.5 49.2/61.2/60.0/61.0/61.0

Table 8.10 *Principle residual stresses in titanium after machining [177]*

Machining Method	Stress Tensor [MPa]
Abrasive water jet p=207 MPa d_P=350 µm	$\begin{vmatrix} -65 & 0 & 0 \\ 0 & -835 & 0 \\ 0 & 0 & -63 \end{vmatrix}$
Abrasive water jet p=138 MPa d_P=350 µm	$\begin{vmatrix} -184 & 0 & 0 \\ 0 & -912 & 0 \\ 0 & 0 & -149 \end{vmatrix}$
Abrasive water jet p=207 MPa d_P=135 µm	$\begin{vmatrix} 48 & 0 & 0 \\ 0 & -474 & 0 \\ 0 & 0 & 26 \end{vmatrix}$
Slab milling	$\begin{vmatrix} -83 & 0 & 0 \\ 0 & -953 & 0 \\ 0 & 0 & -49 \end{vmatrix}$
Ground grinding	$\begin{vmatrix} 661 & 0 & 0 \\ 0 & 276 & 0 \\ 0 & 0 & -37 \end{vmatrix}$

The highest compressive residual stress in abrasive water-jet machining occurs with the use of the largest abrasive-particle size. On the contrary, the lowest residual stresses that were less than half the maximum values, appear with the smallest abrasive particles. The principle stress direction does not correspond with the abrasive water jet penetration-direction. However, the orientation of the maximum shear stress coincides with the jet path along the cut wall.

8.3.3 Micro-Structural Aspects

8.3.3.1 General Aspects of Alteration

Several researchers carry out investigations into the micro-structural alteration of materials machined by abrasive water jets.

In an early investigation, Hashish [331] finds no distortion near abrasive water-jet turned aluminum and boron-carbide surfaces in comparison to the bulk of the materials.

Sano et al. [312] and Hashish [313] check the material deformation near surfaces of thin metal sheets cut by different cutting methods. Hashish [313] compares different machining methods, such as laser, shear cutting, and abrasive water jets. The surface produced by abrasive water jet is free of distortions. Based on SEM-observations of cast iron samples, no evidence of micro-structural changes appear after abrasive water jet cutting. Sano et al. [312] compare etching, blanking, and abrasive water jet cutting of amorphous foil sheets. Based on an x-ray-deflection method, they do not detect any crystallization on the surfaces generated by an abrasive water jet.

Kitamura et al. [330] investigate the conditions of CrMo-steel turbine blades after cutting by abrasive water jet and EDM. Whereas the EDM-cut surface clearly becomes molten, the abrasive water-jet machined surface is unaffected.

8.3.3.2 Surface Cracking in Brittle-Behaving Materials

In brittle-behaving materials, surface cracking is a machining-related problem. Geskin et al. [315] observe cracking in glass samples during piercing by abrasive water jet at the commencement of cutting. Experiments prove that this cracking is eliminated by using a built-in dual compensator that enables a temporary reduction in the pump pressure at the beginning of the cutting. Hashish [313] notices crack formation in abrasive water-jet machined glass samples. He shows that micro-cracks less than one micrometer deep forms when abrasive particles less than $d_P=10$ μm in diameter are used. For metal-matrix composites cut by an abrasive water jet, Savrun and Taya [183] prove a crack-free machined surface.

From Figure 8.33 that shows alumina-ceramic specimens, the amount of cracking in the jet exit area depends on the abrasive-particle diameter. The larger the abrasive diameter, the more severe cracking becomes.

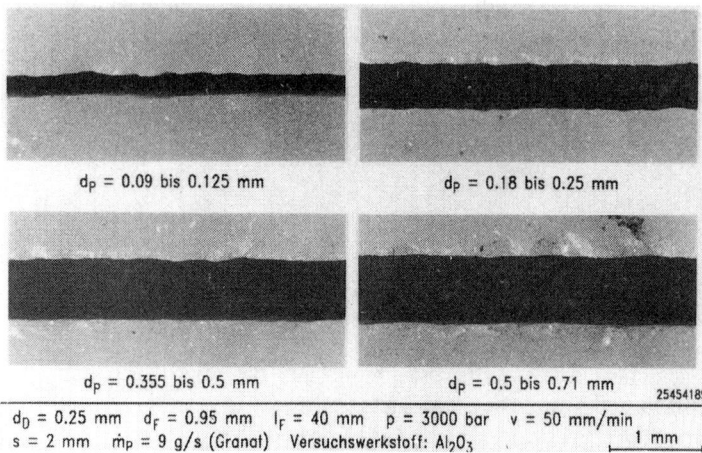

Figure 8.33 *Influence of abrasive diameter on cracking in brittle materials (Univ. of Hannover, IW)*

Nevertheless, if the process parameters are optimized, abrasive water jet offers some qualitative advantages over conventional cutting methods. Figure 8.34 illustrates this aspect for cutting crystal glass. As this example shows, abrasive water jet generates a cut nearly free of chipping. In contrast, abrasive-wire technique produces severe chipping at the bottom of the cut.

Figure 8.35 that shows the lower part of a honeycomb structure cut by an abrasive water jet, presents a contrary case.

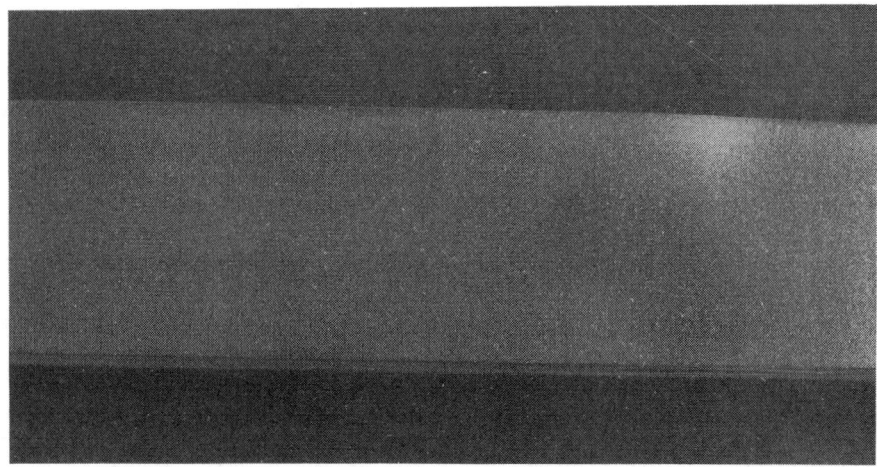

Figure 8.34 *Surface quality of crystal glass cut by an abrasive water jet (Univ. of Hannover, IW)*

8.3 Integrity of Abrasive Water-Jet Generated Surfaces   277

Figure 8.35 *Quality of a honeycomb structure cut by an abrasive water jet (Univ. Hannover, IW)*

In case of abrasive water-jet cutting, the structure is heavily damaged. In this case, each individual cell of the structure leads to a destruction of the structure of the abrasive water jet. This disturbed jet deteriorates the cut quality of the following row.

8.3.3.3 Phase Modifications in Ceramics

An interesting aspect is the chemical alteration of materials due to the water phase in the jet. In magnesia ceramics, the magnesia oxide reacts with the process water. This reaction forms magnesia-hydroxide (brucite). Because this process is connected with a volume expansion, the original material is mechanically damaged (Figure 8.36). Thus, careful drying is recommended after the machining of this material type by an abrasive water jet.

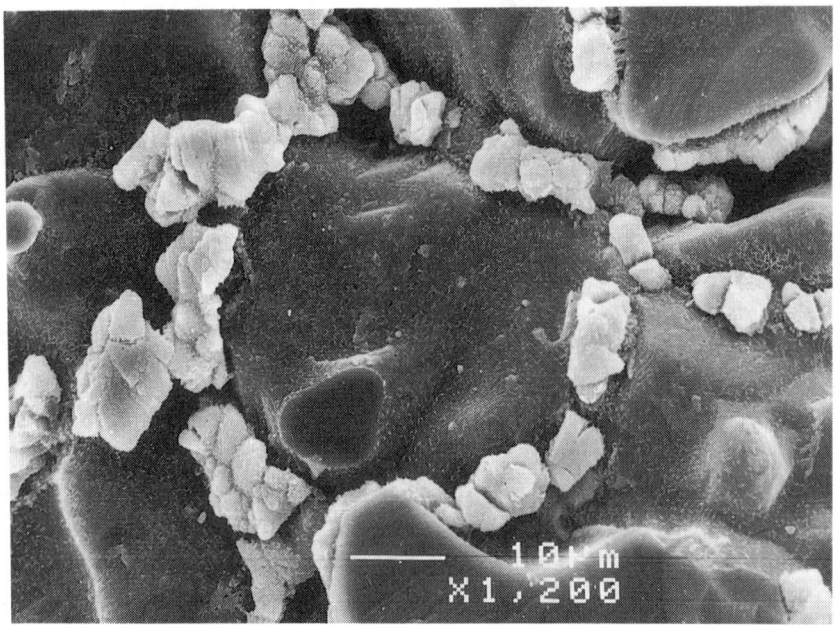

Figure 8.36 *Brucite formation in magnesia-ceramic after abrasive water-jet cutting*

8.3.4 Abrasive-Particle Fragment Embedding

Embedding of fragments from abrasive particles is well known in the solid-particle erosion of soft materials, such as copper and aluminum [332]. Nevertheless, Zu et al. [239] show that abrasive-particle embedding sharply reduces by using water as an acceleration medium. The liquid phase provides a cleaning effect that dislodges the embedded particles.

Under certain conditions, abrasive-particle deposition occurs in abrasive water-jet machining of ductile-behaving materials (Figure 8.37a) as well as in brittle-reacting materials (Figure 8.37b).

Neusen et al. [182] carry out s detailed SEM-based investigation into the abrasive-particle deposition during abrasive water-jet cutting. In metal-matrix composites cut by garnet abrasives, the authors find wear tracks that are plowed in the material by single abrasive particles. The particles themselves are lodged at the end of the wear track, apparently after their kinetic energy reduce below the critical level needed to permit further movement of the particles. Savrun and Taya [183] detect embedded garnet particles in the matrix of metal-matrix composites.

The situation is somewhat different in brittle porous materials, such as ceramics. During the cutting of refractory ceramics, garnet fragment deposition is detected on the bottom of pores that are opened during the cutting process. Probably, the abrasive particles fracture during the mixing-and-cutting and the smaller fragments are swept into the pores by the water flow.

Ramulu et al. [126] find abrasive-embedding in aluminum alloys but not in the same alloy reinforced by SiC-particles. Therefore, the average hardness of a material determines the amount of abrasive-particle embedding.

A special problem is abrasive particles that are wedged within the delamination spaces in fiber-reinforced materials cut by an abrasive water jet. Colligan et al. [322] generally observe this effect during severe delamination.

A recommended practice in using abrasive water jets is to turn off the abrasive supply after a surface has been generated so that the water jet cleans the surface.

8.3.5 Delamination in Composite Materials

Delamination is the collapse in the bounding between two layers of multi-layer materials that generates a crack perpendicular to the cut. In multi-layer and fiber-reinforced materials, delamination generally occurs during the cutting with water jets [333].

Delamination is also observed during the abrasive water-jet cutting. In an early investigation, Ricci et al. [334] detects delamination on the bottom side of graphite-epoxy laminates cut by an abrasive water jet. Ramulu and Arola [187] observe delamination near the cut exit in composite materials. The same authors [183] detect delamination in graphite epoxies; typical regions are 50 µm to 75 µm (or 10-15 fiber heights) from the exit of the workpiece.

Colligan et al. [322] carry out a systematic study into the delamination during the abrasive water-jet machining of graphite-epoxy laminates. They observe the behavior of laminates of different thickness consisting of several plies (ply number 20-144). Colligan et al. detect small fiber-pullout in exit ply, small surface-ply delamination, and severe exit-ply delamination. Singh and Jain [326] find abrasive particles trapped into grooves in the matrix of titanium that act as delamination sources.

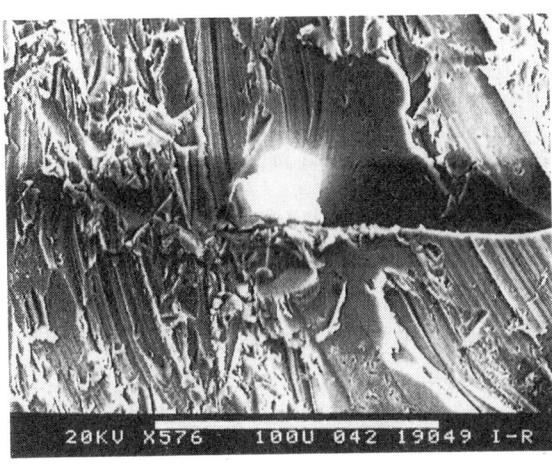

Figure 8.37a *Abrasive-particle embedding in aluminum [320]*

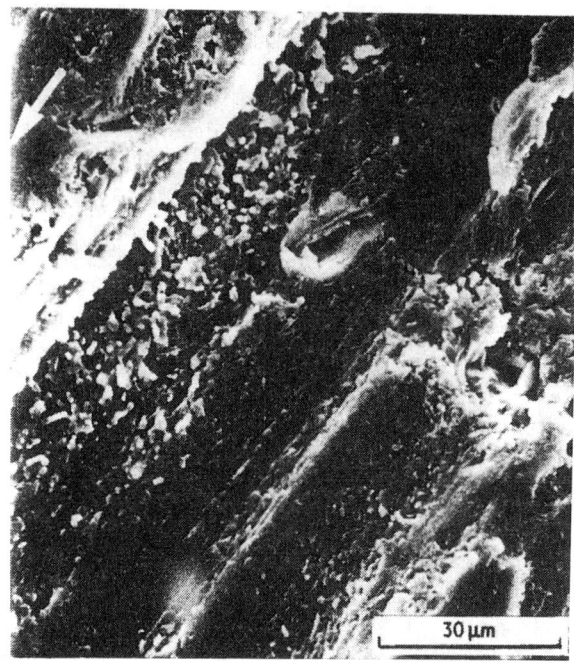

Figure 8.37b *Abrasive deposition in fiber-reinforced plastics (Univ. Washington, Seattle)*

Figure 8.38 shows some results of delamination investigations from Colligan et al. [322]. The type of damage depends on the traverse rate as well as on the abrasive-mass flow rate. For high abrasive-mass flow rates, small scale exit-ply delamination less than 1 mm in length is the only noted damage. A sufficient abrasive-mass flow rate and moderate traverse rate allow the abrasive water jet to easily penetrate the laminate thickness. However, when the abrasive-mass flow rate decreases, the jet has more difficulty cutting through the laminate and, progressively, more exit-ply delamination occurs. The delaminations are much larger in thick laminates, but only the lower three plies are involved in the delamination of the thick laminates. With the thin laminates, only the bottom two plies are involved in delaminations. The traverse rate at which severe delamination begins is independent on the abrasive-mass flow rate. Figure 8.38 shows that in both cases, delamination begins at a traverse rate of about $v=1.77$ m/min.

During drilling of coated materials by an abrasive water jet, a sudden change in the penetration rate results in layer spelling or delamination [316].

8.3 Integrity of Abrasive Water-Jet Generated Surfaces

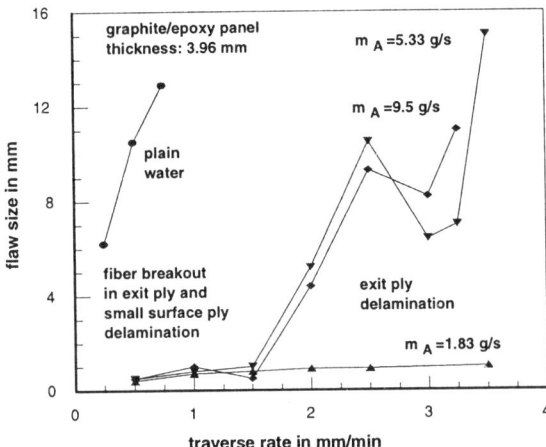

Figure 8.38 *Parameter influence on the delamination size [322]*

8.3.6 Burr Formation

Groppetti and Monno [335] state that burr formation can not be completely avoided in abrasive water-jet cutting of ductile materials. These authors observe the formation of roll-over burrs. Hamatani and Ramulu [184] notice burr formation in metal-matrix composites cut by an abrasive water jet. Generally, burrs are material chips that are not completely separated from the workpiece during the cutting process.

Groppetti and Monno [335] perform a systematic study into the burr formation in brass specimens cut by abrasive water jets. Table 8.11 lists some results. The burr height in this material is between $h_{Burr}=28$ μm and $h_{Burr}=67$ μm. The burr size minimizes at low cutting forces, high pump pressure, low traverse rate, high abrasive-mass flow rates, and small abrasive-particle size. Similar to surface cracking in brittle materials: the larger the particle diameter, the more severe is burr formation (Figure 8.39). Hashish [313] observes a reduction in the burr height even at high traverse rates (Figure 8.40a). The burr heights shown in this figure for inconel are in the same order of magnitude as those observed during the brass cutting.

Machaida et al. [386] find that severe burr formation occurs at large standoff distances (Figure 8.40b). In contrast, burr formation does not take place at small standoff distances not larger than x=5.0 mm. Figure 8.41 illustrates the influence of the target material on the burr formation. Burr formation is very heavy in aluminum; whereas, the generation of burrs is almost prevented in the titanium alloy.

Table 8.11 *Mean burr height of brass specimens cut by abrasive water jets [335]*

Traverse rate in mm/min	Pump pressure in MPa	Mean burr height in μm			
		Abrasive-mass flow rate (g/s)			
		5.0		10.0	
		Abrasive-grain size (mesh)			
		# 80	# 120	# 80	# 120
100	200	55	39	50	45
	300	48	37	41	28
200	200	56	41	50	48
	300	46	39	41	30
400	200	67	60	54	54
	300	55	48	47	36
800	200			66	
	300	59		46	42
1,600	200				
	300				

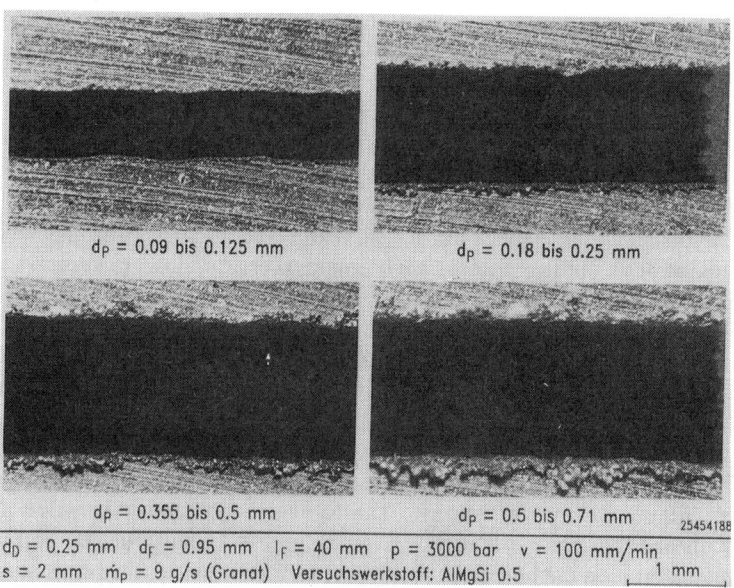

Figure 8.39 *Influence of the abrasive diameter on burr formation (U. Hannover, IW)*

8.3 Integrity of Abrasive Water-Jet Generated Surfaces   283

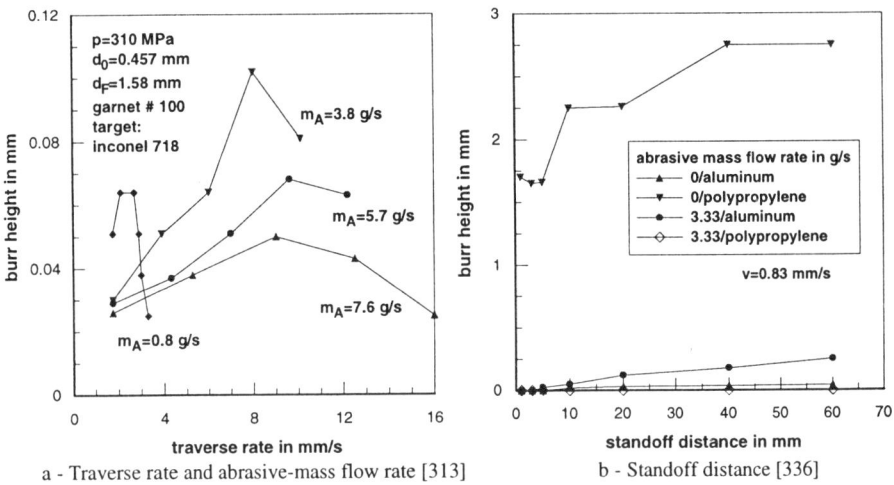

a - Traverse rate and abrasive-mass flow rate [313]   b - Standoff distance [336]

Figure 8.40 *Parameter influence on the burr height*

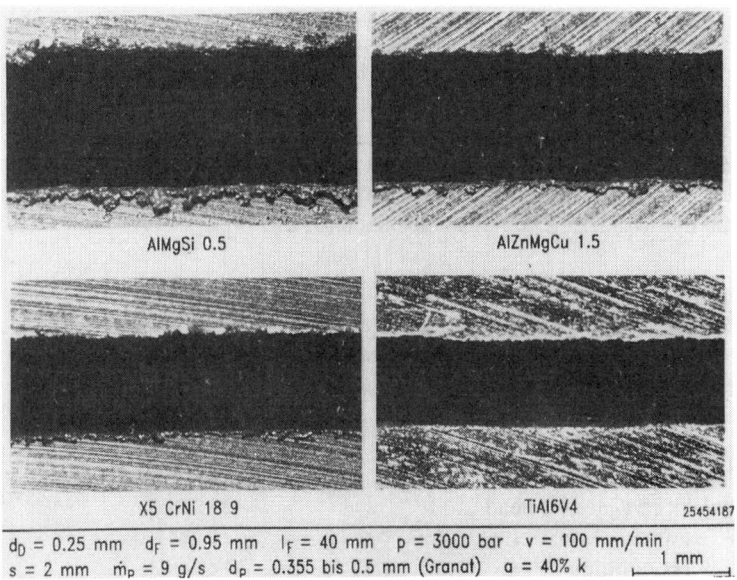

Figure 8.41 *Influence of the target material on the burr formation (Univ. Hannover, IW)*

9 Alternative Machining Operations with Abrasive Water Jets

9.1 Capability of Abrasive Water Jets for Alternative Machining

The versatility of the abrasive water jet in cutting almost any engineering material is a very special feature of this technology. Figure 9.1 shows an example for the machining capability of the abrasive water-jet technique. This figure illustrates several steps of manufacturing a certain complex shape in a hard-to-machine material. Besides cutting, as described in the previous chapters, the manufacturing process includes the following operations:

- milling
- turning
- piercing
- finishing

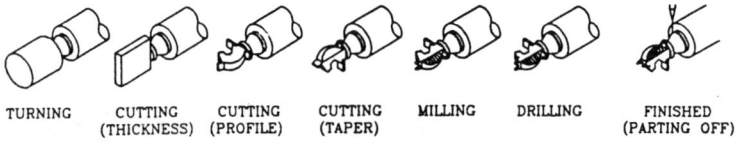

TURNING   CUTTING   CUTTING   CUTTING   MILLING   DRILLING   FINISHED
          (THICKNESS) (PROFILE)   (TAPER)   (PARTING OFF)

Fig. 9.1 *Chess knight manufactured by the abrasive water-jet technique [337]*

Figure 9.2 shows a complex workpiece-shape that is also completely manufactured by abrasive water-jet milling. Even since abrasive water jets just enter these machining areas, several preliminary investigations with promising results are performed that will be the topic of this chapter.

9.2 Milling with Abrasive Water Jets

9.2.1 Concepts of Abrasive Water-Jet Milling

Several concepts of milling with abrasive water jets are developed, such as

- rotary multiple water-jet head
- process-parameter variation principle
- multi-traverse milling
- mask milling
- discrete milling

The principle of rotary water jets inside a single mixing tube is inefficient because of the low material-removal rate. The material-removal rate is more than an order of magnitude less than for conventional abrasive water jets even if the process, especially the rotational speed, is optimized. The main reason for the bad performance of this milling concept is the inefficient mixing process between the abrasive grains and the rotating water jets [338].

An attempt to develop a concept of abrasive water-jet milling that bases on continuously-varying dynamic variables is not successful either. It is known that several process parameters such as the pump pressure, traverse rate, abrasive-mass flow rate, and impact angle significantly affect the efficiciency of the abrasive water jet-milling as well as the geometry of the generated cavities. Nevertheless, a real-time control of the parameters of an abrasive water jet for selective cavity milling is generally impractical with today's hardware technology [338].

Several authors use the multipass linear-traverse cutting as a milling strategy. This principle is based on the superposition of several kerfs to obtain a cavity of defined geometry. The specimen shown in Figure 9.2 is generated by this method. Laurinat [290] and Laurinat et al. [339] develop and investigate this strategy in detail. The key parameter in the multipass-kerfing process is the lateral distance, e, between the single kerfs. Section 9.2.4 discusses the influence of this parameter on the cavity geometry.

Fig. 9.2 *Complex shape, milled in X5CrNi189 by an abrasive water jet (dimensions 50 mmx50mmx20mm, Univ. of Hannover, IW)*

Hashish [338, 340] addresses the abrasive water-jet milling using a rotating workpiece table and masks. Figure 9.3 shows this milling principle. The system contains two axes of motion: the cross feed (linear traverse) of the abrasive water jet, and the specimen rotation.

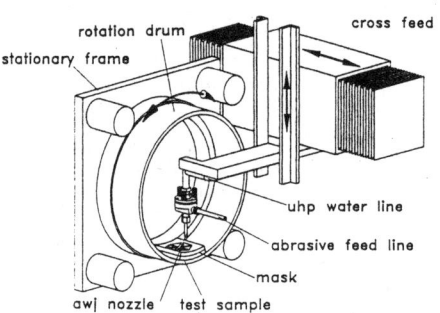

Fig. 9.3 *Setup for abrasive water-jet milling using rotary workpiece and masking [338]*

The cross-feed motion of the abrasive water-jet tool is used to control the amount of overlap of the abrasive water jet from one revolution of the drum to the next. The material samples are placed in the drum and covered with a mask.

Figure 9.4 illustrates the influence of several process parameters on the volume-removal rate. From Figure 9.4a, for the overlap ratio, the highest material-removal rate is obtained with the smallest focus diameter. This result attributes to the fact that small-diameter focus nozzles generate more effective jets (higher power per area). The material-removal rate peaks at a certain overlap ratio that does not correspond to that of the highest milling depth. Varying the standoff distance does not significantly change the material-removal rate or milling depth in the given parameter range (Figure 9.4b). Because the abrasive water jet spreads with an increasing distance from the focus, an increase in the the standoff distance increases the actual degree of overlap. The traverse rate influences both of the target parameters of the milling process. High speeds result in reduced depth of milling and consequently better control over the depth tolerance. Increasing the traverse rate over a certain limit yields a reduction in the volume-removal rate (Figure 9.4c). The milling depth almost linearly increases with the number of passes. From Figure 9.4d, the volume-removal rate drops for fine-gained abrasive particles. The surface uniformity improves by selecting high abrasive-mass flow rates and small-diameter abrasive particles. Garnet # 60 to # 100 result in almost identical material-removal rates, but the surface topography substantially improves by using the finer mesh abrasive-mixture [338].

Hashish [338] performs a practical application - the controlled depth milling of isogrid structures - by using the principles of rotary table and masking. This study indicates the feasibility of milling pockets in an aluminum sample to a uniformity of $\Delta h = 0.025$ mm. It is also possible to mill to 2.36 mm thick skins in aluminum tubes.

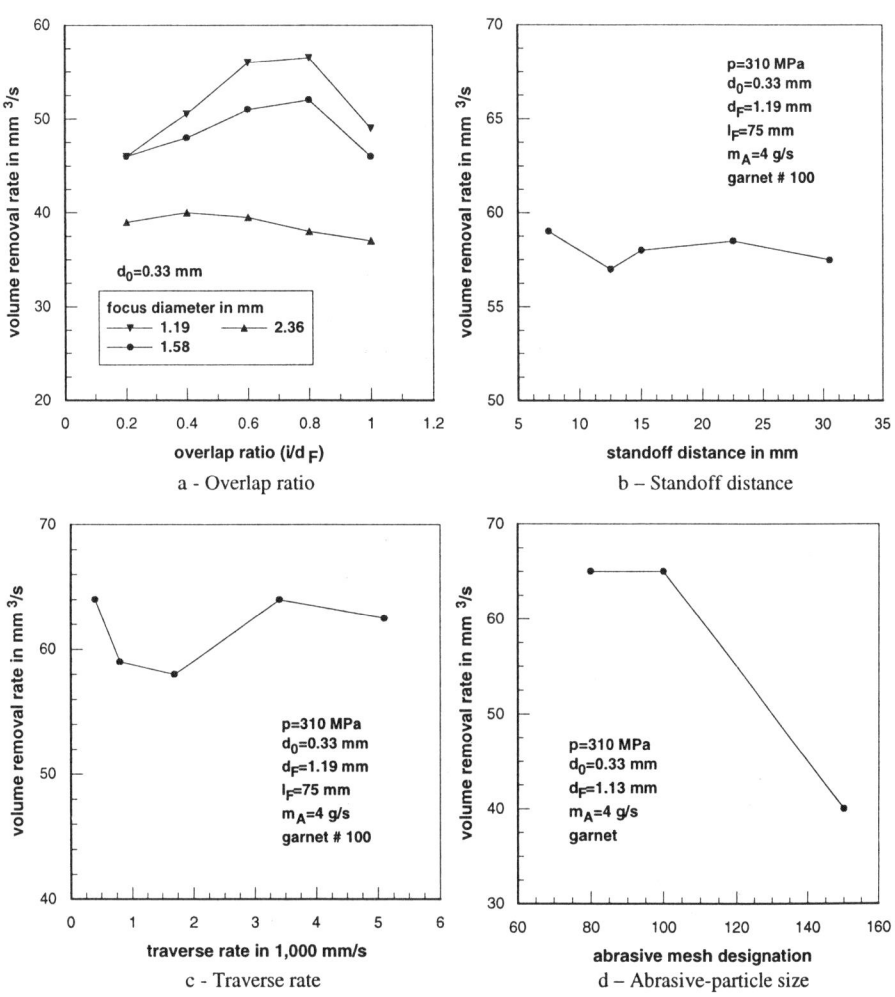

Fig. 9.4 *Parameter influence on the material removal using rotary workpiece and masking [340]*

Öjmertz and Amini [146] introduce a milling concept similar to the multiple-pass cutting. The concept is called discrete milling and basically consists of a system of parallel rows of single cavities. Therefore, discrete milling is not a modified cutting process. Figure 9.5 shows a workpiece structure generated by this milling method.

Moving incrementally a step Δx and Δy, the new position for the next cavity will be $b=b+\Delta b$ and $y=y+\Delta y$. When superposing two cavities that overlap, the abrasive particles in the intersected areas strike a previously cut surface of a certain slope. This slope affects the material-removal rate.

Fig. 9.5 *Workpiece, generated by discrete abrasive water-jet milling (Chalmers University of Technology)*

9.2.2 Parameter Optimization in Abrasive Water-Jet Milling

Unlike cutting, the most important target parameters in milling are the material-removal rate to obtain an effective milling performance, and the geometrical structure of the generated cavities, in particular the uniformity of the milling depth. Another notable difference compared to cutting is the small removal depth that is usually in the range of several millimeters. Thus, energy dissipative processes, especially damping, play a minor role.

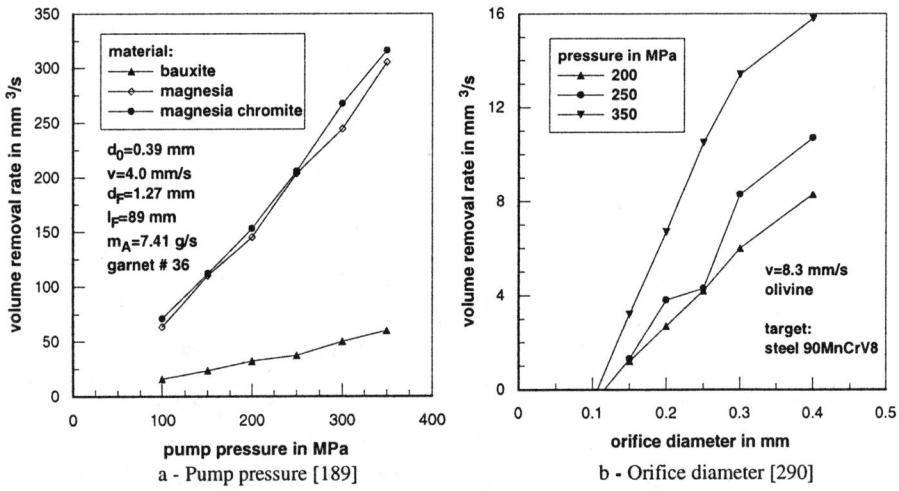

Fig. 9.6 *Influence of hydraulic parameters on the material-removal rate*

Laurinat [290], Laurinat et al. [286], Öjmertz [341], Hochheng and Cheng [314] and Momber et al. [189] perform studies to clarify the influence of the process parameters in abrasive water-jet milling on the material-removal rate. Figure 9.6a shows that in ceramic materials, the material-removal rate linearly increases as the pump pressure increases. As in cutting operations, a threshold pressure can be noticed. Nevertheless, milling with very high pump pressure is associated with an increase in the specific material-removal energy. With an increase in the orifice diameter, the volume-removal rate increases, but the progress drops at a certain diameter of about $d_0=0.3$ mm. Also, a critical threshold diameter of about $d_{0thr}=0.1$ mm is noticed. Figure 9.6b shows these results. Laurinat [290] relates the material-removal rate to the jet power, water-flow rate and abrasive-mass flow rate. In all cases, the relations show typical maximum values at orifice diameters of $d_{0max}=2 \cdot d_{0thr}$. The exceptions are low pressures where the maximum values are located at about $d_{0max}=3 \cdot d_{0thr}$.

Figure 9.7a illustrates the influence of the traverse rate on the material-removal rate. The material-removal rate remains almost constant as a certain traverse rate value is exceeded. The smaller the standoff distance, the higher is this critical traverse rate. Momber et al. [189] find somewhat different relations. For hard-to-machine refractory ceramics, no significant influence on the material-removal rate is observed. The material-removal rate almost linearly increases with an increase in the traverse rate for softer ceramics. For multipass-milling with very high traverse rates, Hashish [289] observes a typical maximum for the material-removal rate.

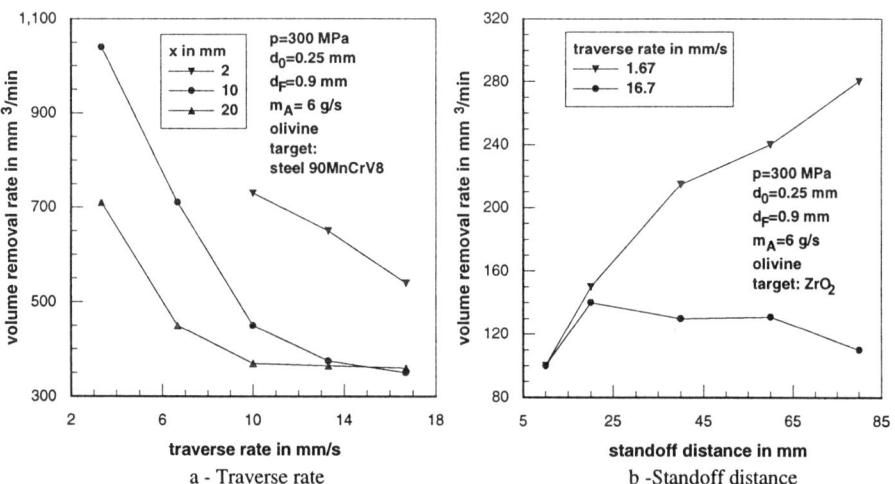

Fig. 9.7 *Influence of cutting parameters on the material-removal rate [290]*

You et al. [138] detect maximum material-removal rates at optimum standoff distances and find that the optimum standoff distance increases with an increase in the focus diameter. Laurinat [290] suggests that this result occurs just for high traverse rates. For low traverse rates, the material-removal rate increases with an increase in the standoff distance (Figure 9.7b). For milling of fiber-reinforced plastics, Hocheng et al. [340] find optimum standoff distances for high pressures and 0°-fiber orientation. For fiber orientations of 90°, the material-removal rate does not notably increase at larger standoff distances. Figure 9.4b shows that the influence of the standoff distance in mask milling is associated with optimum values for maximum material-removal in the range of medium abrasive-mass flow rates.

Figure 9.8 plots typical results for the influence of the impact angle on the material-removal rate. For conventionally ductile-behaving metals, the material-removal rate shows maximum values at impact angles between $\varphi_{opt}=40°$ and $\varphi_{opt}=60°$. In contrast, the ceramic materials exhibit a maximum removal at an angle of $\varphi_{opt}=90°$. For rotating milling-devices, Hashish [296] shows that the optimum impact angle depends on the setup of the other process parameters. For deep-milling applications with rotating abrasive water-jet tools, Hashish [296] finds an improvement in the process by a slight angling ($\varphi=3°-5°$) of the tool. At an optimized angle of $\varphi=5°$, it is possible to mill glass samples over 300 mm thick.

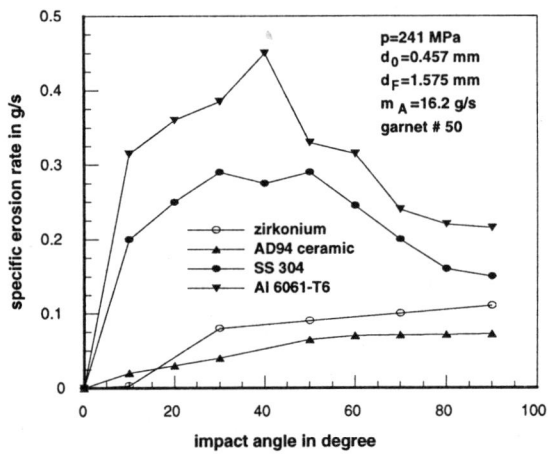

Fig. 9.8 *Influence of the impact angle on the material removal [230]*

Figure 9.9a illustrates the influence of the focus diameter. Notice a maximum for the material-removal rate. Nadeau et al. [285] report a similar behavior for material-removal processes with abrasive water jets. Blickwedel [69] finds an optimum focus length for material-removal processes. For low abrasive-mass flow rates, Figure 9.9b confirms this result.

Figure 9.10 illustrates the material-removal performance for varying abrasive-mass flow rate*s*. In the range of low abrasive-mass flow rates, the material-removal rate increases as the abrasive-mass flow rate increases. However, depending on the pump pressure, there exists an optimum abrasive-mass flow rate for milling. For milling, this optimum is at somewhat larger values compared to cutting [290]. For short focus length and for difficult-to-machine materials, this optimum switches to higher values [189]. Hu et al. [293] show for ductile materials and Momber et al.

[189] for brittle materials that the material-removal rate almost linearly increases with an increase in the abrasive-mass flow rate.

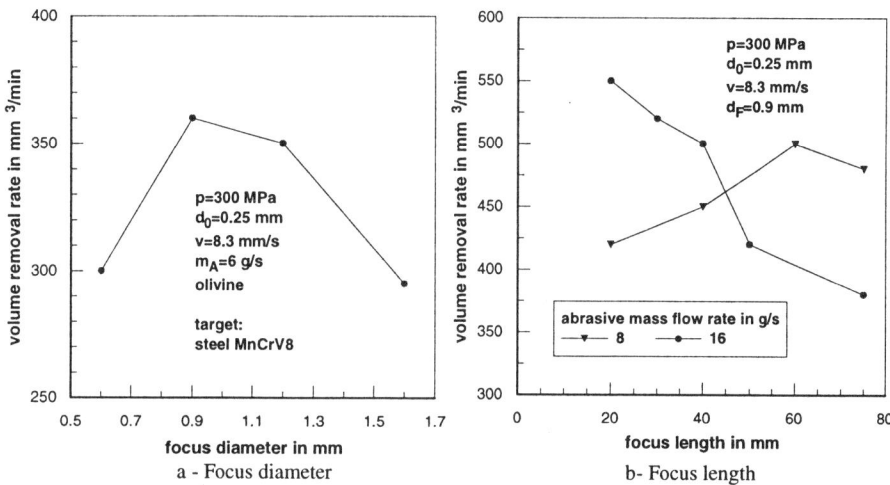

Fig. 9.9 *Influence of mixing parameters on the material-removal rate [290]*

The ratio between the material-removal rate and water-flow rate increases with an increase in the abrasive-mass flow rate. From the point of view of abrasive exploitation, comparatively high abrasive-mass flow rates can be used for milling operations [286]. For the ratio between the material-removal rate and abrasive consumption, typical optimum values exist at low abrasive-mass flow rates. The specific material-removal energy drops with an increase in the abrasive-mass flow rate [189].

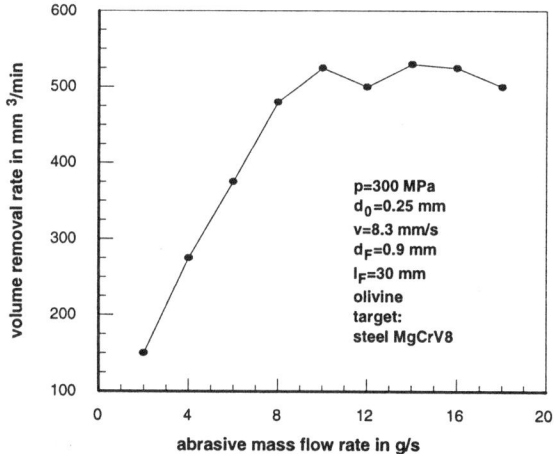

Fig. 9.10 *influence of the abrasive-mass flow rate on the material removal [290]*

For multi-pass milling, Öjmertz [341] finds that the material-removal rate slightly increases as the lateral increment between the milling passes increases.

9.2.3 Quality of Abrasive Water-Jet Milling

Two parameters estimate the quality of surfaces generated by abrasive water-jet milling: the depth of cut tolerance and the bottom roughness. Öjmertz [341] and Laurinat [290] investigate the influence of several process parameters on these features in detail. Öjmertz notices that the tolerance of the milling depth linearly increases as the pump pressure increases (Figure 9.11).

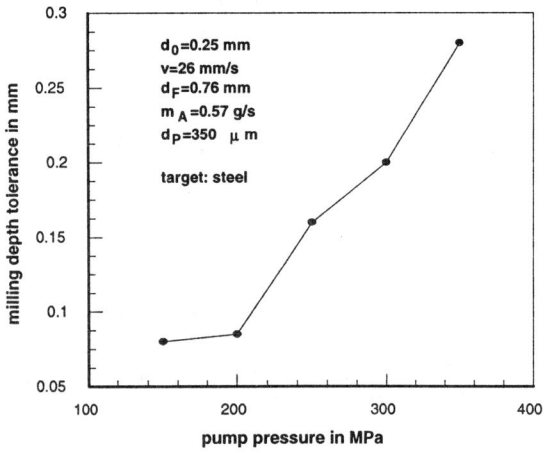

Fig. 9.11 *Influence of the pump pressure on the milling-depth tolerance [341]*

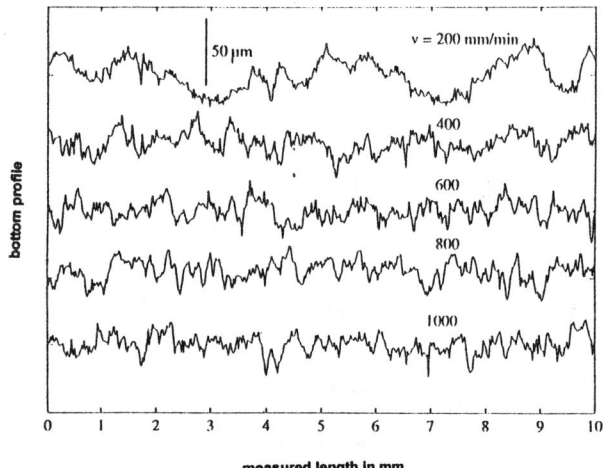

Fig. 9.12 *Influence of the traverse rate on the bottom topography [290]*

Figure 9.12 illustrates the relation between the ground topography and traverse rate. As the traverse rate increases, the profile becomes more uniform. In case of high traverse rates that are associated with low depths, the transition depth that determines the introduction of an unsteady material-removal process that is cyclic (section 5.4.2) is not achieved. Figure 9.13a illustrates this relation for multi-pass milling. As the number of milling passes increases, the bottom contour becomes more regular. Nevertheless, in multi-pass milling, the lateral increment influences the surface topography. As Öjmertz [341] shows, a decrease in the increment improves the milling-depth tolerance. For fiber-reinforced plastics, Hocheng et al. [342] recommend a lateral increment of 50 % of the top width for acceptable plain surfaces.

The bottom roughness also benefits from a jet impact angle that is inclined (Figure 9.13b), especially if the abrasive type is selected properly.

Figure 9.14 shows the influence of the abrasive-mass flow rate on the bottom roughness. In tendency, the bottom roughness increases as the abrasive-mass flow rate increases. This effect is dramatic for high pump pressures.

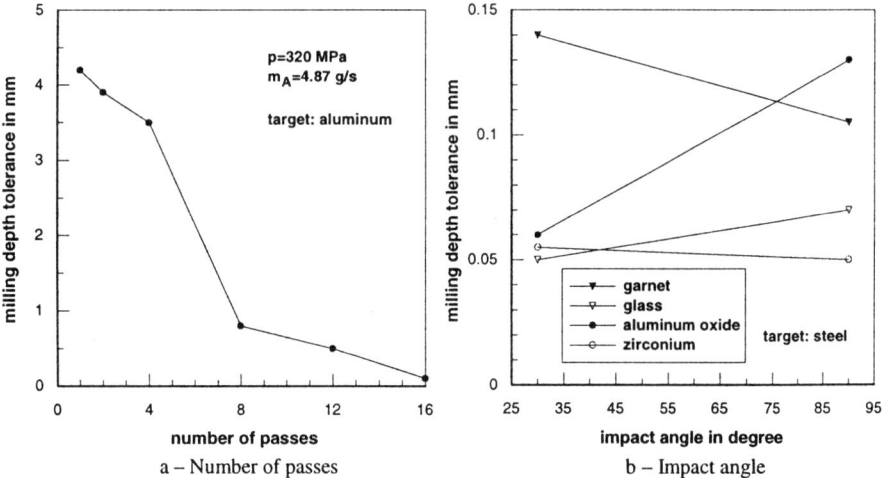

Fig. 9.13 *Influence of cutting parameters on the milling-depth tolerance [341]*

Hashish [289] investigates the influence of the abrasive type and abrasive-grain size on the surface finish in abrasive water-jet milling. He finds that combinations of different abrasive types ore different mesh-sizes do not improve the surface structure. If steel is completely machined with garnet as an abrasive material, the deviation in the removal depth is $\Delta h=2.5$ mm. A combination of garnet for the material-removal part and silica sand or glass beads for surface finishing yields $\Delta h=3.0$ mm. If garnet # 20 is used for the material-removal part and garnet # 150 for finishing, the tolerance in the removal depth is $\Delta h=0.51$ mm. In contrast, as garnet # 80 alone is applied for the material removal as well as for the finish operation, a surface uniformity of $\Delta h=0.203$ mm is achieved. Therefore, irregularities that are produced by one milling step can not be corrected in the subsequent passes.

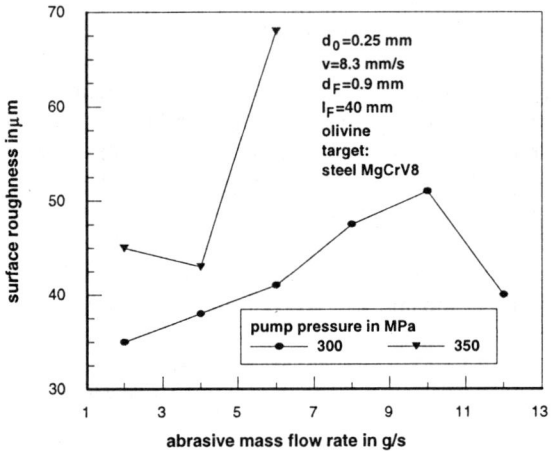

Fig. 9.14 *Influence of the abrasive-mass flow rate on the bottom roughness [290]*

Momber et al. [42] show that the uniformity of the depth significantly improves if the material-removal process is performed with an optimum abrasive-grain size distribution (Figure 9.15).

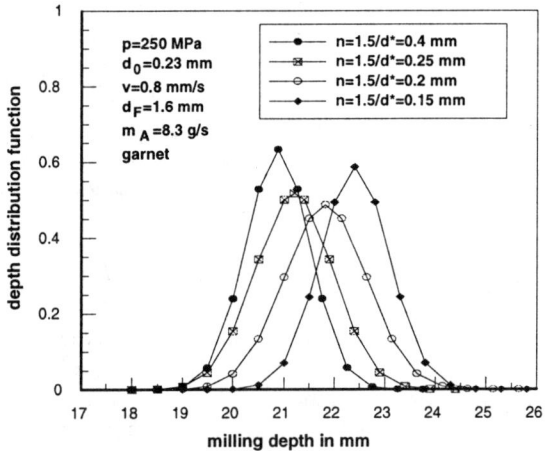

Fig. 9.15 *Influence of particle-size distribution parameters on the milling-depth variation [42]*

9.2.4 Modeling of Abrasive Water-Jet Milling

9.2.4.1 General Milling Model

Several attempts have been made to model the abrasive water-jet milling. A key problem in abrasive water-jet milling is the geometry of the generated cavities. A preliminary investigation by Hashish [289] proofs that the geometry of the generated cavities mainly depends on the traverse rate and standoff distance. Laurinat [290] develops a milling model that covers the cavity profile, material-removal rate, and multiple-step milling.

Figure 9.16 illustrates the influence of the lateral distance, e, on the cavity geometry for multiple-step milling. For e=0, the depth of the original cavity simply increases,

$$h = n_P \cdot h_0. \qquad (9.1)$$

In the equation, n_P is the number of passes over the kerf. The width of the top of the kerf is not influenced. For values of $0 < e < b_T/2$, and $n_P > b_T/e$, the depth of the cavity as well as its width are influenced. By superposition, the depth of the generated profile is [286, 290]

$$h = \frac{h_0 \cdot b_T}{2 \cdot e}. \qquad (9.2)$$

For $e = b_T/2$, the depth of the cavity is equal to the depth of the single cut, $h=h_0$. If the lateral distance exceeds a value of $e=b_T/2$, the depth of the cavity is not influenced and

$$h = h_0. \qquad (9.3)$$

In these cases, just a certain number (n_P) of separated cavities is generated.

For ceramic materials, Freist et al. [343] perform investigations into the shape of cavities milled by abrasive water jets. These authors approximate the geometry of a cavity generated by single pass kerfing by a cosine function (Figure 9.17),

$$h(b) = \frac{h_{max}}{2} \cdot (\cos b - 1), \qquad (9.4a)$$

$$b_T = 2 \cdot \pi. \qquad (9.4b)$$

This relation is especially applied for high values of b_T/h and for profiles with low surface roughness.

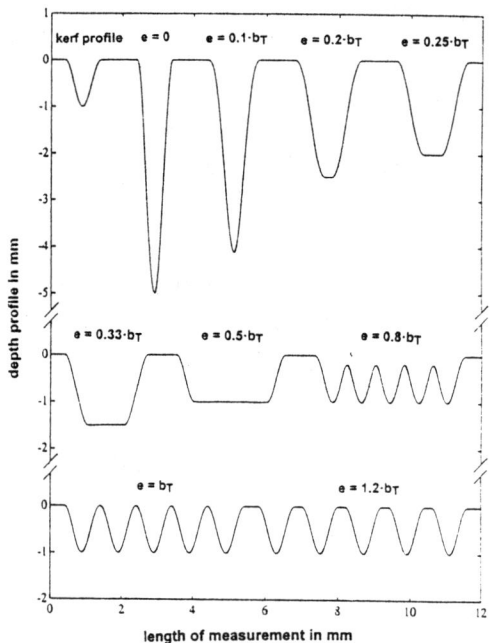

Fig. 9.16 *Influence of the lateral distance on the cavity geometry [290]*

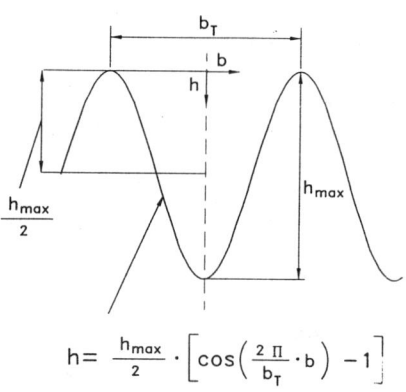

$$h = \frac{h_{max}}{2} \cdot \left[\cos\left(\frac{2\pi}{b_T} \cdot b\right) - 1\right]$$

Fig. 9.17 *Approximation of a kerf geometry for abrasive water-jet milling [286]*

Laurinat [290] modifies Eq. (9.4),

$$h = \frac{h_{max}}{2} \cdot \left[\cos\left(\frac{2 \cdot \pi}{b_T} \cdot b\right) - 1\right]. \tag{9.5}$$

Eq. (9.5) is suitable to describe a variety of different cavity profiles. Nevertheless, the cosine function fails for certain geometries, such as sharp contours and deep profiles. For those geometries, other mathematical relations need to be chosen, as for example a high-order polynomial. Laurinat et al. [286] link the cavity-geometry parameters with the traverse rate and standoff distance. A regression analysis gives

$$h_{max} = \frac{2 \cdot R_{Ec}}{v \cdot (c_1 \cdot x + c_2)}, \tag{9.6a}$$

for metallic materials, and

$$h_{max} = (A_1 \cdot v + A_2) \cdot x + \frac{2}{A_3 \cdot v^{A_4}} \tag{9.6b}$$

for ceramics. For both material groups, the width of the cavity is

$$b_T = a_1 \cdot x + a_2. \tag{9.7}$$

Tables 9.1 and 9.2 list the corresponding regression parameters for different materials. Figure 9.18 presents some comparisons between Eqs. (9.6) and (9.7) and experimental results.

Table 9.1 *Regression parameters for metal milling [286, 290]*

Regression parameter	Material		
	90MnCrV8	X5CrNi189	C 45k
c_1	0.918	1.027	1.223
c_2	0.080	0.081	0.075
R_{Ec}	1.00	1.96	1.78
a_1	0.0798	0.0814	0.0754
a_2	0.918	1.027	1.223

Table 9.2 *Regression parameters for ceramic milling [290]*

Material	A_1 min/mm	A_2 -	A_3 min/mm^2	A_4 -
Al_2O_3	$5.32 \cdot 10^{-6}$	$-3.68 \cdot 10^{-3}$	0.0604	0.744
ZrO_2	$7.11 \cdot 10^{-7}$	$-1.48 \cdot 10^{-3}$	0.0528	0.791
SiC	$-4.49 \cdot 10^{-6}$	$-4.62 \cdot 10^{-4}$	0.1769	0.782
Si_3N_4	$5.52 \cdot 10^{-6}$	$-9.42 \cdot 10^{-4}$	0.5643	0.863

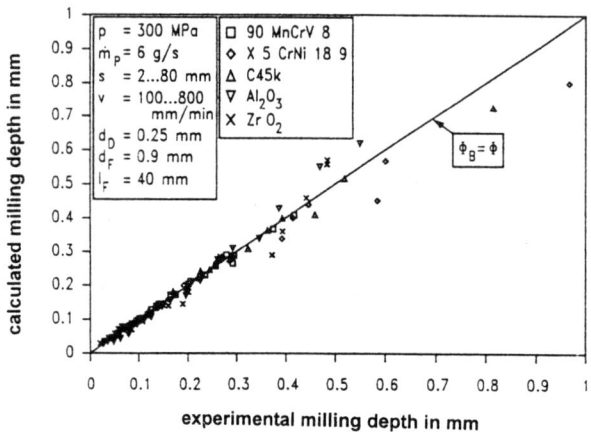

Fig. 9.18 *Verification of Laurinat's milling model [290]*

The material-removal rate in milling is [290]

$$\dot{V}_M = A_M \cdot v = \int_{-b/2}^{b/2} h(x) dx \ . \tag{9.8a}$$

Solving the integral for a cosine-function, the material-removal rate is

$$\dot{V}_M = h \cdot b_T \cdot v \ . \tag{9.8b}$$

With the regression parameters given in Tables 9.1 and 9.2, the material-removal rate can be calculated for different materials.

9.2.4.2 Milling Model for Fiber-Reinforced Plastics

Hocheng et al. [342] develop a set of non-dimensional equations for milling fiber-reinforced plastics. For several target parameters, they obtain

$$TP = c_1 \cdot \left(\frac{v_P - v_{thr}}{v}\right)^{c_2} \cdot \left(\frac{v \cdot x^2 \cdot \rho_P}{\dot{m}_A}\right)^{c_3} \cdot \left(\frac{d_P}{x}\right)^{c_4} \cdot \left(\frac{\sigma_f \cdot x^2}{v \cdot \dot{m}_A}\right)^{c_5} . \tag{9.9}$$

In this equation, TP is a target parameter according to Table 9.3. Table 9.3 lists the regression parameters for these equations for the different target parameters. The regression parameters deliver additional information about the influence of the process parameters. The pump pressure and traverse rate count the maximum for the material-removal rate, milling depth, and width-depth-ratio.

9.2 Milling with Abrasive Water Jets

Table 9.3 *Regression parameters for Hocheng's model [342]*

Target relation	Regression parameters					
	c_1	c_2	c_3	c_4	c_5	R^2
h/x	0.12	2.4	0.8	0.9	-0.9	0.90
$V_M/(v \cdot x^2)$	0.018	1.9	0.3	0.5	-0.9	0.98
b/h	1.4	-2.3	-0.95	-0.9	0.9	0.89

9.2.4.3 Model for Discrete Milling

For the discrete abrasive water-jet milling, Öjmertz and Amini [146] adapt a weight function of the expected volume removal, $w(\nabla_Z)$. In the function, ∇_Z, is the surface gradient. The weight function is determined by first fitting a fairly accurate surface model to a unit cavity. The geometry of two superposed model cavities is then compared with the print of two superposed true cavities. This procedure yields a discrepancy in each point of the surface. These discrepancies form a residual function. Finding a correlation between the residuals and the surface gradient at each location yield the weight function. This function depends on the target material. Figure 9.19 shows a comparison between an experimentally generated cavity and a simulated cavity. In this case, the milling procedure consists of parallel rows of cavities with an increment of e=0.5 mm. As a limitation, the model is valid as long as the material is not substantially eroded by the water drops in an abrasive water jet.

Öjmertz and Amini [146] assume that the shape of the removed cavity corresponds to the distribution of the abrasive particles in the abrasive water jet at least in the stadium of initial penetration. This leads to a saddle-shaped cavity which is in fact obtained during discrete milling-experiments (Figure 9.20).

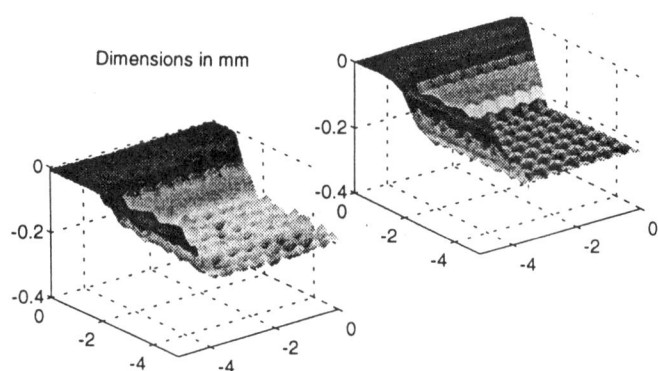

Fig. 9.19 *Computer simulation using the discrete-milling model [146] left: experiment, right: simulation*

The profile of the cavity generated by discrete milling is mathematically described by a set of exponential functions [146]. The functions are

$$h(r) = \sum_{i=1}^{N_S} \left[\alpha_i \cdot e^{-\beta_i \cdot r^2} \right] \qquad (9.10a)$$

in polar coordinates, and

$$h(b, y) = \sum_{i=1}^{N_S} \left[\alpha_i \cdot e^{-\beta_i \cdot (b^2 + y^2)} \right] \qquad (9.10b)$$

in Cartesian coordinates, respectively. In the equations, N_S is the number of sub-functions superposed, and α and β are shape parameters of the sub-functions.

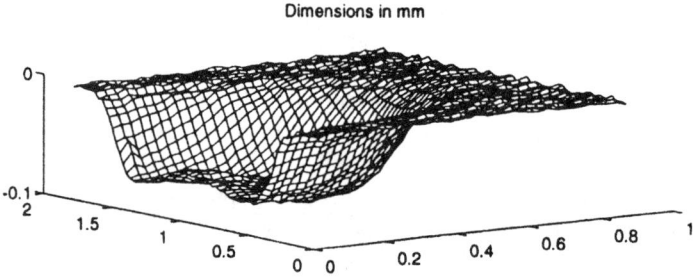

Fig. 9.20 *Cavity milled by an abrasive water jet in WC-Co (6%) carbide[146]*

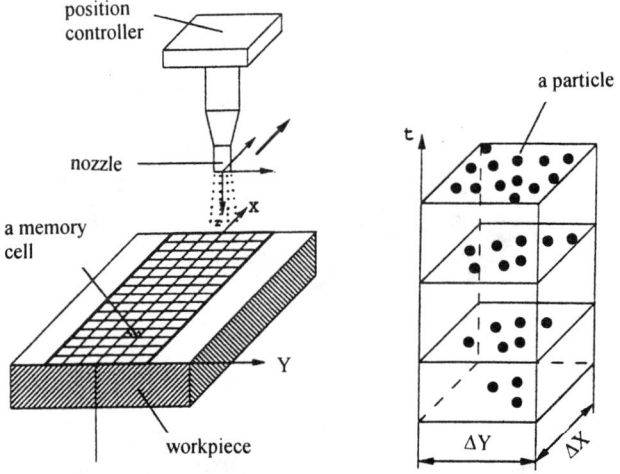

Figure 9.21 *Machining simulation with a memory-cell network [344]*

9.2.4.4 Numerical Milling Model

Section 6.7 briefly addresses Yong and Kovacevic's [344] modeling concept for abrasive water-jet machining. Figure 9.21 shows the simulation procedure for a planar milling process. Figure 9.22 illustrates a cutting view of a cone-shaped cavity from the 3D-simulation algorithm.

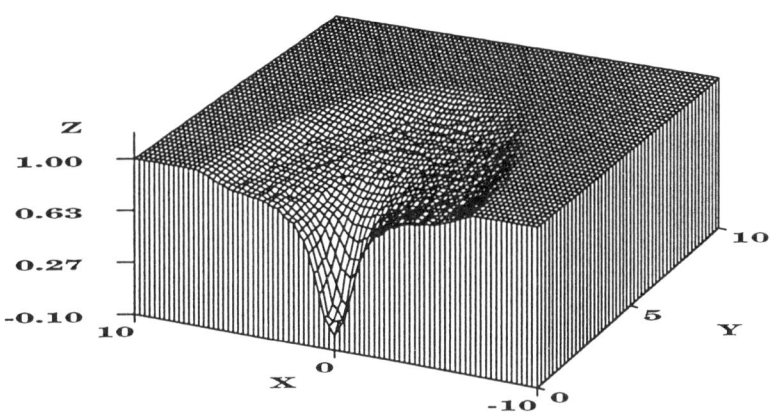

Figure 9.22 *Cone-shaped cavity from a 3D-simulation [342]*

9.3 Turning With Abrasive Water Jets

9.3.1 Macromechanism of Abrasive Water-Jet Turning

Abrasive water jet-turning simply consists of spinning a workpiece around the center of rotation while simultaneously traversing the abrasive water jet in the desired contour over the specimen to achieve an axis-symmetric shape. Figure 9.23 conceptually illustrates the process. Ansari [287] performs a visualization study in abrasive water jet-turning using high-speed camera technique. This author finds that the material removal takes place at the face of the workpiece rather than at the circumference. Only at high traverse rates, the location of the material removal moves to the circumference of the specimen. Therefore, almost no radial deflection of the abrasive water jet is detected. In contrast, the abrasive water jet deflects axially with cyclical changing.

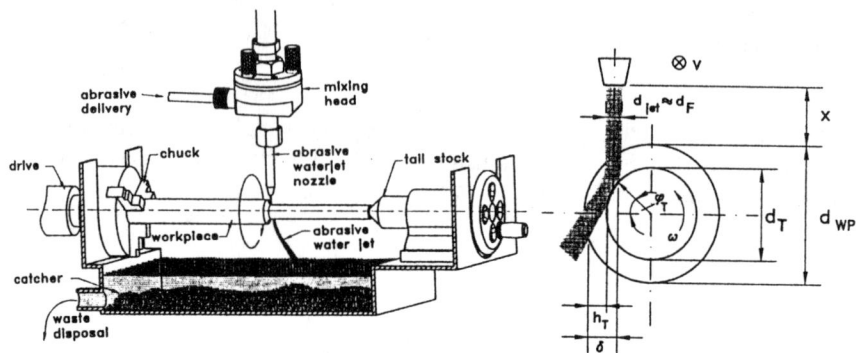

Fig. 9.23 *Principle and geometry of the abrasive water-jet turning*

9.3.2 Parameter Optimization in Abrasive Water-Jet Turning

Figure 9.23 depicts the pertinent parameters in abrasive water-jet turning. From the geometry of the process, the volume-sweep rate is

$$\dot{V}_S = \frac{\pi}{4} \cdot \left(d_{WP}^2 - d_T^2\right) \cdot v . \qquad (9.11)$$

The volume-sweep rate is the material volume swept by the combined specimen rotation and abrasive water jet traverse in unit time. In contrast, the true volume-removal rate is

$$\dot{V}_M = \pi \cdot h_T \cdot (d_{WP} - h_T) \cdot v . \qquad (9.12)$$

Generally, the volume-removal rate is less than the volume-sweep rate, though in a few exceptional cases (during finish turning) they are equal. The depth of cut, h_T, in Eq. (9.12) is a function of several process parameters, very similar to simple cutting processes. A third major target parameter in abrasive water-jet turning is the surface quality.

Figure 9.24a shows the influence of the pump pressure on the volume-removal rate with the abrasive-mass flow rate as parameters. The principal trend is that under most conditions, the volume-removal rate increases as the pump pressure increases. However, the shape of the volume-removal curve declines with an increase in the pump pressure and asymptotically approaches the corresponding volume-sweep rate. This behavior is more pronounced for high ratios of h_T/d_1. The increase in the volume-removal rate with higher pump pressure is due to the higher velocity of the impacting abrasive particles that are capable of generating higher erosion rates. However, as the pump pressure increases, it reaches a point where is no more material that is to be removed. The volume-removal rate has a limit that is set by the volume-sweep rate. Under these conditions, both the diameter deficit and radial deflection of the abrasive water jet reduce to zero.

Figure 9.24b illustrates the effect of the orifice diameter and abrasive-mass flow rate. In the figure, the volume-removal rates of the two larger orifices are almost identical at low abrasive-mass flow rates, but significantly higher than for the small-diameter orifice. For the small-diameter orifice, an increase in the abrasive-mass flow rate beyond a certain value does not affect the volume-removal rate. The hydraulic momentum of this water jet is just able to effectively accelerate comparatively small abrasive-mass flow rates. The same is for the larger orifice diameters, but the critical abrasive-mass flow rate appears at a higher value.

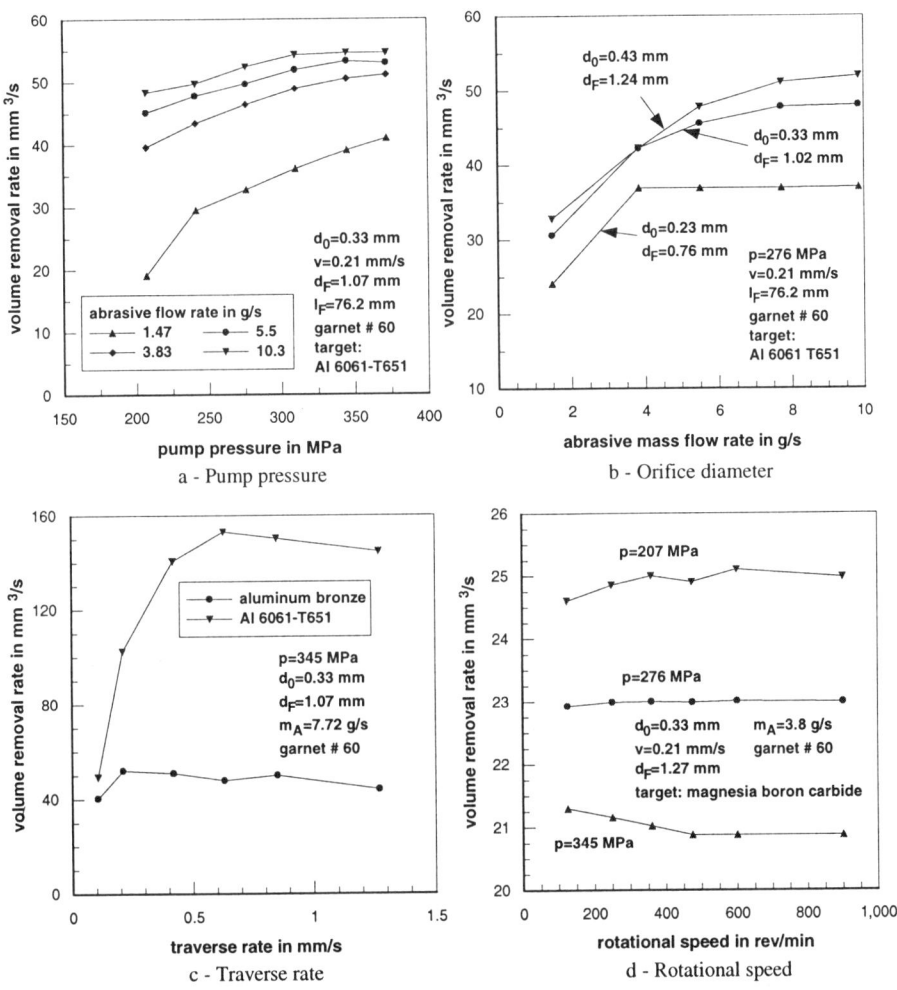

Fig. 9.24 *Parameter influence on the material removal in turning [287]*

Figure 9.24c shows the traverse rate as a parameter. The typical functions in this figure are explained using similar arguments as for the influence of the pump pressure. This figure also expresses the influence of the target material on the material-removal process.

Figure 9.24d illustrates the influence of the rotational speed on the turning process. There is no significant relationship between this parameter and the material-removal rate. This result is explained due to the fact that the relative velocity between the abrasive particles and workpiece is not significantly affected by the rotational speed. Considering, for example, a rotational speed of $\omega=2,500$ min^{-1} and a workpiece diameter of $d_1=50.8$ mm, the corresponding tangential speed is about 6.6 m/s. This spped is two orders slower than typical abrasive-particle velocities.

Figure 9.24a plots abrasive-mass flow rate curves as a function of the pump pressure. Because these tests are performed at identical h_T and v, the volume-sweep rate is the same for all conditions. If the abrasive-mass flow rate is plotted directly against the volume-removal rate, the volume-removal rate increases as the abrasive-mass flow rate increases (Figure 9.24b). Nevertheless, the slopes of the curves decrease. This decrease indicates inefficient abrasive suction and acceleration in the system. Section 7.5.1 addresses these effects.

Figure 9.25 illuminates the influence of several abrasive parameters on the turning process. Figure 9.25a indicates the effect of the initial abrasive-particle size on the volume-removal rate. The general trend is that at a given pump pressure, the difference in the volume-removal rate between the two mesh sizes is negligible.

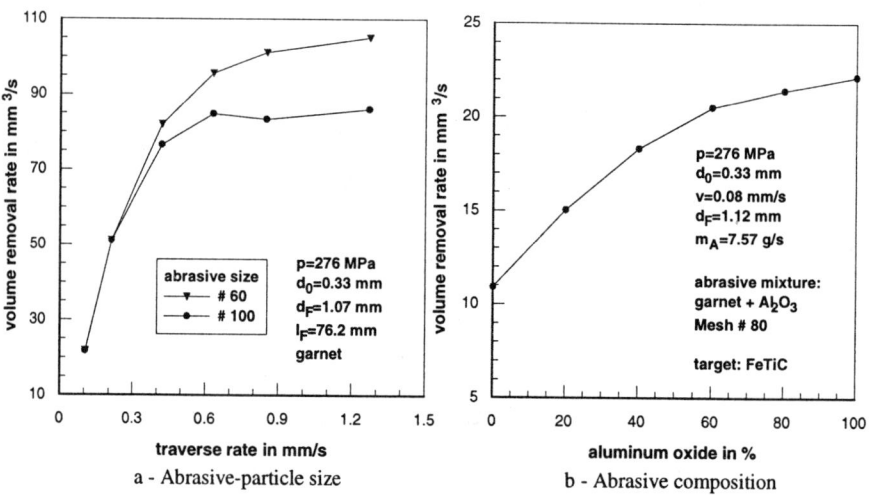

Fig. 9.25 *Influence of abrasive parameters on the material removal in turning [287]*

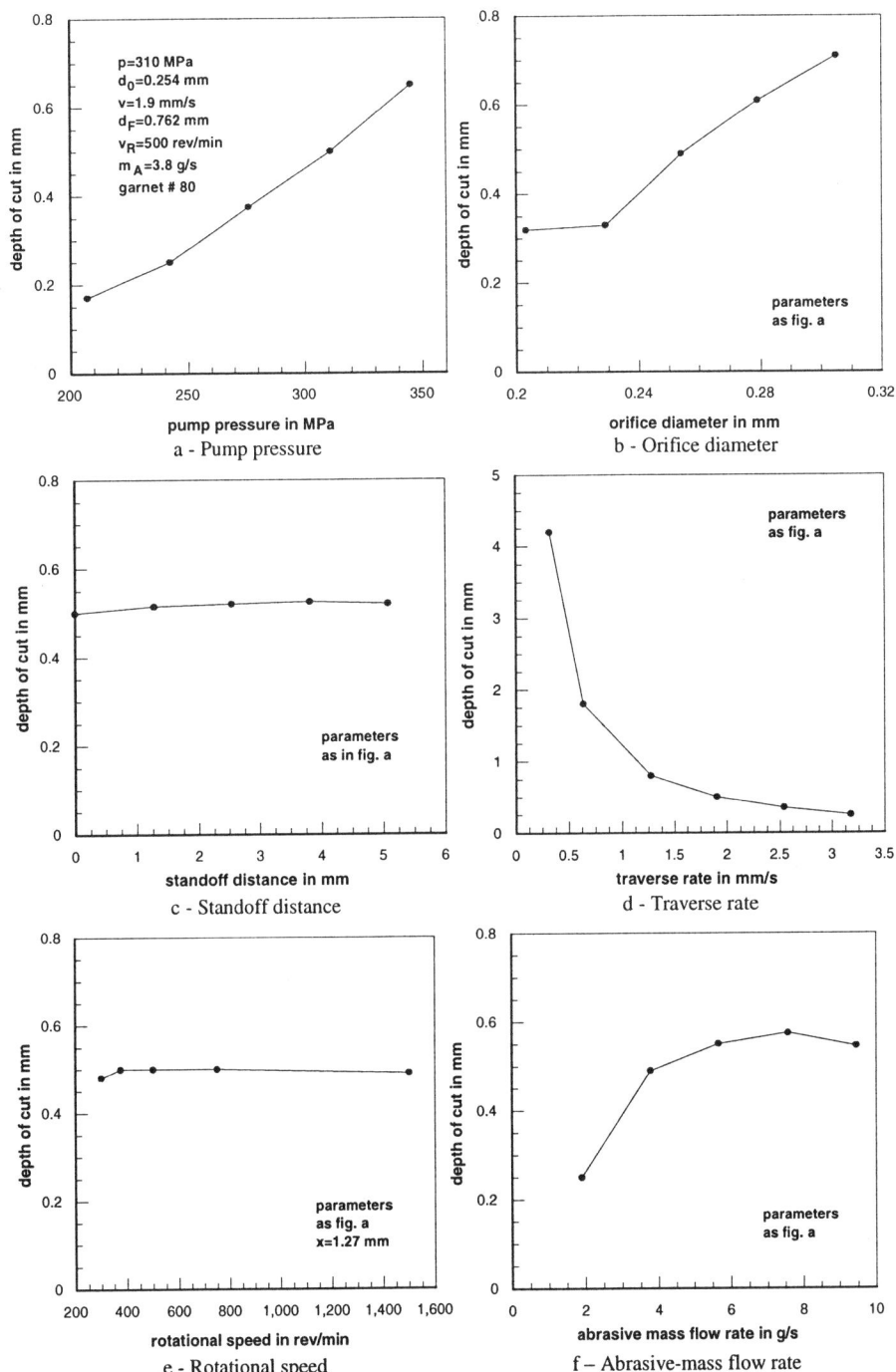

Fig. 9.26 *Parameter influence on the depth of cut in abrasive water-jet turning [345]*

Figure 9.25b shows that the volume-removal rate with 100 % aluminum-oxide abrasive is twice that with 100 % garnet. The incremental improvement in the volume-removal rate decreases with an increase in the aluminum-oxide concentration. However, mixtures with more than 60 % aluminum oxide do not offer any substantial improvement of the material-removal process.

Zeng et al. [345] perform a systematic investigation into the depth of cut in abrasive water-jet turning. Figure 9.26 shows some results.

The depth of cut almost linearly increases with an increase in the pump pressure. The relation between the orifice diameter and depth of cut shows the same tendency. The depth of cut and, so, the final turned diameter of the workpiece are nearly independent on the standoff distance. The same is with the rotational speed. An increase in the traverse rate decreases the achievable depth of cut. All these relations are almost identical to the trends shown in the abrasive water jet-cutting (chapter 7). Excluding the traverse rate, these trends are qualitatively similar to those discussed above for the process parameter impact on the material-volume removal rate. For the influence of the abrasive-mass flow rate, the depth of cut decreases as the abrasive-mass flow rate becomes comparatively high.

Figure 9.27 reveals three different data trends for the relationship between the depth of cut and the ratio $\tau = \delta/d_F$. For $\tau \leq ^4/_3$, the depth of cut significantly increases with an increase in the ratio τ. For $^4/_3 < \tau < ^{10}/_3$, the slope reduces. For $\tau \geq ^{10}/_3$, the depth of cut is almost independent on the ratio K. Zeng et al. [345] show that for $\tau \leq ^4/_3$, the depth of cut is a function of the lateral feed as well as of the focus diameter. For $^4/_3 < \tau < ^{10}/_3$, the depth of cut is a function of the lateral feed alone. For $\tau \geq ^{10}/_3$, the depth of cut is independent of both the lateral feed and focus diameter.

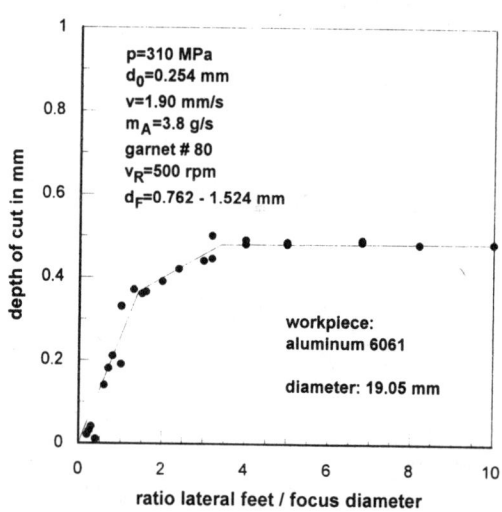

Figure 9.27 *Influence of the geometry on the depth of cut in turning [345]*

9.3.3 Quality of Abrasive Water-Jet Turning

Hashish [331] investigates the quality of surfaces generated by the abrasive water-jet turning in detail. Figure 9.28a illustrates the influence of the pump pressure on the surface. For aluminum, the surface waviness not significantly reduces as the pump pressure reaches values beyond p=200 MPa. Therefore, there is a pump pressure limit for an efficient, high-quality abrasive water-jet turning.

Generally, a high traverse rate increases the surface waviness. However, no changes in the surface waviness occur when the traverse rate reduces below a critical value. This value depends on the other process parameters as well as on the target material.

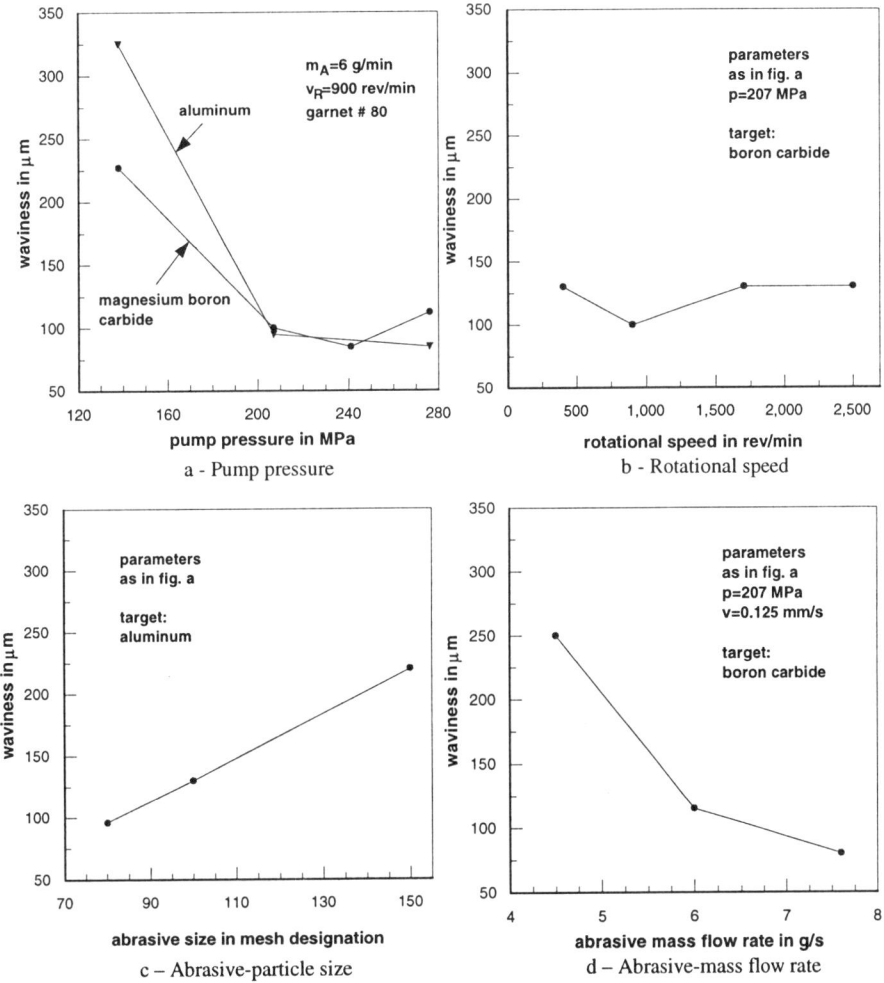

Figure 9.28 *Parameter influence on the surface waviness in turning [331]*

Figure 9.28b shows the effect of the rotational speed on the surface waviness. From this figure, varying the rotational speed, at least under the given process conditions, does not affect the textural characteristics of the surface. A possible explanation is the large velocity slip between the target rotational velocity and the velocities of the impacting abrasive particles. Therefore, even an increase in the rotational speed by a factor of 6 does not influence the mechanics of the material removal.

Hashish [331] finds that the direction of rotation affects the surface quality. The surface waviness is greater when the sample is turned so that the relative velocity between abrasive water jet and specimen increases.

Figure 9.28c shows the effect of the abrasive-particle size on the waviness. The surface quality reduces as the abrasive particles become finer. This effect relates to the worsen material removal capacity of small-grained abrasives (section 7.5.2).

Hashish [331] discusses the effect of the impact angle on the surface waviness. Surface waviness improves when the jet is angled into the workpiece in the traverse direction.

Figure 9.28d shows roughness values of specimens turned by abrasive water jets under different abrasive-mass flow rates. An increase in the abrasive-mass flow rate improves the surface texture. Nevertheless, the surface improvement is not very pronounced at high abrasive-mass flow rates. Hashish [331] even observes an increase in the surface waviness for very high abrasive-mass flow rates. Experimental results from this author also suggest that the influence of the abrasive mass-flow rate on the surface structure becomes more significant if the input process energy is comparatively low.

Hashish [331] briefly addresses the effect of the abrasive material. The author finds that olivine produces rougher surfaces than garnet. Although the material-removal rates are just slightly lower for olivine and the cost of olivine are about 25% of that for garnet, this result suggests that when the surface waviness is not a critical feature, use olivine for a cost effective abrasive water-jet turning. Hashish also performs some surface finish improvement tests by applying multiple-pass turning steps. But the surface waviness not significantly improves due to a second turning step.

9.3.4 Modeling of Abrasive Water-Jet Turning

9.3.4.1 Analytical Turning Model

The final turned diameter, d_T, after an abrasive water-jet turning is (Figure 9.23)

$$d_T = d_{WP} - 2 \cdot h_T. \tag{9.13}$$

In this equation, the unknown parameter is the depth of cut. Ansari [287] and Zeng et al. [345] develop models for the estimation of this parameter.

The model of Ansari is an adaptation of Hashish's [260] model for linear abrasive water-jet cutting (section 6.2). The model bases on the assumption that the depth of cut consists of two parts,

$$h_T = h_{Tc} + h_{Td}. \tag{9.14}$$

In this equation, in analogy to Hashish's cutting model, h_{Tc} is the 'cutting-wear' depth of turning, and h_{Td} is the 'deformation-wear' depth of turning. For the estimation of h_{Tc}, Ansari derives

$$h_{Tc}^2 - 2 \cdot \sqrt{d_{WP} \cdot \delta - \delta^2} \cdot h_{Tc} + \frac{v_P^2 \cdot \dot{m}_A}{\pi \cdot v \cdot \sigma_f \cdot \psi \cdot K} \cdot \left[\sin(2 \cdot \varphi_T) - \frac{6}{K} \cdot \sin^2(\varphi_T) \right] = 0 \tag{9.15}$$

$$\text{with} \quad \varphi_T = \cos^{-1}\left(1 - \frac{2 \cdot \delta}{d_{WP}}\right). \tag{9.15a}$$

This equation is quadratic in the argument an can be solved for a set of given parameters. For K and ψ, see section 5.1. Ansari approximates the 'deformation-wear' depth as,

$$(h_{Td} - h_{Tc})^2 - 2 \cdot (h_{Td} - h_{Tc}) \cdot \sqrt{d_{WP} \cdot \delta - \delta^2} + \frac{\dot{m}_A \cdot (v_P - v_{cr})^2}{\pi \cdot d_{WP} \cdot v} = 0. \tag{9.16}$$

This equation is also quadratic in the argument. Eqs. (9.15) and (9.16) properly reflects some qualitative features of the abrasive water-jet turning. Figure 9.29 shows an example.

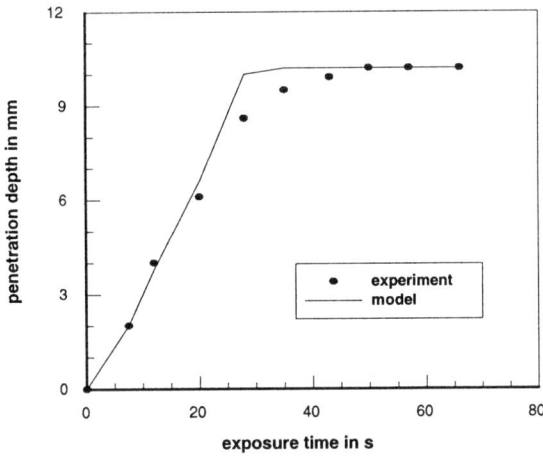

Figure 9.29 *Numerical results from Ansari's turning model [287]*

9.3.4.2 Regression Turning Model

Zeng et al. [345] apply a regression method for the estimation of the depth of cut in abrasive water-jet turning. Their model is

$$h_T = a_0 \cdot N_T \cdot p^{a_1} \cdot \dot{m}_A^{a_2} \cdot v^{a_3} \cdot d_{WP}^{a_4} \cdot d_0^{a_5} \cdot d_T^{a_6} \cdot \delta^{a_7}. \tag{9.17}$$

In this equation, the exponents a_0, a_6, and a_7 depend on the ratio τ. Table 9.4 illustrates these relations. The parameter N_T, defined as 'machinability number for abrasive water-jet turning' [345], characterizes the target material response in the abrasive water-jet turning. This parameter is estimated by back-calculations from the data of two turning tests. Generally, this number plays the same role as the 'machinability number', N_m, for the abrasive water-jet cutting (section 5.9.2.1). Table 9.5 lists some values of of the parameter N_T. Eq. (9.17) does not include any quality index. In general, use the condition $\tau \leq 3/4$ for finishing processes, and the condition $\tau \geq 3/4$ for rough material-removal processes [345].

Table 9.4 *Regression parameters for Zeng's et al. model [345]*

Regression parameter	Ratio τ		
	$\leq 4/3$	4/3 to 10/3	$\geq 10/3$
a_0	$1.08 \cdot 10^7$	$6.82 \cdot 10^6$	$5.19 \cdot 10^6$
a_1		2.673	
a_2		0.247	
a_3		1.240	
a_4		1.371	
a_5		1.808	
a_6	0.250	-	
a_7	1.605	0.292	-

Table 9.5 *Machinability numbers for turning [345]*

Material	Machinability number
Al 6061	213
Copper	56.7
Glass	1,648
Granite	1,370
Plexiglass	919
SS-304	40
Titanium	71.5

To avoid threading during the abrasive water-jet turning, the traverse distance of the abrasive water jet during one rotation of the workpiece must be equal or less than the jet (focus) diameter. Thus,

$$v = q_O \cdot d_T \cdot v_{rot}.\qquad(9.18)$$

In this equation, q_O is an overlapping index. The smaller the q_O, the more is the overlapping. From Figure 9.30a, values of $q_O \leq 0.5$ avoid threading. However, if q_O is reduced by decreasing the traverse rate, deeper cuts are generated (Figure 9.30b) that yields a poorer surface finish. Since the depth of cut is independent on the rotational speed (Figure 9.26e), apply a higher rotational speed to achieve a higher surface quality. For $q_O<0.5$, a further reduction of q_O by increasing the rotational speed does not improve the surface finish any further. Therefore, $0.3<q_O<0.5$ is recommended [345].

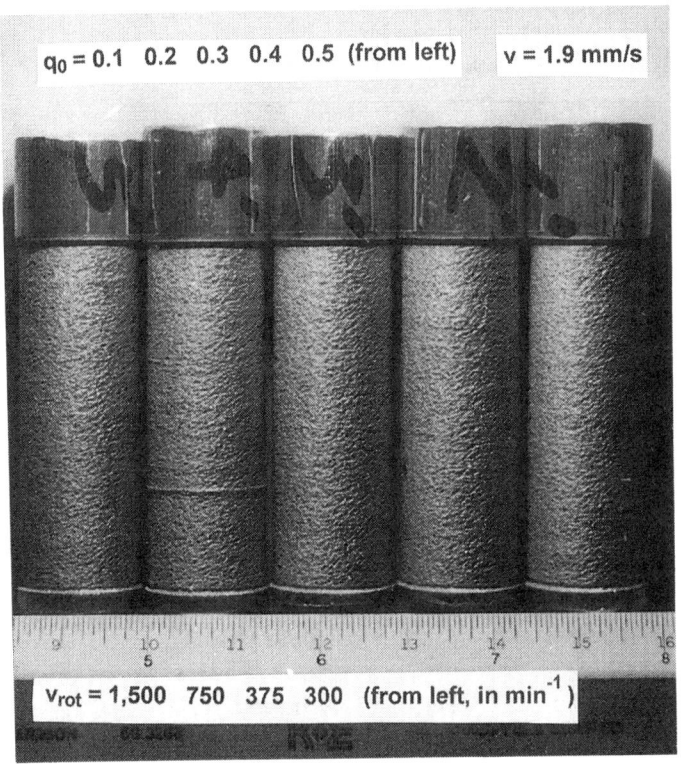

Figure 9.30a *Influence of q_O on the surface finish in abrasive water-jet turning (Univ. of Rhode Island, Kingston)*

Figure 9.30b *Traverse rate influence on the surface finish in abrasive water-jet turning (Univ. of Rhode Island, Kingston)*

9.4 Piercing with Abrasive Water Jets

9.4.1 Macromechanism of Abrasive Water-Jet Piercing

The generation of holes in a material by an abrasive water jet is realized by three methods: piercing, drilling and trepanning. In piercing, neither the jet nor the specimen perform rotational movements. The jet just penetrates the material in its axial direction until it leaves the workpiece.

Figure 9.31 illustrates the general structure and features of a hole that is generated by piercing. This process distinguishes into three phases:

- water jet impact
- abrasive water jet penetration
- abrasive water jet dwelling

Therefore, for the piercing time,

$$t_P = t_W + t_{JP} + t_D . \tag{9.18}$$

Figure 9.32 that bases on acoustic-emission measurements, visualizes this temporal development. Several features associate with these certain phases:

- The water-jet impact phase as a highly dynamic, non-localized process determines the quality of the hole-top surface.
- The penetration phase that is characterized by an almost perpendicular abrasive impact angle at the hole bottom and by rebound flow determines the efficiency of the process (e.g., penetration rate, drilling time).
- The dwelling phase, where the abrasive particles impact the hole wall at shallow angles is responsible for the hole geometry.

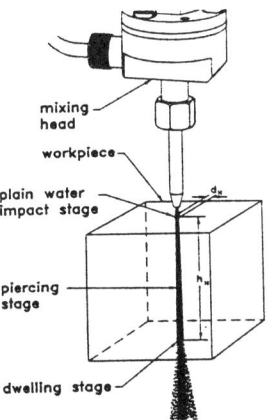

Fig. 9.31 *Piercing process and hole geometry*

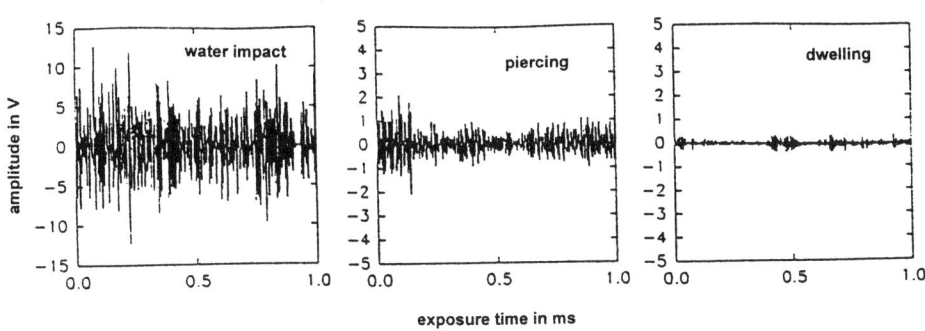

Fig. 9.32 *Acoustic-emission signals acquired from different piercing-stages [346]*

9.3.2 Parameter Optimization in Abrasive Water-Jet Piercing

Kim et al. [299], Hunt et al. [247], and Yanagiuchi and Yamagata [317] report the first systematic investigations into the piercing by an abrasive water jet. Kim et al. [299] and Hunt et al. [247] demonstrate that abrasive water jets effectively pierce alumina and high-strength ceramics. The authors find that the pump pressure, abrasive-mass flow rate, and abrasive type influence the piercing time.

Figure 9.33 shows some relations. Figure 9.33a plots the influence of the pump pressure on the piercing time for different materials. As the pump pressure increases, the piercing time decreases. Kim et al. [299] show that the effect of softer abrasive particles (garnet) on the influence of the pump pressure is more pronounced than that of harder abrasives. Liu et al. [127] use suspension-abrasive jets at low pump pressures (p=9 MPa) for piercing of glass, and find the relation

$$t_P = \frac{C_1}{p} + C_2 \cdot p + C_3. \tag{9.19}$$

Generally, there exists a critical pressure above which no further decrease in the piercing time occurs. As Figure 9.33a illustrates, the parameters C_1 to C_3 in Eq. (9.19) depend on the target material. They also depend on the abrasive type [247].

Figure 9.33b expresses the influence of the abrasive-mass flow rate on the piercing process in ceramics. The piercing time decreases for comparatively high abrasive-mass flow rates. But the particular trend significantly depends on the abrasive type. For soft abrasives (garnet), the piercing time initially increases with an increase in the abrasive-mass flow rate. In contrast, a hard abrasive (silica carbide) reduces the piercing efficiency at high abrasive-mass flow rates [247]. Hashish and Walen [347] do not find a general trend for the influence of the abrasive-mass flow rate on the piercing process in ceramic-coated components. But in their study, the piercing time is always comparatively high for the highest abrasive-mass flow rate. These authors also find that a medium orifice diameter (d_0=0.23 mm) is most efficient for piercing.

Liu et al. [127] investigate the influence of the standoff distance. For soft and ductile-behaving materials, the penetration rate generally decreases with an increase in the standoff distance. In contrast, the penetration rate exhibits a maximum at an optimum standoff distance for brittle-behaving glass.

Mazurkiewicz et al. [67] investigate the influence of the mixing-chamber geometry on the piercing process in aluminum and find that the piercing time substantially reduces with a proper mixing-chamber design (section 3.2).

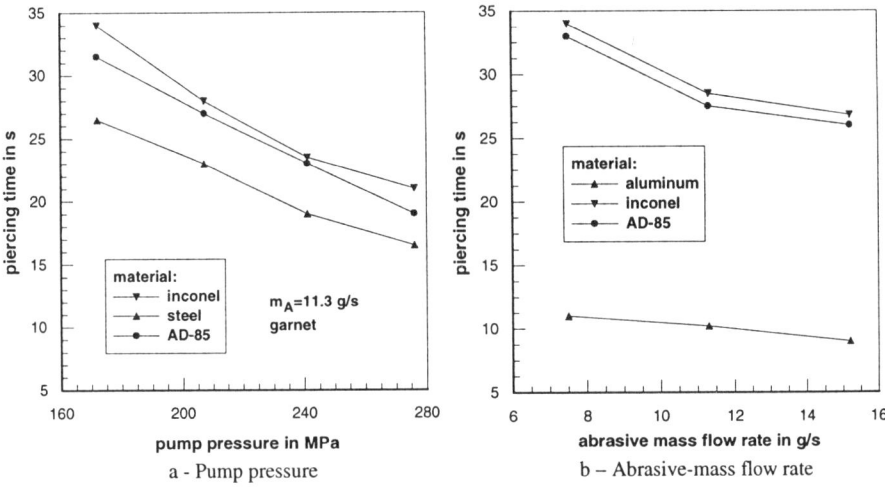

Fig. 9.33 *Parameter influence on the piercing time [247]*

9.4.3 Geometry and Quality of Holes Pierced by Abrasive Water Jets

9.4.3.1 Hole Geometry

Yanagiuchi and Yamagata [317] first consider geometry aspects of the abrasive water-jet piercing. They detect tapering effects as illustrated in Figure 9.34a. Also, the authors find a relation between the hole diameter and focus diameter,

$$d_H \cong 2 \cdot d_F. \tag{9.20}$$

Figure 9.34a shows that the diameter on top as well as on the bottom of the hole almost linearly increases with an increase in the pump pressure. The progress is much more severe for the bottom diameter. This observation indicates that the pump pressure dominates the dwelling period. Therefore, the hole taper increases at high pump pressures.

Hamatani and Ramulu [184] observe in composite materials that the hole taper increases as the standoff distance increases. Also, they find that the taper is more severe for the softer target material. For the softer target material, the increase is almost linear; whereas, the taper digressively augments with an increase in the standoff distance for the hard ceramic composite. The authors also observe a reduced taper if the abrasive-mass flow rate increases (Fig. 9.34b).

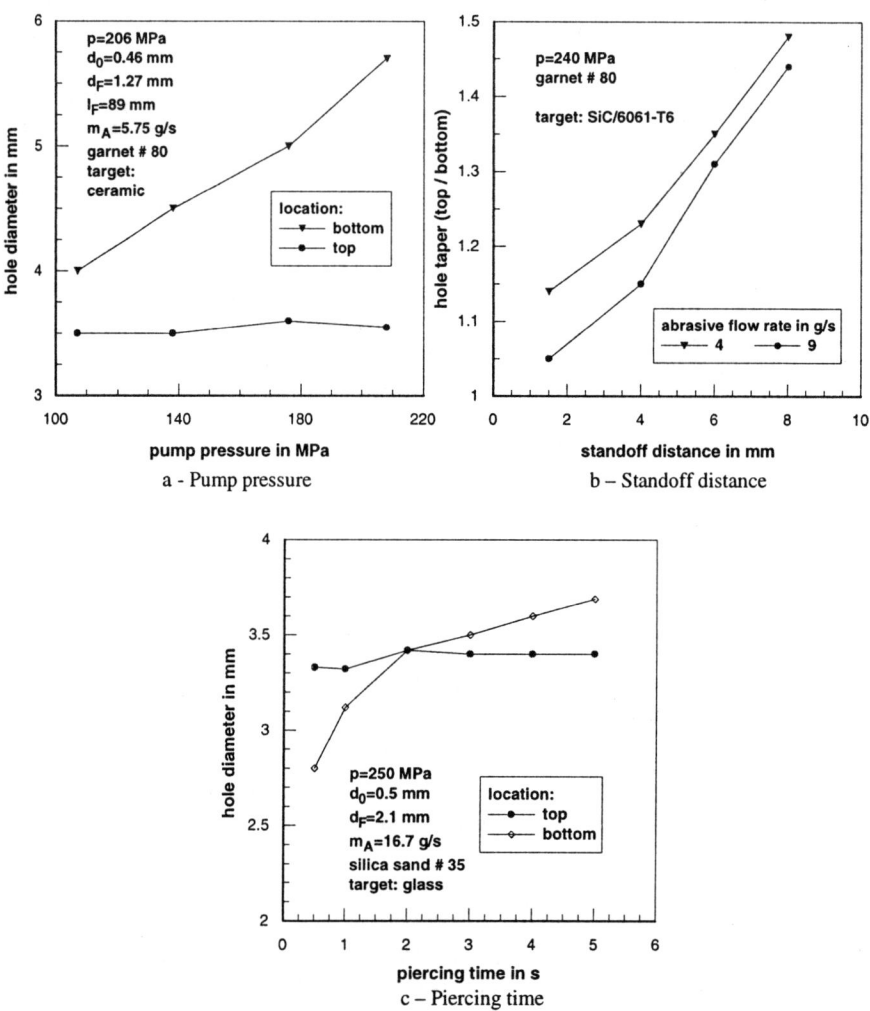

Fig. 9.34 *Parameter influence on the hole geometry [184, 317]*

It is also observed that the hole geometry for soft ceramics pierced by abrasive water jets at different standoff distances illustrates more complex relations. In this case, the direction of the taper changes as a function of the standoff distance. The top diameter increases almost linearly with an increase in the standoff distance; whereas, the bottom diameter remains more or less constant. There is an optimum standoff distance with no taper effects. The relation between the hole diameter and abrasive-mass flow rate is of similar complexity. Whereas, the top-hole diameter does not depend on the abrasive-mass flow rate, the bottom-hole diameter is very sensitive to changes in the abrasive-mass flow rate in the range of small abrasive-mass flow rates. For high abrasive-mass flow rates, the exit diameter remains almost constant. Generally, the hole taper is small for small abrasive-mass flow rates as well as for high abrasive-mass flow rates. The latter tendency is in agreement with observations from Hamatani and Ramulu [184].

Figure 9.34c illuminates the influence of the piercing time on the hole-exit diameter in a composite material. With an increase in the piercing time, the diameter almost linearly increases.

Figure 9.35 illustrates the precision of the abrasive water-jet piercing. In this case, 21 holes are drilled in several shrouds by an abrasive water jet [347]. Another investigation, where more than 1,000 holes are pierced in a ceramic-coated metal (d_H=0.5 mm) with a standard-deviation of σ_{dH}=0.025 mm, indicates the precision capability of abrasive water jets for hole piercing [316]. Figure 9.36 shows holes pierced in plastic boards by an abrasive water jet.

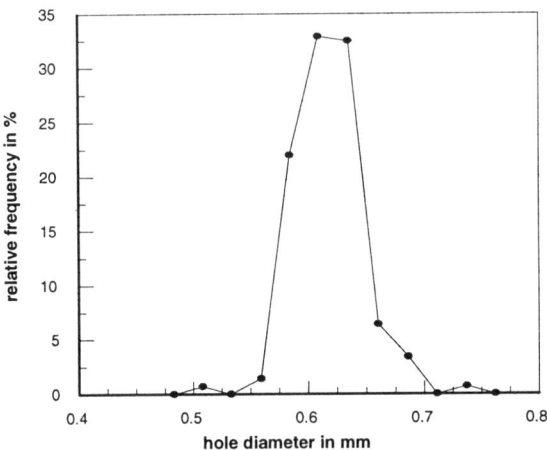

Fig. 9.35 *Precision of material piercing by an abrasive water jet [347]*

9.4.3.2 Hole Quality

The piercing by an abrasive water jet includes several quality aspects. Figure 9.37 summarizes these aspects.

Surface chipping is evident for brittle materials. Hashish [348] observes this damage in glass piercing. Shipping occurs especially at high pump pressures and also due to lags in the abrasive-material delivery. Hashish and Walen [347] states that the chipping reduces if the abrasive-line length reduces. During the treatment of coated workpieces, piercing from the metal side produces chipping as the jet breaks through.

318   9. Alternative Machining Operations with Abrasive Water Jets

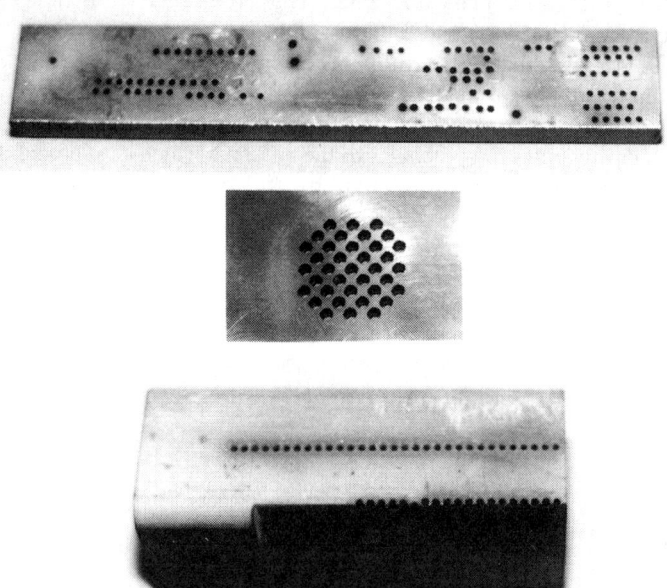

Figure 9.36 *Holes pierced into engineering materials by abrasive water jets (Univ. of Texas at Arlington, Arlington)*

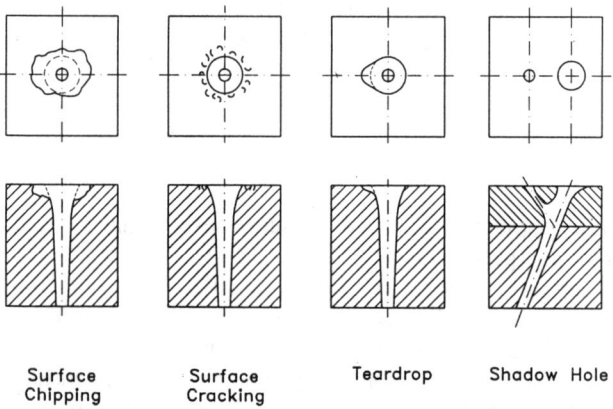

Surface Chipping   Surface Cracking   Teardrop   Shadow Hole

Fig. 9.37 *Quality aspects of piercing by abrasive water jets*

Cracking on top as well as on the exit of a pierced hole actually emanates from sub-surface pressurization or from wave loading under high shock loads, such as the impact of plain water jets (Figure 9.32). In piercing of ceramic coated metal parts, most cracks are circumferential and appear to emanate from the interface between the ceramic and the metal [347]. Yanagiuchi and Yamagata [317] do not find any cracking in soda-lime glass pierced by an abrasive water jet at a comparatively low pump pressure (p=50 MPa). In contrast, Geskin et al. [315] observe cracking during the piercing of glass as a result of the sudden impact of the abrasive water jet. The authors prove that impact cracking can is eliminated by using a built-in dual compensator. This technique enables a temporary reduction of the water pressure up to p=95 MPa at the beginning of the piercing process. Later, Hashish and Walen [347] adapt this concept for the precision piercing in advanced aircraft engine components without any chipping or cracking.

Hamatani and Ramulu [184] use high-pressure abrasive water jets (p=240 MPa) to pierce composite materials and detect damaged zones around the entry of the holes. The surface damage zone appears to be more random in nature in metal-matrix composites than in ceramic composites. However, in ceramic composites, the severity of damage is higher near the pierced hole. The size of this damage zone is about 500 µm.

In the case of teardrop, the hole is not round or elliptical in shape. Teardrop is actually a jet rebound effect.

Shadow holes are the result of the jet deflection in layered materials. The rebound of the jet off the basic component interface produces another 'shadow' hole. This effect becomes more severe for piercing at shallow angles.

Laminar composites, such as graphite epoxy, kevlar, and fiber components, delaminate during the abrasive water jet piercing [313]. Limited tests on supporting delamination-sensitive materials by using steel plates with holes drilled in them for the exiting jet flow are not successful in eliminating delamination at high pump pressures. This result suggests that hydrodynamic pressurization plays a role in delamination [349].

9.4.4 Modeling of Abrasive Water-Jet Piercing

9.4.4.1 Phenomenological Piercing Model

Hashish [207] conducts the first investigation into the mechanism of material piercing by an abrasive water jet using high-speed photography. He assumes that in piercing, the abrasive particles impact the target material at large impact angles and that the efficiency of the piercing process reduces at deeper depths due to backflow effects. Figure 9.38a shows that the penetration rate for piercing decreases as the piercing time increases. Ramulu [192] shows that this deceleration in the penetration rate occurs even at very short exposure times smaller than one second. For these reasons, the piercing depth digressively increases with an increase in the piercing time. Several authors observe this phenomnena for glass [207, 315],

320   9. Alternative Machining Operations with Abrasive Water Jets

photoelastic materials [192], and ceramics [346]. These relations can be described by

$$h(t) = h_{max} \cdot \left(1 - e^{-C \cdot t}\right). \tag{9.21}$$

In this equation, the parameter C is sensitive to the workpiece-material properties. For ceramics, the parameter C increases as the compressive strength decreases. In contrast, the parameter h_{max} is a constant value for a certain material group, as for example refractory ceramics [350]. From Figure 9.38b, the material-removal rate is almost constant over the piercing time. This result suggests that the hole diameter increases with an increase in the piercing time. Therefore, assume the predominance of a pressurization process in the blind hole that associates with a circumferential material removal at a certain depth or time, respectively.

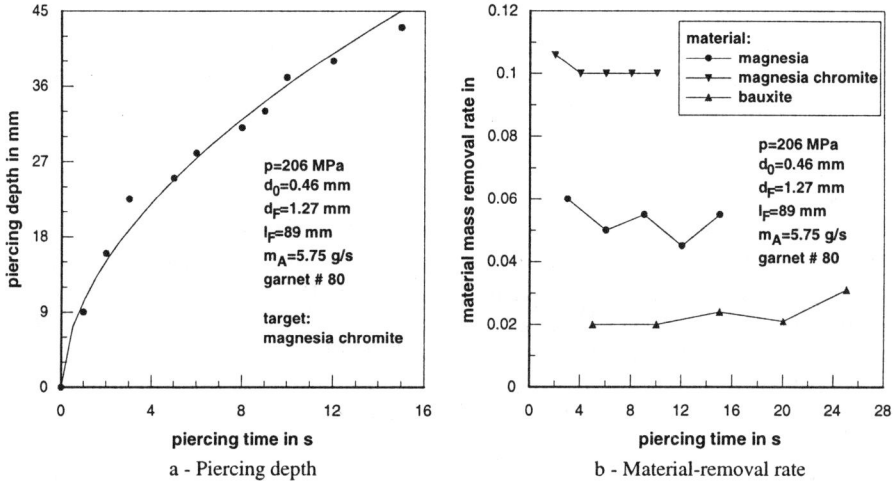

Fig. 9.38 *Piercing-time influence on the piercing of ceramics [350]*

In Figure 9.38a, this critical time is between $t_{Cr}=3$ s and $t_{Cr}=10$ s. These relations imply a non-linear relation for the penetration rate,

$$v_{PR} = \frac{\Delta h}{\Delta t_P} \neq const. \tag{9.22}$$

Figure 9.39 shows the maximum shear stress generated in a hole pierced by abrasive water jets. With increasing penetration, the stress field exponentially abates due to energy-dissipative processes, such as damping and back flow. Similarly, from Figure 9.40 that shows the relation between the drilling time and RMS-voltage of acoustic-emission signals acquired during the abrasive water-jet piercing, the signal intensity almost linearly drops with an increase in the depth of the hole. This result indicates damping effects on the hole bottom.

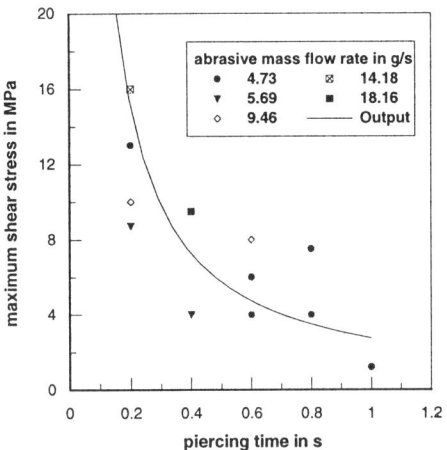

Fig. 9.39 *Relation between the the penetration of an abrasive water jet and shear stress [192]*

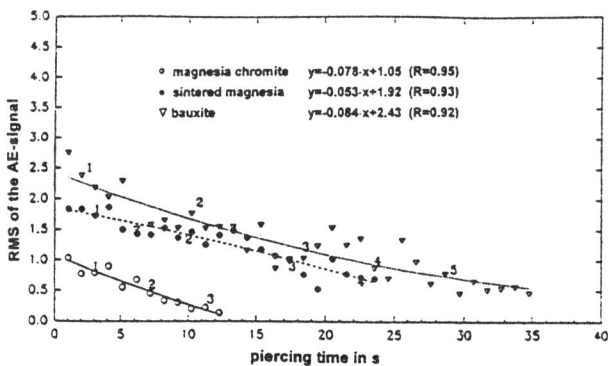

Fig. 9.40 *Acoustic-emission signals acquired at different piercing times [346]*

9.4.4.2 Analytical Piercing Model

Raju and Ramulu [351] introduce an analytical model of abrasive water-jet piercing. Based on an energy balance for the net-energy that is available in the workpiece (Figure 9.41a),

$$\xi \cdot \frac{1}{2} \cdot \dot{m}_A \cdot (v_A^2 - v_e^2) = \pi \cdot \frac{d_H^2}{4} \cdot \frac{dh}{dt}, \qquad (9.23)$$

the authors derive a linear differential equation for the piercing rate:

9. Alternative Machining Operations with Abrasive Water Jets

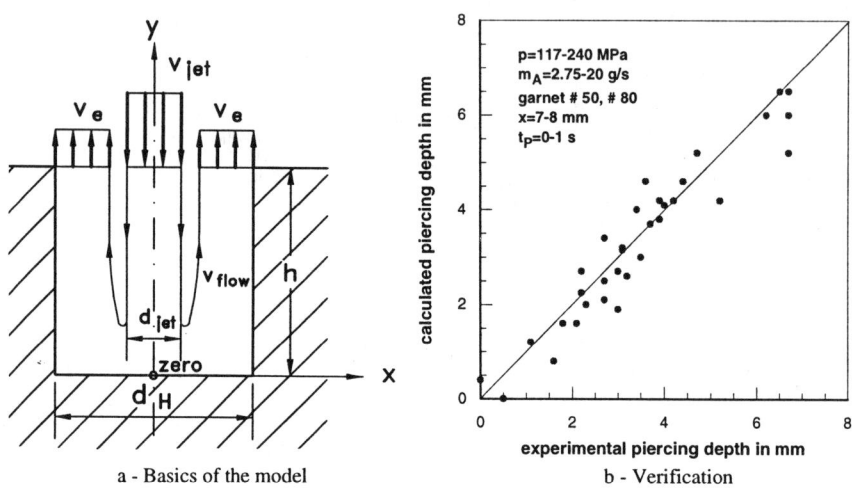

a - Basics of the model   b - Verification

Fig. 9.41 *Raju and Ramulu's model for the abrasive water-jet piercing [351]*

$$\frac{dh}{dt} = K_3 \cdot \left[v_A^2 - \left[\frac{K_2}{K_1} \left(\frac{1}{a} - \frac{1}{a + K_1 \cdot h} \right) - \frac{dh}{dt} \right]^2 \right], \quad (9.24)$$

$$a = \frac{1}{v_A} - K_1 \cdot h, \quad (9.24a)$$

$$K_1 = \frac{c_f \cdot \rho_{jet} \cdot \left[2 + \frac{d_H}{d_{jet}} \right]^2 \cdot (\pi \cdot d_{jet})}{2 \cdot (\dot{m}_A + \dot{m}_W)}, \quad (9.24b)$$

$$K_2 = \frac{c_f \cdot \rho_{jet} \cdot \left[2 + \frac{d_H}{d_{jet}} \right]^2 \cdot (\pi \cdot d_{jet} + \pi \cdot d_H)}{2 \cdot (\dot{m}_A + \dot{m}_W)}, \quad (9.24c)$$

$$K_3 = \frac{2 \cdot \xi_D \cdot \dot{m}_A}{\pi \cdot d_H^2}. \quad (9.24d)$$

This equation is numerically solved. The parameters d_H, c_f, and ξ_D are estimated by experiments. For garnet as abrasive material, Raju and Ramulu [351] find: c_f=0.00032 and ξ_D=2.5·10^{-11} for garnet # 50, and c_f=0.00051 and ξ_D =5·10^{-11} for garnet # 80.

As Figure 9.41b shows, the model reflects several features of the abrasive water-jet piercing and exhibits a good correlation to experimentally estimated results.

9.4.4.3 Regression Piercing Model

Zeng and Munoz [352] present a regression model for the estimation of the piercing time that bases on Zeng and Kim's [262] model for abrasive water-jet cutting. The time for piercing a hole of a depth h is

$$t_P = \frac{c_0 \cdot h^{c_1} \cdot d_0^{c_2}}{N_{mP} \cdot p^{c_3} \cdot d_F^{c_4} \cdot \dot{m}_A^{c_5}} \, . \tag{9.25}$$

In this model, the constants c_0 to c_5 are determined from regression analysis. The 'machinability number for piercing', N_{mP}, is evaluated for engineering materials by a reference test.

9.4.4.4 Simulation Model for Piercing

Yong and Kovacevic [353, 354] develop a general model for the simulation of piercing by abrasive water jets. Basically, the authors extend the conventional equation for solid-particle erosion

$$h_P = C \cdot v_P^2 \tag{9.26}$$

for multiple particle-impact phenomena. The model considers damping effects as well as the chaotic behavior of impacting abrasive particles. For turbulent flow, Eq. (9.26) modifies

$$\Delta h_{Pj} = \frac{1}{h_{Pj-1}^2 + 1} \cdot [1 - r_i]^{(0.2 \cdot h_{Pj-1} + 2)/7} \, . \tag{9.27}$$

The final piercing depth in each element is the accumulation of the erosion of all the particles. Figure 9.42a illustrates the assumed surface network and the 'memory cells' that cover the drilling surface. Figure 9.42b shows simulated holes. Figure 9.43 presents comparisons between the model and experiments.

9. Alternative Machining Operations with Abrasive Water Jets

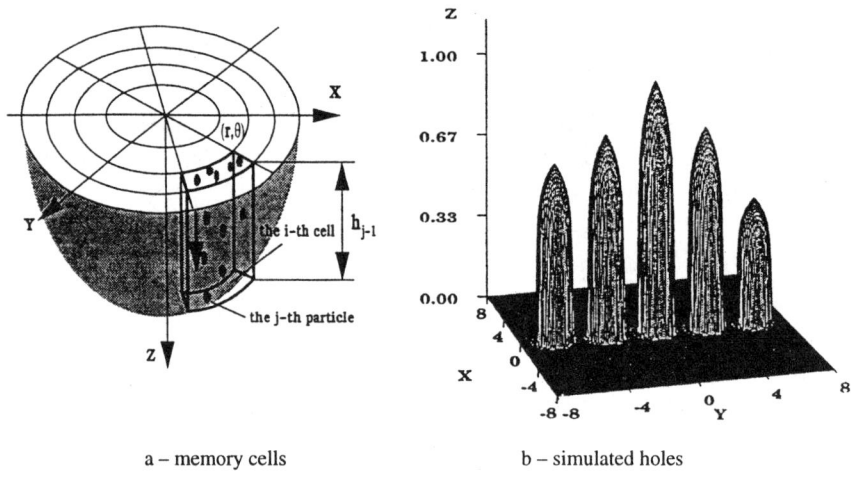

a – memory cells   b – simulated holes

Fig. 9.42 *Simulation model for the piercing by abrasive water jets [353, 354]*

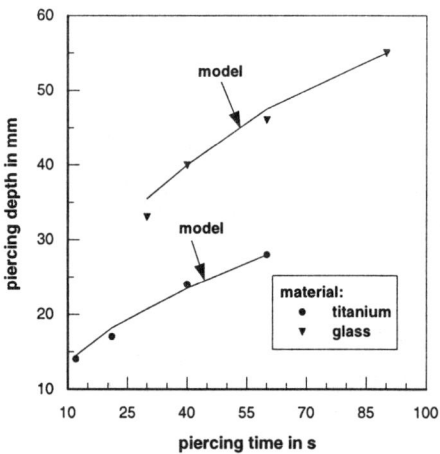

Fig. 9.43 *Verification of piercing simulations [353, 354]*

9.5 Hole Trepanning and Deep-Hole Drilling with Abrasive Water Jets

9.5.1 Hole Trepanning with Abrasive Water Jets

Figure 9.44 shows holes that are generated in ceramics due to trepanning by abrasive water jets. Hole trepanning is simply a non-linear cutting process. However, due to the curvature of the cut, the geometry of the jet-material interface is more complex than that of a straight cut. During trepanning, due to the jet trailback, the diameter of the hole increases as the workpiece thickness increases. Jet spreading produces an opposite effect [316]. Ohlsson et al. [355] conduct a comparative study between the abrasive water jet-piercing, and linear and circular movement of the abrasive water jet. Figure 9.45a illustrates the positive effect of the movement of the abrasive water jet on the penetration time. Figure 9.45b shows the influence of the diameter of the circular movement on the drilling time. There is an optimum diameter range, $1.5 \cdot d_F < D_{Opt} < 2 \cdot d_F$, that does not depend on the traverse rate. For steel, the authors recommend traverse rates between 5 mm/s<v<10 mm/s.

Figure 9.44 *Holes in refractory ceramics due to trepanning by abrasive water jets (WOMA Apparatebau GmbH, Duisburg)*

Table 9.6 contains results of hole drilling using the method of abrasive water jet-trepanning in an inconel plate. The accuracy of the traverse mechanism and the pattern of motion from the piercing location to the wall of the hole primarily control the roundness of the top-side of the holes.

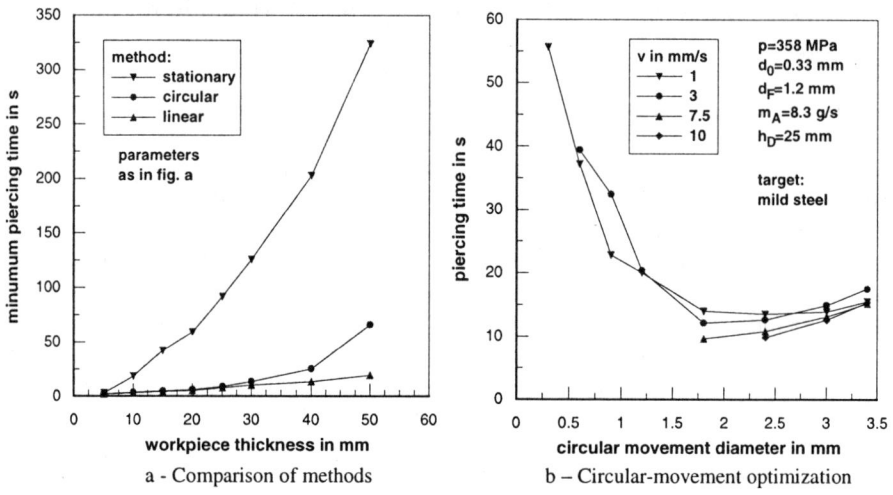

Fig. 9.45 *Comparison of abrasive water-jet based drilling methods [355]*

Table 9.6 *Hole trepanning in a 16 mm thick inconel plate by abrasive water jets [316]*

Parameter	Standard deviation	Average value	Minimum value	Maximum value	Variance
Piercing time [s]	0.790	37.92	35.80	42.10	0.62
Total drilling time [s]	1.289	48.91	45.60	54.10	1.66
Hole top-diameter [mm]	0.010	2.16	2.13	2.18	-
Hole bottom-diameter [mm]	0.025	2.16	2.06	2.29	-

9.5.2 Deep-Hole Drilling with Abrasive Water Jets

Hashish [356] conducts a preliminary study to determine the overall feasibility of deep-hole drilling applications. The author introduces and tests three drilling concepts (Figure 9.46a).

As a first exploring step, he performs computer simulations for studying the flow-pattern-coverage over the workpiece. For oscillating and rotary jets jets, the rotational speed shows a range of high sensitivity. Altering the rotational speed by only 5 %, the flow pattern covers the entire surface and the hole drilling starts.

Hashish [356] reports about the drilling of tungsten by an exclusively designed mixing tube for forming rectangular abrasive water jets. Multiple parallel water jets accelerate the abrasives. The drilling rate is 115 mm/h.

9.5 Hole Trepanning and Deep-Hole Drilling with Abrasive Water Jets

With a rotary water jet, holes are drilled in several materials up to 15 cm deep. The surface roughness is comparatively low, a typical value is 2,2 µm. A rotary abrasive water jet-device is the most efficient. Table 9.7 lists cross sections of holes that are drilled in several materials by this method.

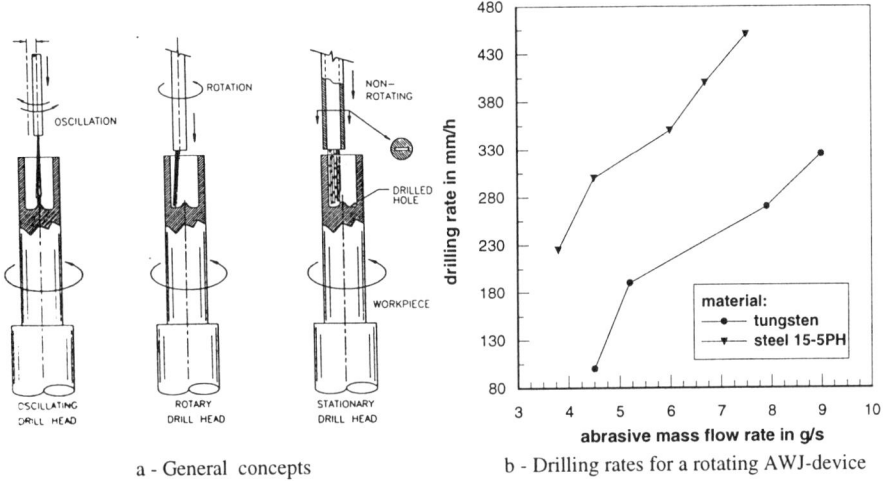

a - General concepts   b - Drilling rates for a rotating AWJ-device

Fig. 9.46 *Deep-hole drilling by abrasive water jets [356]*

Table 9.7 *Cross sections of holes drilled by rotary abrasive water jets [356]*

Material	d_F [mm]	d_0 [mm]	m_A [g/s]	v [mm/s]	Offset [mm]
Stainless steel	1.19	0.33	3.8	0	4.57
					2.79
					4.57
					3.81
Molybden	1.19	0.33	3.8	0	3.81
Tungsten	1.19	0.33	3.8	0	3.81
Stainless steel	10.16x3.05				3.81
	1.19				3.81
Tungsten	3.18	0.46	7.5		4.57

From the point of view of parameter optimization, the drilling rate almost linearly increases as the pump pressure and abrasive-mass flow rate increases (Figure 9.46b). In steel, the drilling rate reaches up to 457 mm/h. In tungsten rods, the maximum observed drilling rate is 305 mm/h.

9.6 Polishing with Abrasive Water Jets

9.6.1 Abrasive Water-Jet Polishing Concepts

Table 9.8 and Figure 9.47 list several concepts for using abrasive water jets for polishing. Nevertheless, these concepts are still in the status of exploration.

For polishing with an external injection of abrasives in a high-speed water jet, the water jet is shallow angled (Table 9.8). The abrasives, in a slurry flow, are fed between the water jet and the workpiece under high pump pressures. In this method, when the sample rotates under the stationary jet, a ring of material is exposed to the jet effect. Most of the abrasive particles deflect off of the material surface [357]. This deflection results in slow polishing rates.

Table 9.8 *Abrasive water-jet polishing investigations [357, 358]*

Method	Target material	Abrasive material	Process parameters	Results
External injection	Silica carbide		p=60 - 207 MPa φ = 10 - 15°	R_a=4.5 - 5.6 µm
Flat surface	Zerodur®	Garnet d_P=25 µm	p= 35 - 345 MPa φ= 2 - 20° d_0 = 0.33 mm d_F= 1.52 mm	Estimating and fixing the grinding regime
Tapered and curved suspension nozzle	Glass Zerodur®		v= 0.31 mm/s	Curved nozzle is advantageous
Radial flow nozzle	Glass Silica carbide Diamond film	Silica carbide mesh # 600	v = 0.31 mm/s p = 35 MPa $\dot{m}_A$ = 10 g/s	R_a=0.13 µm polishing rate: 2.7 micro/s/mm^2

Hashish [357] briefly discusses the polishing of flat surfaces by abrasive water jets. The author subject rotating workpiece samples to the abrasive water jet under various process conditions. After processing, the micro-structure of the polished surfaces is inspected and evidence of brittle fracture, ductile flow, or no effects are recorded. Hashish [357] constructs a chart that illustrates the influence of the pump pressure and impact angle of the grinding regime. The shallower the impact angle, the higher the pump pressure that can be applied before the fracture-mode grinding starts.

9.6 Polishing with Abrasive Water Jets

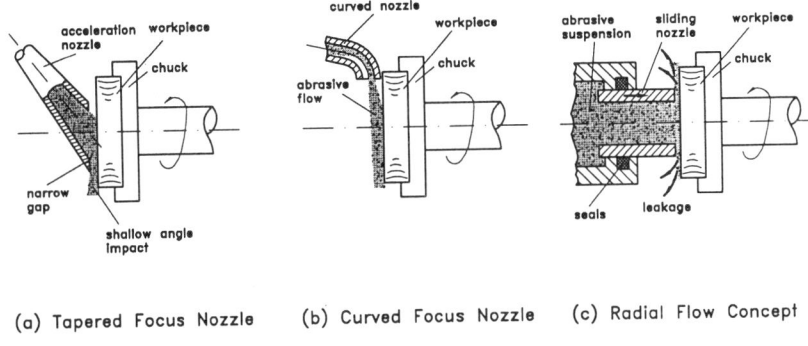

(a) Tapered Focus Nozzle   (b) Curved Focus Nozzle   (c) Radial Flow Concept

Fig. 9.47 *Concepts for polishing with abrasive water jets [357, 358]*

Figure 9.47 shows the polishing with a suspension-abrasive water jet with tapered and curved nozzles. The special designs allow the application of shallow impact angles. From preliminary tests [357], higher grinding rates occur with the curved nozzle design compared to a straight nozzle at identical process conditions. Also, for short periods of exposure and without rotating the workpiece, the effect on the material is not even.

Hashish [358] discusses the concept of a radial flow nozzle. Figure 9.47c illustrates the basic idea. A nozzle made from soft materials such as nylon or copper delivers the suspension to the workpiece. This nozzle is mounted inside a supply chamber such that it flows axially under the pressure inside the suspension chamber. The friction between the nozzle and the wall of the supply tube, in addition to the pressure force under the nozzle, balances the nozzle extrusion force due to the chamber pressure. This method establishes a boundary layer under the nozzle in which the abrasive particles flow at a nearly zero-degree impact angle. Glass polishing is very sensitive to the nozzle material. As, for example, polyethylene is used, the results significantly improve over those obtained with metal and nylon nozzles. In contrast, the polishing of reaction-bonded silicon-carbide indicates effective polishing with metallic nozzles. Hashish [358] reports about the polishing of diamond films with a suspension-abrasive water jet. The surface finish improves from about 5 μm to 1μm in about 40 minutes polishing time.

9.6.2 Quality Aspects of Abrasive Water-Jet Polishing

Li et al. [359] further investigate the feasibility and limitations of material polishing. They find that abrasive water jets can polish materials as ceramics, stainless steel and alloys. The surface quality strongly depends on the size and impact angle of the abrasive grains. Figure 9.48 illustrates these effects. An initial surface roughness of 2 μm to 12 μm reduces down to 0.5 μm at a reasonable productivity by using garnet with $d_P=100$ μm at moderate pressure levels. As Figure 9.48a shows, the impact angle plays a predominant role in the angle range up to φ=20°. As this angle is exceeded, changes in the surface quality are small. Also, as the abrasive-particle size decreases, surface roughness improves. This improvement is very significant int the size range up to mesh # 250. Beyond this limit, the importance of the abrasive size decreases.

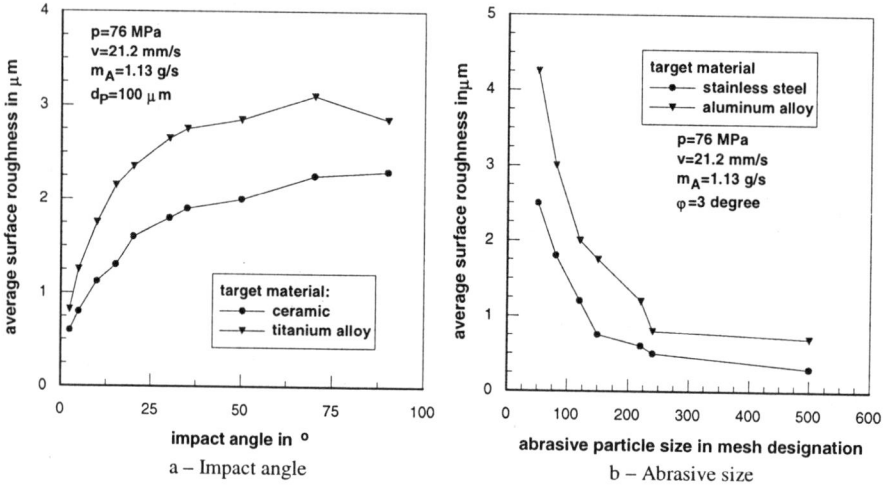

Fig. 9.48 *Parameter influence on the surface roughness [359]*

Microhardness measurements show that the polishing process does not cause obvious hardness modification (Figure 9.49) or micro-defects at the sample subsurface.

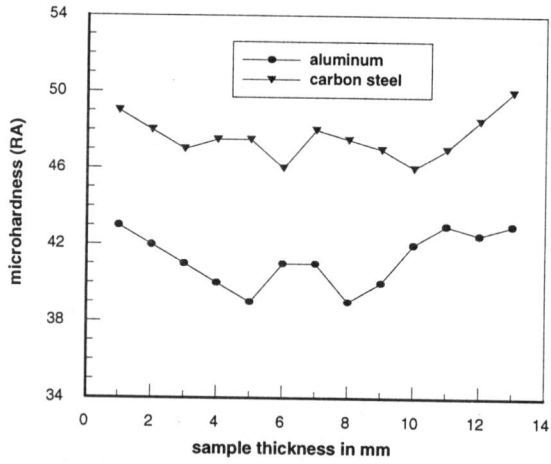

Fig. 9.49 *Microhardness of various materials polished by an abrasive water jet [359]*

9.7 Screw-Thread Machining with Abrasive Water Jets

Figure 9.50a illustrates the principle of thread machining with an abrasive water jet. This process is actually a turning process. Sheridan et al. [360] show that by carefully controlling the process parameters, abrasive water-jet cutting provides an effective method for machining screw threads in composite materials.

Figure 9.50b shows a typical thread profile machined by an abrasive water jet. The profile is not identical to a desired ACME-profile that has sharp corners and a level root. These characteristics are impossible to achieve with abrasive water jets because of the jet geometry. The machined profile possesses a curved root due to the circular jet. The thread root curvature depends on the effective jet diameter. Another important feature is the asymmetry of the thread profile. Examination of the thread walls reveals that one side (the left side in Figure 9.50b) exhibits a greater thread angle than the opposite side. This difference is a result of the jet deflection away from both the cutting direction and the rod axis. This bi-directional jet deflection is associated with normal jet deflection due to energy dissipation and jet deflection due to turning. Also, the thread microstructure consists of smooth root and side wall regions and relatively rough regions at the corner between the side wall and the root and at the corner between the side wall and the crest.

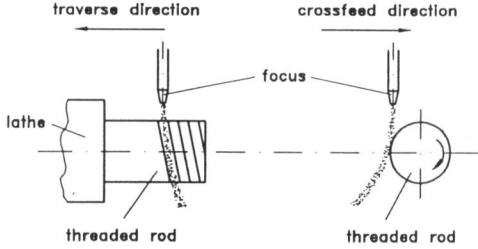

a - Basic setup and principle

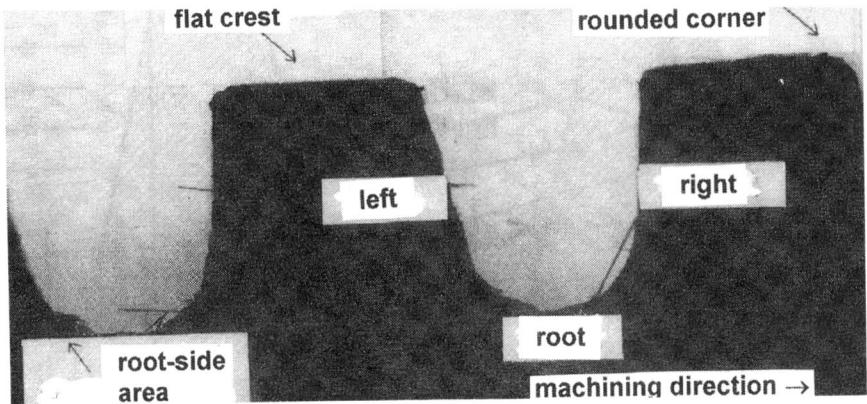

b - Thread profile machined by an abrasive water jet

Fig. 9.50 *Screw-thread machining with an abrasive water jet [360]*

SEM-microscopy reveals that this rough surface is associated with exposed fibers and cylindrical holes. These results indicate a fiber-pullout failure mechanism.

Figure 9.51 shows the impact of several process parameters on the thread width. Very low traverse rates result in relatively high thread widths (Figure 9.51a). At those low traverse rates, the thread width becomes so large that adjacent threads begin to merge. This process eliminates the crest region and reduces the major diameter of the thread. Also, at low traverse rates, an increase in the crossfeed-rate does not greatly affect the crest width or thread width dimensions. Figure 9.51b illuminates the impact of the focus/orifice combination on the machining process. The influence of the nozzle combination is much more pronounced than that of the crossfeed-rate. For the nozzle combination 3, the thread profile consists of relatively wide crest regions, narrow thread widths and diminished thread depth.

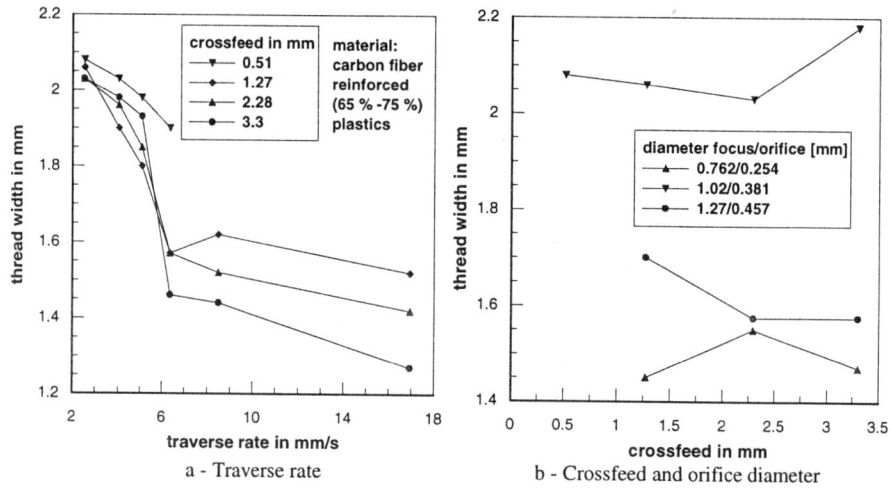

Fig. 9.51 *Parameter influence on the thread geometry [360]*

Sheridan et al. [361] subject composite shaft/steel nut assemblies that are machined by an abrasive water jet to a modified tensile test. This test is designed to pull the nut off of the shaft. The authors find a shear mode failure with shear strengths between $\tau_S=28.2$ MPa (undirectional fiber-reinforced) and $\tau_S=47.4$ MPa (woven-fabric composite). In addition, the authors perform a finite-element simulation for the mechanical behavior of the machined threads. Sheridan et al. [359] discover single contact points between the nut and the thread that generate stress concentrations and finally the initiation of the shear failure. Failure is also initiated at the rounded corner between the thread root and the thread side wall. Since SEM-observation show several matrix damage and fiber-pullout at these locations, they act as flaws and sources of localized damage.

10 Control and Supervision of Abrasive Water-Jet Machining Processes

10.1 General Aspects of Abrasive Water-Jet Process Control

As in any other machining method, control and supervision of the abrasive water-jet machining improves both the efficiency and quality of the process. Figure 10.1 shows a block diagram of a possible new generation of an abrasive water-jet machining system.

Mazurkiewicz and Karlic [363] make an early attempt to develop an appropriate control algorithm for linking abrasive water jetting to automated manufacturing equipment. In detail these authors find:

- The abrasive water-jet machining is a strongly non-linear process.
- The gain parameters can change sign and vary over a wide range.
- The time constants strongly depend on many parameters and on disturbances.
- It is possible to control depth of cut, width of cut and material-removal rate using the two control variables standoff distance and traverse rate.
- Important in the design of the control system is the methods used to monitor all process outputs as well as the use of an automated compensation control system.

Since that early approach, several methods are developed to measure, supervise, and control the major parameters of the process.

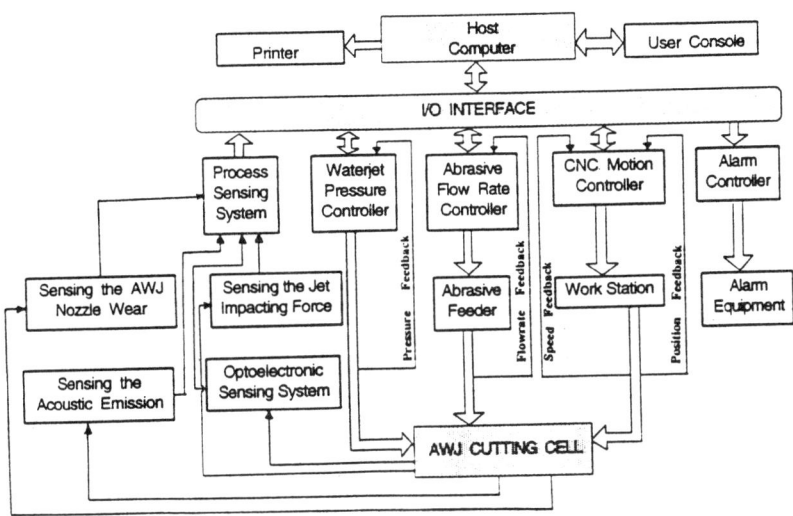

Figure 10.1 *Block diagram of a new generation of an abrasive water-jet cutting system [362]*

10.2 Control of the Abrasive-Particle Suction Process

10.2.1 General Demands

There exist several metering methods to fix the maximum abrasive-mass flow rate that is available for the abrasive water-jet cutting. Nevertheless, these methods do not give any information about the actual abrasive-mass flow rate, because this value depends on several process parameters (section 3.3). Therefore, in order to control and supervise the abrasive-suction process, some possibilities are investigated. Basic demands on a successful control unit are an accuracy of ±5 %, invariance of the abrasive-particle size and type, as well as of the ambient temperature and humidity [364].

10.2.2 Acoustic Sensing

Louis and Meier [76] use acoustic signals to control the suction performance of an abrasive water jet-system. Figure 10.2 shows time domain signals for three different suction conditions for a submerged abrasive water-jet system. In the experiment, just air is sucked without abrasives. For normal conditions (air sucked in), the sound intensity is very low. In case of sucked in water (that could be due to a broken suction hose), the sound level rapidly increases. Also, when the transport hose is clogged (in that case the cutting head produces a vacuum), the sound level is very high, which is due to cavitation effects [76].

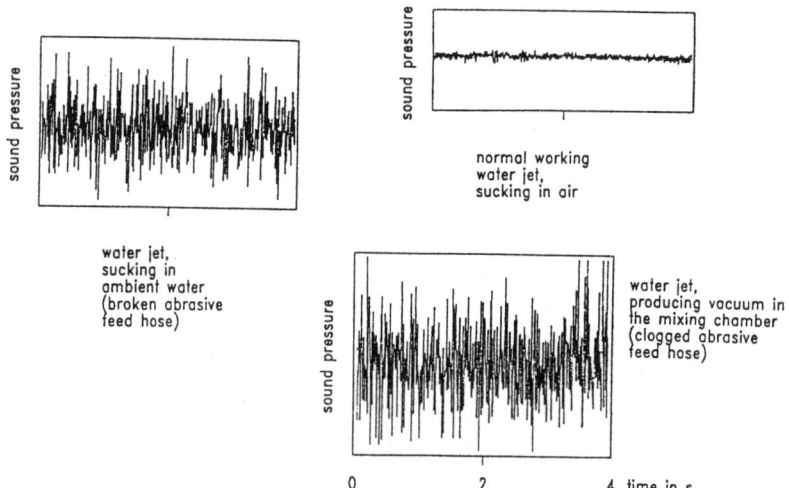

Figure 10.2 *Time-domain acoustic signals, indicating the abrasive-suction performance of a submerged injection-abrasive water jet [76]*

10.2.3 Workpiece Reaction-Force Measurements

Hunt et al. [365] suggest the application of the workpiece reaction-force to control the abrasive-suction process. Figure 10.3a shows a typical example. In this case, the abrasive hopper actually run out of abrasives. A clogged abrasive-particle stream generates the same effect. As this happens, the force-signal intensity suddenly rises to a higher level and becomes stable. This stabilized force corresponds to a force measured during the piercing of the material when no further penetration occurs.

10.2.4 Vacuum Sensor

Zeng and Munoz [366] use a vacuum sensor to control the abrasive-suction process. As Figure 10.3b shows, this concept is able to detect plugged suction hoses as well as plugged focus nozzles.

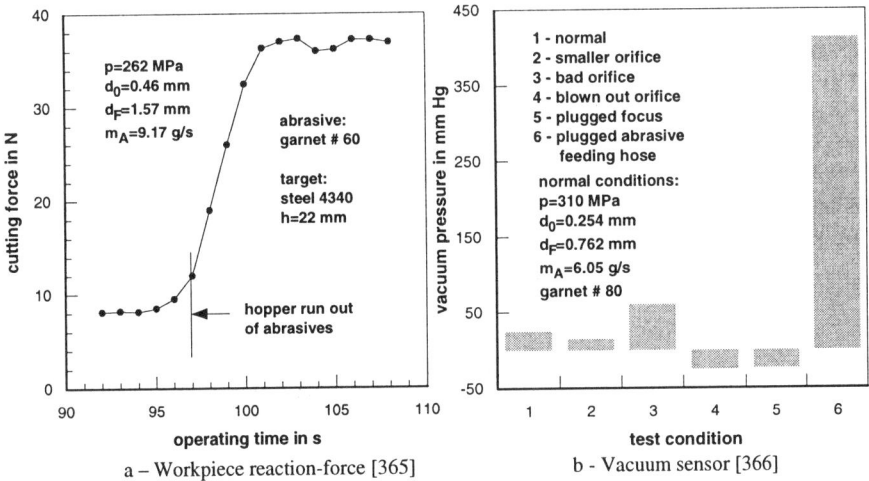

a – Workpiece reaction-force [365]
b - Vacuum sensor [366]

Figure 10.3 *Process-control methods for the abrasive suction*

10.2.5 Actual Abrasive-Mass Flow Rate

Hashish et al. [364] discuss several methods to control and supervise the actual abrasive-mass flow rate. An opacity sensor measures the optical transmittance through the abrasive-feed tube. The abrasive flow blocs the light transmission in proportion to the abrasive-mass flow rate. Another possible way is to measure the amount of electrical charge passing through the abrasive-feed tube per unit time. The electrical charge is carried by the abrasive particles, so the charge-transfer-rate is be proportional to the abrasive-mass flow rate. A simple method is the estimation of weight change of the abrasive hopper with cutting time. Brandt et al. [125] exploit this method to control the abrasive-mass flow rate of a suspension-abrasive water-jet system. A similar way is placing a weight sensor in the path of the abrasive stream [364]

10.3 Control of Water-Orifice Condition and Wear

10.3.1 Optical Jet Inspection

The basics of this approach is that any change in the fluid conditions in the orifice influences the jet structure. As the orifice gets worn or damaged, the jet coherence worses. Knaupp [367] reports about an attempt to optically monitor water-jet structures on-line. Using a video camera, this author records shadowgraphs of the water jets. As Figure 10.4 shows, differences in jet structure can clearly be monitored. As the orifice is worn, the jet width significantly increases. Additional information can be obtained by digitalization and PC-aided pixel analysis.

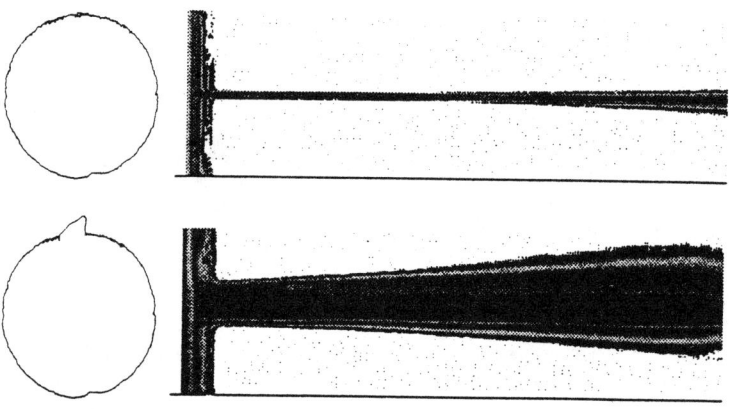

Figure 10.4 *Shadographs of water jets from different orifice conditions [367]*

10.3.2 Vacuum-Pressure Measurements

This approach comprises the monitoring of the vacuum level in the mixing chamber, located just below the orifice [368]. The water-jet coherence length is selected as a measure of the orifice condition. Preliminary sensor readings show good correlation between the coherence length and vacuum pressure in the mixing chamber (Figure 10.5).

10. 4 Control of Focus Condition and Wear

10.4.1 General Comments

The wear of the focus is a main feature of the abrasive water-jet machining. To supervise and control this wear process is therefore a major condition for a successful application of abrasive water jets.

Kovacevic [369] gives a compressed review on principles and methods for the detection of the focus-nozzle wear. The methods used to detect the focus wear categorizes as either direct or indirect. Qualitative methods consist of direct visual inspection of the tip of the nozzle and indirect observation of any deterioration in the quality of the cut surface or the undesirable changes in dimensions of the workpiece. Direct quantitative methods consist of assessing the focus wear by either measuring the inside focus diameter at its tip, or measuring the material loss of the focus by radiometric techniques. Both methods suffer from serious drawbacks. The former technique requires the interruption of the cutting process by turning the jet off and thus is unsuitable for on-line wear measurement. Radiometric technique on the other hand requires special preparation of the focus and also poses potential hazards due to radioactivity. Indirect methods are based on the measurements of some parameters, such as the change in the abrasive-water jet diameter at the focus exit or the normal force on the workpiece, noise, vibration, etc., that correlate to the focus wear.

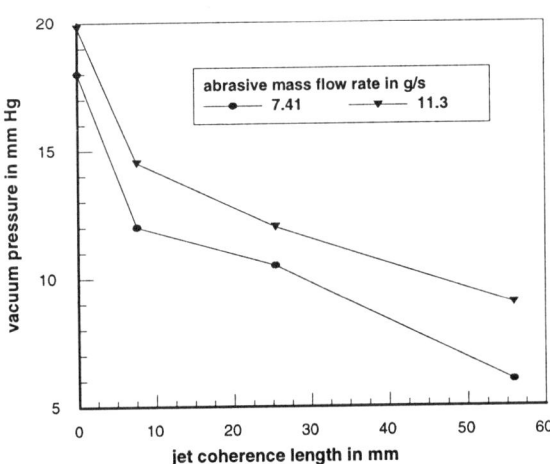

Figure 10.5 *Pressure-based sensor response for orifice control [368]*

Systems for sensing the nozzle wear divides into three groups with respect to the measured object [367]:

- Sensing systems that have the focus as the measured object.
- Sensing systems that have an abrasive water jet as the measured object.
- Sensing systems that have the workpiece as the measured object.

10.4.2 Direct Tracking

A direct tracking of the focus wear is possible with a sensing unit that is embedded at the tip of the focus. Two such sensing units are possible [369]. One sensing unit is based on a conductive loop designed to detect a predetermined threshold that represents the focus life. The sensor system consists of a wear-sensor probe and logic unit connected to a PC. Figure 10.6 shows a simplified circuit diagram of a wear-sensor probe. It consists of a ceramic substrate and the wear probe. The wear probe divides into four quadrants with a hole in the center of the diagram equal to the inside diameter of the new focus. Within each quadrant lies a number of conductive loops. The sensor is attached to the tip of the focus where it is subjected to the same wear modes as the focus itself. Each conductive loop is cut when the nozzle has worn to its position and since the location of the loop is known in advance, the focus wear status is determined. The last conductive loop in the wear probe determines the maximum allowed increase in the focus diameter. In the arrangement proposed by Kovacevic [370], a total of eight loops, the maximum allowed focus diameter is $d_F=2.0$ mm, which represents a 66 % increase with respect to the original one. It is evident that the number of conductive loops in the wear probe varies depending on the maximum allowed increase of the focus diameter.

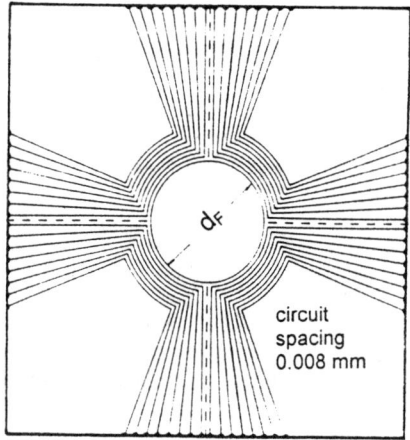

Figure 10.6 *Focus-wear sensor probe [370]*

10.4 Control of Focus Condition and Wear

In order to monitor the increase in the focus diameter, a digital logic-device is needed. The main function of the digital logic-device is to monitor the continuity status of each wear-sensor probe. Kovacevic [370] develops a software program for a logic device.

A problem for practical applications of this method is the sensor erosion due to reflected abrasive particles. Hamatani and Ramulu [184] observe this effect of multiple reflection of abrasive particles between workpiece and focus front.

10.4.3 Jet-Structure Monitoring

The abrasive water-jet diameter near the focus is used as a variable for monitoring the focus wear. In that case, the increase in the focus diameter directly relates to the increase in the abrasive water-jet diameter. Holland [120] makes a first attempt in this direction. A monitoring system, as developed by Kovacevic [371] and shown in Figure 10.7, is based on a video-capture system. This PC-based system consists of three main components: a CCD matrix-array camera, a frame-grabber, and a video-interface board. The matching between the object and the sensor is due to the camera.

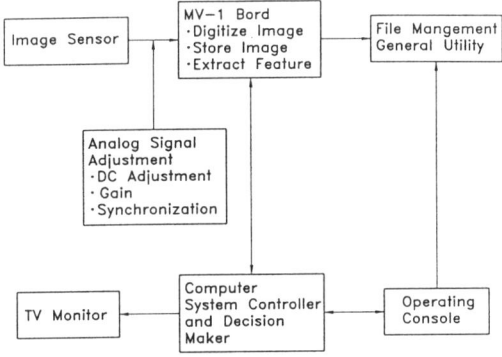

Figure 10.7 *Block-diagram of a video-capture system to monitor the focus diameter [371]*

Once this matching is achieved, the optical image is then converted into digital signals. The TV-monitor makes the visual display of the abrasive water jets.

Kovacevic [371] carries out a number of tests to verify the system performance. Table 10.1 summarizes some results. The actual diameter of the abrasive water jet is larger than the actual focus diameter. This difference increases as the actual focus diameter and pump pressure increase. The increase in the abrasive-mass flow rate has a negligible effect on this difference. The increased wear of the focus nozzle (this is larger diameter) makes the clearance larger between the abrasive water jet and focus.

Table 10.1 *Optically estimated abrasive water-jet diameters versus actual focus diameters [371]*

$\dot{m}_A$ [g/s]	d_F [mm]	Estimated jet diameter d_{jet} [mm]							
		Pump pressure p [MPa]							
		207			241			275	
7.5	1.15	1.18	1.16	1.19	1.22	1.20	1.23	1.25	1.28	1.24
	1.50	1.58	1.55	1.59	1.59	1.60	1.63	1.63	1.68	1.68
	1.90	1.98	2.01	2.02	2.08	2.05	2.04	2.08	2.14	2.07
	2.25	2.34	2.37	2.40	2.42	2.45	2.48	2.50	2.52	2.54
11.3	1.15	1.16	1.19	1.20	1.24	1.32	1.21	1.25	1.30	1.29
	1.50	1.60	1.59	1.57	1.64	1.59	1.61	1.65	1.70	1.69
	1.9	2.01	1.99	2.03	2.03	2.07	2.09	2.10	2.16	2.09
	2.25	2.41	2.35	2.38	2.49	2.45	2.48	2.51	2.53	2.56
15.2	1.15	1.19	1.20	1.21	1.26	1.25	1.25	1.25	1.30	1.30
	1.50	1.58	1.61	1.61	1.63	1.63	1.60	1.65	1.71	1.70
	1.9	2.02	2.04	1.99	2.05	2.10	2.06	2.17	2.13	2.08
	2.25	2.35	2.40	2.42	2.47	2.49	2.49	2.57	2.55	2.54

10.4.4 Air-Flow Measurements

Louis and Meier [76] suggest the measuring of the volume-flow rate of the sucked in air to control the condition of the abrasive water jet-focus. Their basic idea is that any change in the geometry of the focus influences its suction capability. Figure 3.19a shows the relation between the air-volume flow rate and focus diameter for different pump pressures. More air is sucked in the focus as the focus diameter increases. After constructing a calibration curve it is possible to detect changes in the focus diameter even in the range Δd_F=0.1 mm to 0.2 mm [83]. Nevertheless, the measure of the air-flow rate does not give any information about what causes the change in the air-flow rate. An increase in the focus diameter is only one possible reason; other reasons could be: a change in the alignment between the orifice and focus, jamming of abrasive particles, a leakage in the suction hose. Also, on-line measurements of the air-flow rate can only be performed for suction processes without abrasives because of the possible destruction of the flowmeter.

To overcome this limitation, Louis and Meier [76] recommend measuring the pressure loss in a certain length of the suction hose. Figure 10.8 shows some test results. The curves are very similar to those in Figure 3.15, but the pressure loss is measured by a bypass.

The almost linear relation between the focus diameter and vacuum pressure is analytically derived by Zeng and Munoz [366],

$$\frac{\Delta \dot{p}_V}{\Delta d_F} = C_0. \tag{10.1}$$

Therefore, the pressure loss in the abrasive-transport hose offers the possibility to measure the focus diameter on-line.

10.4 Control of Focus Condition and Wear   341

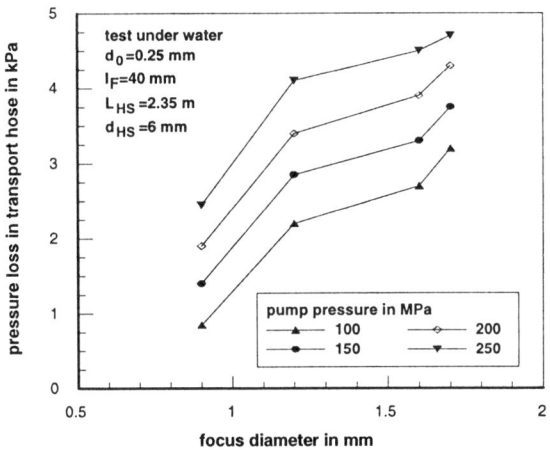

Figure 10.8 *Relation between the focus diameter and pressure loss in the abrasive-suction hose [76]*

10.4.5 Infrared Thermography

Kovacevic et al. [208] and Mohan et al. [246] study the temperature distribution in an abrasive water-jet focus to develop an alternative method for focus-wear monitoring through infrared sensing (section 5.9). The focus is subjected to a gradual wear till the inside diameter increased by 70%. The focus temperature is measured for three typical focus diameters.

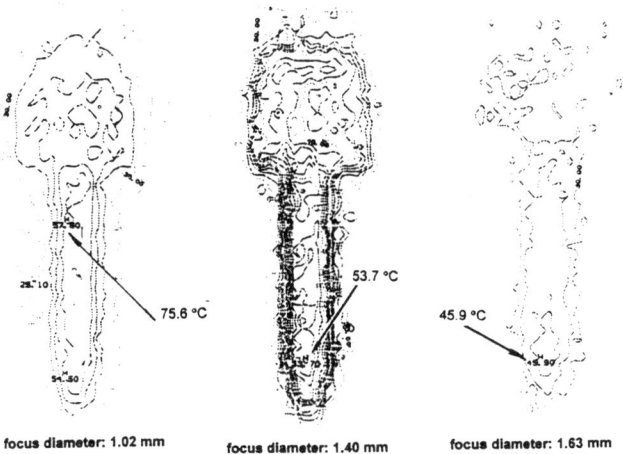

Figure 10.9 *Influence of the focus-diameter on the thermal energy distribution [208]*

Figure 10.9 shows typical temperature distributions in partially worn-out focuses. Figure 10.10a gives the relationship between the focus diameter and the measured focus peak-temperature. The relation very reasonably fits into a second-order polynomial. Figure 10.10b depicts the shift in the position of the peak temperature as the focus wear progresses. In the figure, L/l_F, is plotted in the y-axis; whereas, L is the distance between the location of the peak temperature and the top of the focus.

From these figures, notice two essential features: first, as the focus diameter increases, the peak temperature drops down and, second, its position shifts towards the focus exit as the focus diameter increases. Both observations indicate the progress of radial and axial wear. In the axial direction, for a new (small sized) focus, the wear due to impacting abrasive particles concentrates at the entry zone (top) of the focus (section 3.6), Figure 10.9a illustrates this process by the presence of a hot spot in this zone. However, with continuous use (an increase in the focus diameter), the inner walls of the focus at the entry zone gets a fluid-mechanical stable shape (section 3.6) and the wear process gets shifted towards the focus exit. This process is well documented by the corresponding movement of the peak temperature (Figure 10.9c). In the radial direction as the focus diameter increases, more space is available for the abrasive water-jet stream to flow. This process leads to a reduction in the frictional force between the jet and the focus walls. As a result, the average focus temperature reduces. The quadratic relationship between the focus diameter and the peak temperature as well as the L/l_F-ratio indicates that infrared-thermography is a viable alternative technique for the focus-wear monitoring.

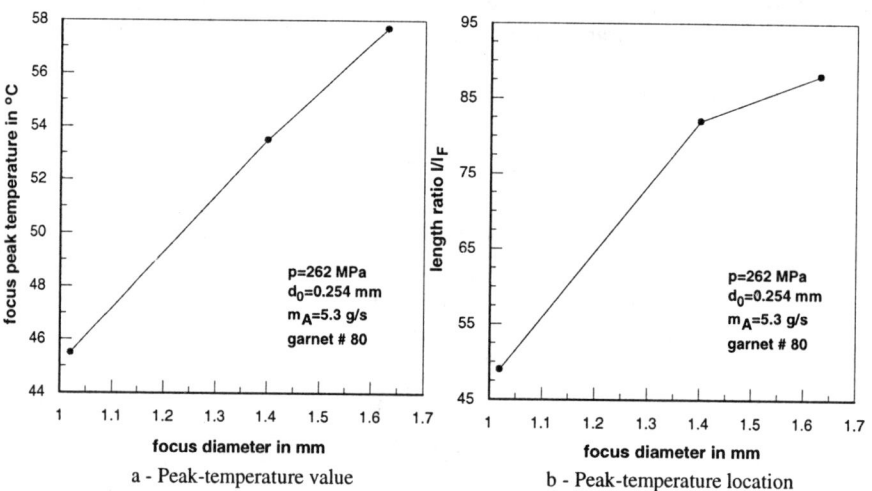

Figure 10.10 *Relation between the focus peak-temperature and focus diameter[208]*

10.4.6 Acoustic Sensing

Acoustic sensing methods are used to detect the focus wear. These methods base on the fact that a change in the focus diameter affects the sound generated by the abrasive water jet. Barker et al. [372] show by for plain water jets that the sound significantly increases as the surface of the jet increases. Depending on the geometry of the nozzle, the sound generated is tonal (generated by periodic vortex shedding) or random noise (generated by turbulence). In order to investigate the feasibility of using the acoustic sensing technique, Kovacevic and Evizi [373] and Kovacevic et al. [374, 375] conduct systematic experiments.

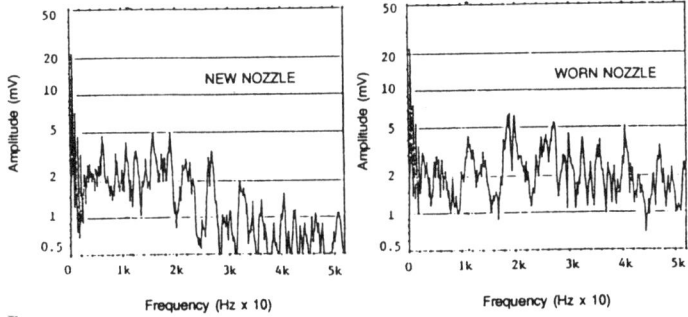

Figure 10.11 *Sound spectra for different focus conditions [373]*

Figure 10.11 shows frequency-domain spectrums of sound produced by an abrasive water jet when it flows through a new and a worn focus. As the focus wear increases, the level of the noise increases, especially at frequencies over 20 kHz. Kovacevic et al. [374, 375] show in a more detailed investigation by advanced signal processing methods that there are two dominant frequencies in the acoustic signature (Figure 10.12a). With an increase in the focus wear, the two dominant frequencies shift to higher frequency values while the amplitudes of the two dominant frequencies decrease (Figure 10.12b). Also, the amplitude of the acoustic signal in the high frequency range (above 20 kHz) is a good indicator for monitoring the focus wear. Kovacevic et al. [375] propose a method for the identification of the focus wear based on an on-line monitoring of the acoustic signal.

In an advanced study of this problem, Mohan et al. [376] develop a system for monitoring and compensating the focus wear that is based on a neural network. A closed-loop controls the position of the focus by compensating the wear. The acoustic signal is measured, and the averaged frequency domain signal is inputted to the trained artificial neural network. The neural network instantaneously determines the corresponding focus diameter, and this information is used to control the position of the cutting head using a piezo-electric actuator. Once the network is trained, it is used to determine the focus diameter while the machining operation takes place.

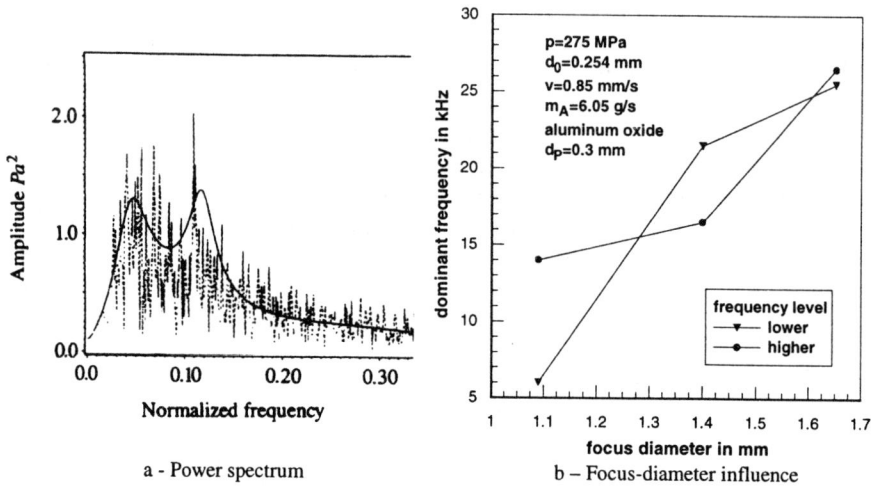

Figure 10.12 *Acoustic-signal behavior during abrasive water-jet flow through a focus [374]*

10.4.7 Workpiece Reaction-Force Measurement

A number of studies [114, 377, 378] shows that the normal force that is generated by an abrasive water jet on a workpiece increases as the focus diameter increases. Figure 10.13 presents examples. Based on a multiple regression analysis, Kovacevic and Beardsley [114] derive

$$F_W = \frac{1}{e^{19.278}} \cdot p^{2.114} \cdot \dot{m}_A^{0.243} \cdot d_F^{0.508}. \tag{10.2}$$

Kovacevic and Chen [376] and Kovacevic [378] develop this concept further by detecting and analyzing the dynamic component of the reaction force signals and relate them to the focus wear. This approach bases on the time-series analysis technique in order to characterize the signal dynamics of the workpiece reaction-force with an auto-regressive model. This model tracks only the dynamic properties but not the mean level of the signal. The current value of the measured workpiece normal force can, for example, be expressed using a n-th order auto-regressive model that is simply a linear combination of n previous values [379],

$$F_W(t) = a(t) \cdot F_W(t-i) + n(t). \tag{10.3}$$

In order to establish the relative importance of the model parameters, a discrimination-index parameter is used that is capable of separating different cutting conditions. Figure 10.14 shows 2-D parameter planes that provide a relationship between the two most significant parameters under selected focus-wear conditions for two different cutting conditions. There exists a distinct difference between the focus-wear conditions in both operations. Under practice conditions, focus wear is monitored by the magnitudes of the parameters of the auto-regressive model.

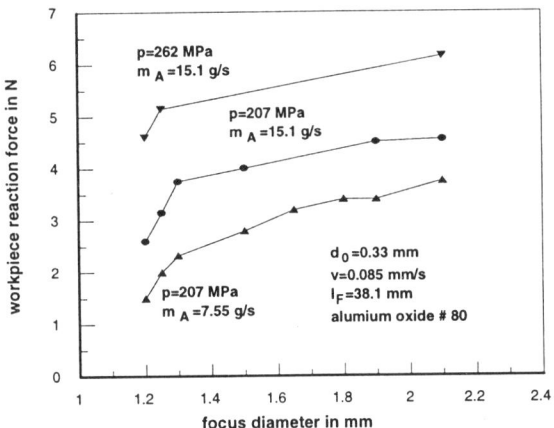

Figure 10.13 *Relation between the focus diameter and average workpiece reaction-force [376]*

Kovacevic and Zhang [380] propose an additional approach for monitoring the status of the focus wear. The proposal bases on an on-line fuzzy recognition-technique that is applied to workpiece reaction-force measurements. In the approach, several states of focus wear (initial, small, high, severe) are defined that correspond to grades of the measured normal force. The values between two neighboring measures are produced through simulation based on a fuzzy algorithm.

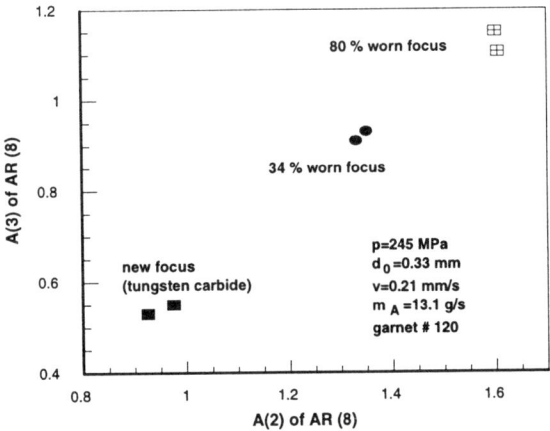

Figure 10.14 *2-D parameter planes for different focus-wear conditions [379]*

10.4.8 Off-Line Focus-Diameter Measurement

Hashish [362] et al. briefly address a system for estimating the focus diameter off-line. When the measurement needs to be done, the CNC control unit brings the focus to an enclosed measurement station where the end of the focus is imaged onto either a one- or two-dimensional solid-state-camera. By an image processing technique, the diameter and shape of the focus is determined. The sensor suggested for this approach consists of an imaging lens, a one-dimensional CCD-camera, and control and data processing-software.

10.5 Measurement and Control of Abrasive Water-Jet Velocity

10.5.1 Inductive Methods

Swanson et al. [381] use an inductive measurement method that bases on magnetic abrasive particles. As shown in Figure 10.15a, the abrasive water jet is encircled by two small coils of wire that are connected to sensing-electronics and tinning circuitry. Magnetic particles of the same general size as the abrasive particles are mixed with the abrasives and accelerated in the same water jet. When one of the magnetic particles passes through the coil, it induces a small electric signal that is repeated at short time later as the particle passes through the second coil. The transit time between the fixed coil spacing distance is accurately measured. This value is used to calculate the individual abrasive-particle velocity,

$$v_P = \frac{x_{coil}}{\Delta t}. \tag{10.4}$$

In the experiments, a certain amount of steel shot is mixed into the garnet abrasives. Figure 10.15b shows that the percentage of the steel shot influences the measured average particle velocities. Also, the deviations of the measured velocities are very high, sometimes about 100 %. Later [78], ferrite particles with a density closer to the abrasive material are used as tracers. As Figure 10.15c indicates, the measured particle velocity depends on the density of the used tracer material. In contrast, the density of the material does not significantly influence the final particle velocity.

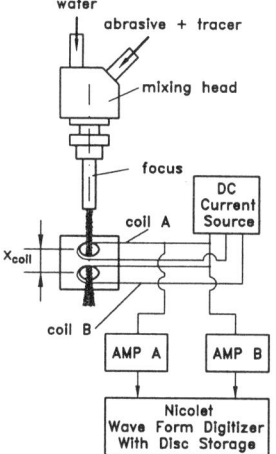
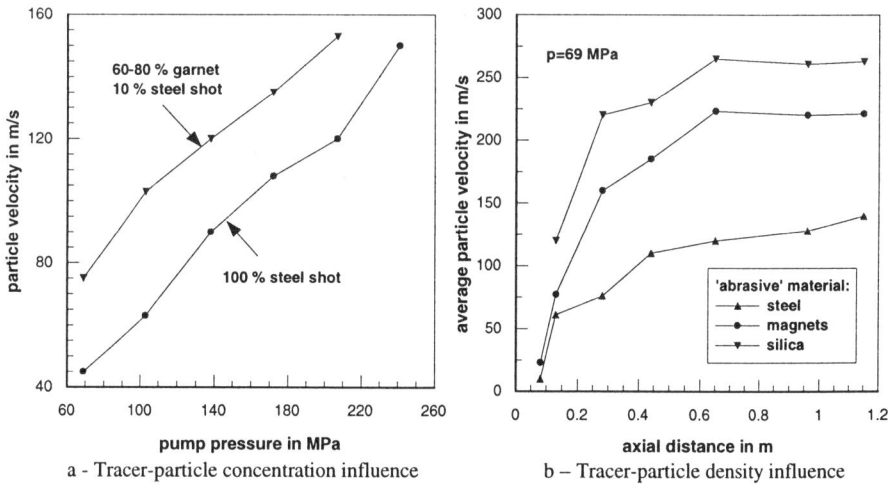

a - General experimental setup

a - Tracer-particle concentration influence

b – Tracer-particle density influence

Figure 10.15 *Measurement of abrasive-particle velocities by the inductive method [78, 381]*

10.5.2 Impact-Crater Counting

Isobe et al. [77] develop a method to indirectly measure the abrasive-particle velocity by counting impact craters. Supposing an inspection plane perpendicular to the central axis of the abrasive water jet, as shown in Figure 10.16a, the number of abrasive particles that passes this plane in a unit time is

$$N_{P1} = \int_{A_{jet}} E(r) \cdot v_P(r) dA . \qquad (10.5)$$

Similarly, assuming a second inspection plane parallel to the central axis, the horizontal component of the abrasive-particle velocity relative to the surface is constant. Therefore,

$$N_{P2} = \int_{A_{jet}} E(r) dA \cdot L . \qquad (10.6)$$

From Eqs. (10.5) and (10.6), the abrasive-particle velocity is

$$v_P = \frac{N_{P1}}{N_{P2}} . \qquad (10.7)$$

In their experiment, Isobe et al. [77] direct an abrasive water jet at aluminum plates with two inspection faces as shown in Figure 10.16a. Based on the impact craters caused by the abrasive particles, the distributions $E(r)$ and $v_P(r)$ are obtained.

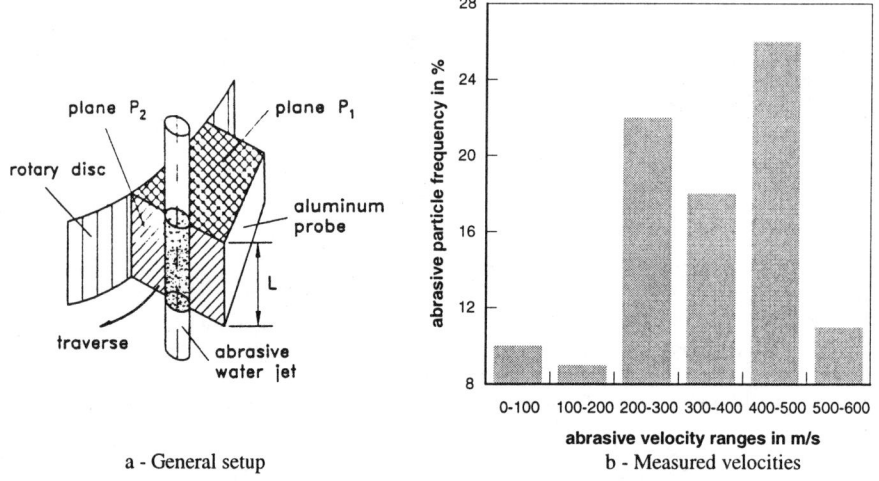

Figure 10.16 *Abrasive-particle impact-crater counting experiments [77]*

Figure 10.16b shows calculated velocities of a steel shot abrasive (d_P=550 μm-650 μm) distributed over the jet radius. A notable result from these measurements is that the ratio between the measured average particle velocities and the calculated abrasive particles, Eq. (3.2b), is about η_T=0.57, which is low compared to other measurement methods. This low value is because of the unusually high abrasive-mass flow rate used in this study.

10.5.3 Laser-Based Methods

10.5.3.1 Laser-2-Focus-Velocimeter

Himmelreich [47] and Himmelreich and Rieß [382] describe the structure and function of a Laser-2-Focus Velocimeter in detail. The L-2-Focus measure-volume is made up by the two focal spots of the upstream (start) beam and the downstream (stop) beam that are emitted from the optical head. As a particle passes through both focal spots, it emits two scattered light pulses with a characteristic time shift (Figure 10.17).

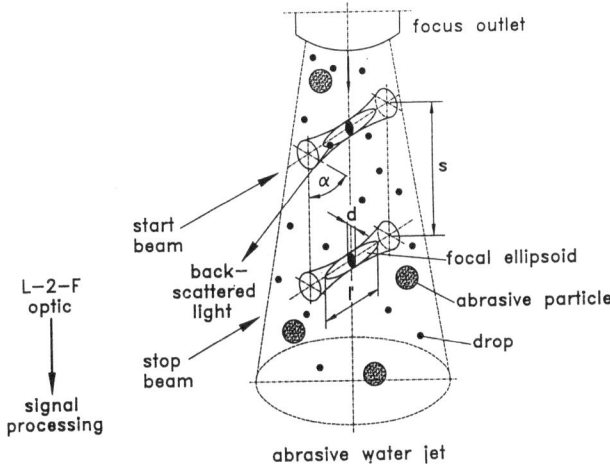

Figure 10.17 *Particle-velocity measurement by Laser-2-Focus-Velocimeter [47]*

This time shift is measured and is called 'time of flight', t_F. At each measurement point, a sample of t_F-measurements is recorded and statistically evaluated. From this processed time-of-flight distribution, the mean value is estimated. The mean particle velocity is, then,

$$\bar{v}_P = \frac{x_{Beam}}{\bar{t}_F}. \tag{10.8}$$

In the equation, x_{Beam} is the beam distance, given by the laser equipment.

Figure 10.18 shows characteristic time-of-flight distributions for a water jet and an abrasive water jet, respectively. For the water jet, the maximum in the function appears at $t_F=0.68$ μs. This time corresponds to a mean velocity of $v_0=409$ m/s. The clear narrow peak suggests a comparatively low jet turbulence (ca. 3.8 %). As abrasive particles are added to the water jet, a dual peak in the time-of-flight distribution function is generated. This clearly separated dual peak is caused by significant phase-velocity differences after an inefficient mixing process. The left peak that attributes to the water phase, keeps its maximum at $t_F=0.86$ µs. The right peak that attributes to the abrasive particles, exhibits a maximum at $t_F=1.06$ µs. The corresponding velocity is $v_P=247$ m/s. Therefore, the average velocity of the jet severely decreases to $v_{jet}=279$ m/s.

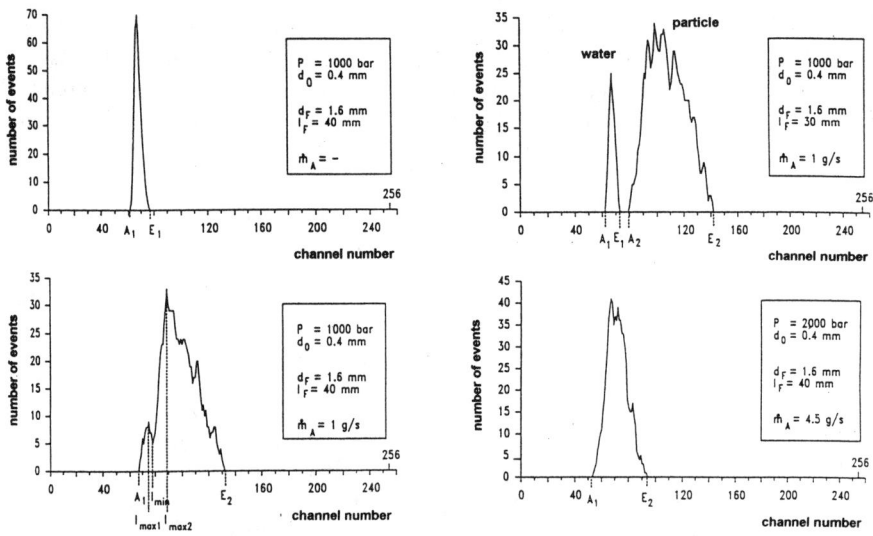

Figure 10.18 *Time-of-flight images from L-2-Focus-measurements [47]*

The main advantages of laser-2-focus-velocimeters are the capability of processing high data-rates, particles of different size, and velocity ranges from less than 1.0 m/s to 1,000 m/s without a change in the optical arrangement. Disadvantages are the increase in the noise level as the turbulence increases and the limitation to turbulent degrees of about 30 %.

10.5.3.2 Laser-Transit-Velocimeter

Chen and Geskin [46, 84] use a Laser-Transit-Velocimeter with a He-Ne-laser as a light source for measuring the water and abrasive velocities in an abrasive water jet. The velocimeter has a lens system that splits the laser beam into two equal intensity beams and focuses them in a small region at two closely-spaced focal points. Thus, as shown in Figure 10.19, a particle passing through the focal point of either of these split beams generates a scattering light that is detected and converted to a voltage signal. The particle velocity is computed by dividing the known distance between the two focal points by the time difference between the successive signals. The distance between the two foci for measurements in abrasive water jets is about

0.45 mm [46]. With the laser-transit-anemometer technique, the velocity of plain water jets is also measured since many sub-micron particles are normally contained in the real fluid. Large size differences between multi-micron abrasive particles and sub-micron particles contained in the same water can be easily distinguished in laser-transit-velocimeters by the difference in the intensity of the scattered light.

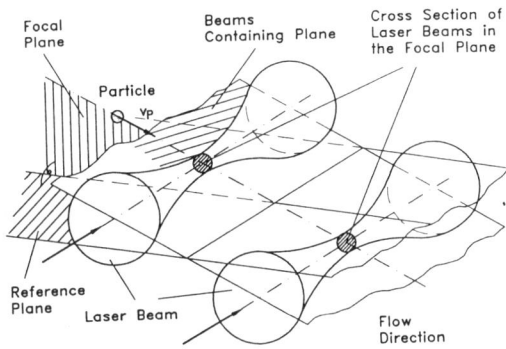

Figure 10.19 *Particle-velocity measurement by Laser-Transit-Velocimeter [46]*

10.5.3.3 Laser-Doppler-Velocimeter

A Laser-Doppler-Velocimeter, as used by Neusen et al. [48, 58] consists basically of a laser, transmitting optics, receiving optics, a photo detector, and some side equipment, such as signal processor and data acquisition device. In their experiments, Neusen et al. [48] use a forward scatter-light collection. A polarized laser beam enters a beam splitter that separates the beam into two parallel beams by use of a partial mirror. After the beam splitter, the two parallel beams pass through a focusing lens that redirects the beams, and causes them to converge at the focal point of the lens. In the focal point, there exists an intersection volume referred to as the probe volume (Figure 10.20a). The particle velocity is predicted by measuring the frequency of the scattered light. The relation between the interference frequency and the particle velocity is [48]

$$v_P = \frac{f_D \cdot \lambda_L}{2 \cdot \sin \varphi_D}. \tag{10.9}$$

In the equation, $\varphi_D = 2.415°$, and $\lambda_L = 632.8 \cdot 10^{-9}$ m. The scattered light passes through the collecting optics on to a photo multiplier where the optical signal is transformed into an electrical signal that is processed, including statistical analysis, by standard software. Figure 10.20b shows a typical histogram for abrasive water-jet velocities estimated by a laser-doppler-velocimeter. These measurements correspond to an average abrasive velocity of $v_P = 250$ m/s with a standard deviation of $\sigma_{VP} = 41.8$ m/s for 1,018 points collected. Laser-doppler-velocimeters are also used to measure velocities of water jet and water drop.

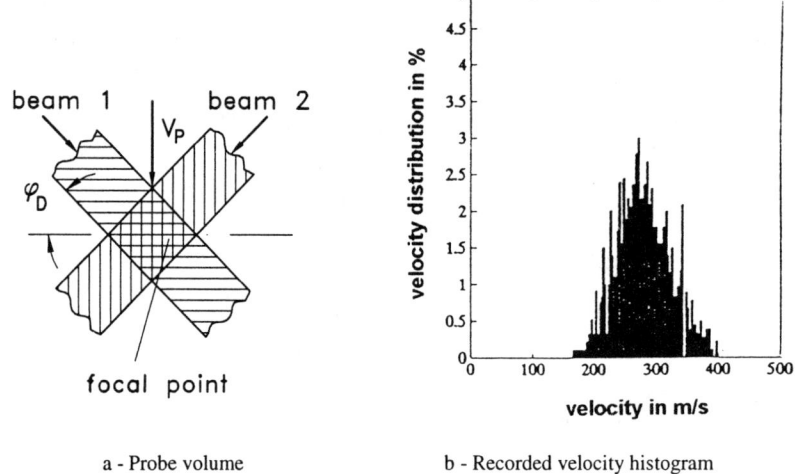

a - Probe volume   b - Recorded velocity histogram

Figure 10.20 *Particle-velocity measurement by Laser-Doppler-Velocimeter[48]*

10.5.3.4 Laser-Light-Section-Procedure

Blickwedel [47] suggests the Laser-Light-Section-Procedure to investigate velocity and structure of abrasive water jets. Figure 10.21 illustrates this technique. Theoretically, the velocity of a particle entrained in a jet is estimated by this technique,

$$v_P = \frac{l_0 \cdot \left(\frac{\Delta x}{\Delta x_{ref}}\right) - d_P}{t_B}. \tag{10.10}$$

No attempt is known so far to use Eq. (10.12) for the estimation of velocities for water jets and abrasive water jets.

10.5.4 Other Optical Methods

10.5.4.1 Schlieren-Photography

Another method that briefly is addressed, is the photographic characterization of jet velocities. Cheng and Geskin [46] use Schlieren-Photography to estimate the velocity of an abrasive water jet. If the shock angle is measured from the photograph, the velocity of the flow is

$$v_{jet} = \frac{c_{Air}}{\sin \theta_S}. \tag{10.11}$$

In this equation, $c_{Air}=340$ m/s is the speed of sound in air. From Eq. (10.11), Chen and Geskin [46] estimate velocities between $v_{jet}=430$ m/s to 660 m/s. Comparative values estimated on the same jets by laser-transient-anemometer are in the range of $v_P=400$ m/s to 500 m/s. Because Schlieren-photography does not distinguish between the velocity of the liquid phase and the entrained abrasive particles, the higher velocity values measured by this method are probably those of the water phase in the abrasive water jet.

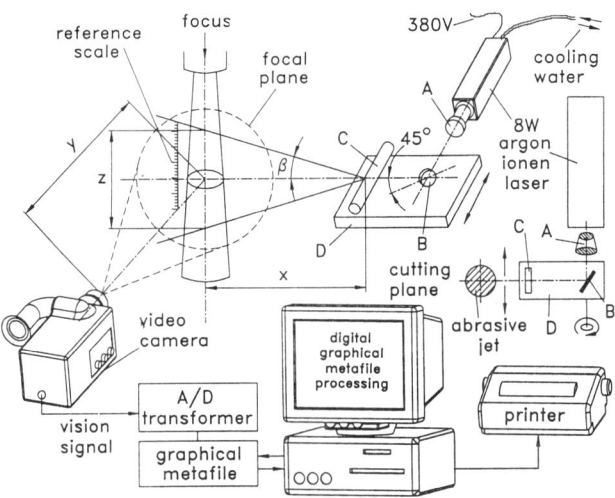

Figure 10.21 *Particle-velocity measurement by Light-Section-Procedure [47]*

10.5.4.2 High-Speed-Photography

Shimizu [142] applies High-Speed-Photography for estimating the velocity of suspension-abrasive water jets. This technique is used to record irregularities of the air-liquid interface of the jet and liquid lumps. The velocities of the irregularities and the liquid lumps are calculated from the high-speed photographs (Figure 10.22). Typical time intervals between the individual frames are about 36 µs. The recorded velocities have values of about 75 % up to 95 % of the theoretical maximum velocity.

354   10. Control and Supervision of Abrasive Water-Jet Machining Processes

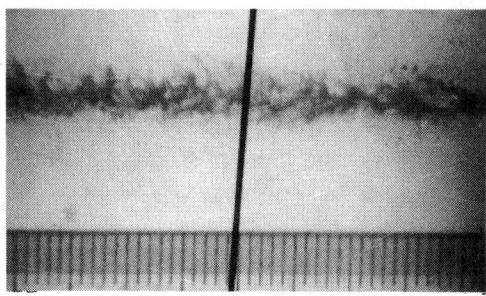

Figure 10.22 *High-speed-photograph of a suspension-abrasive water jet [142]*

10.5.5 Jet Impact-Force Measurements

As already pointed out in chapter 3, the velocities of water jets and abrasive water jets can be estimated by jet impact-force measurements. The underlying relation is the jet momentum, Eq. (3.3), that gives

$$v_{jet} = \sqrt{\frac{4 \cdot F_{jet}}{\pi \cdot \rho_{jet} \cdot d_{jet}^2}} \;. \tag{10.12}$$

Practically, jet forces are measured by piezoelectric transducers (Figure 10.23a). During the measurement, usually the vertical (z-) component of the jet is considered. Figure 10.23b plots typical force signals for an impacting water jet and an abrasive water jet.

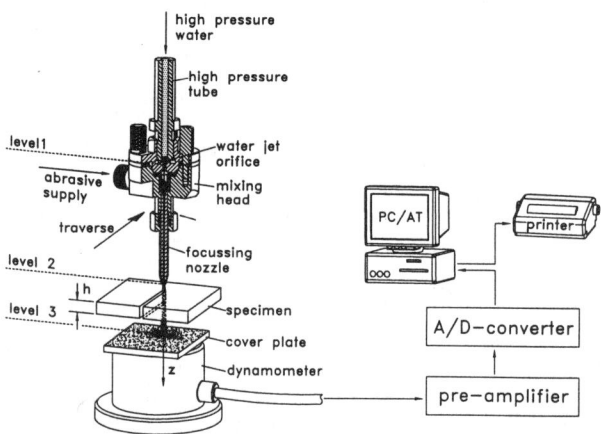

Figure 10.23a *Jet impact-force measurement by a piezoelectric force transducer [49]*

10.5 Measurement and Control of Abrasive Water-Jet Velocity 355

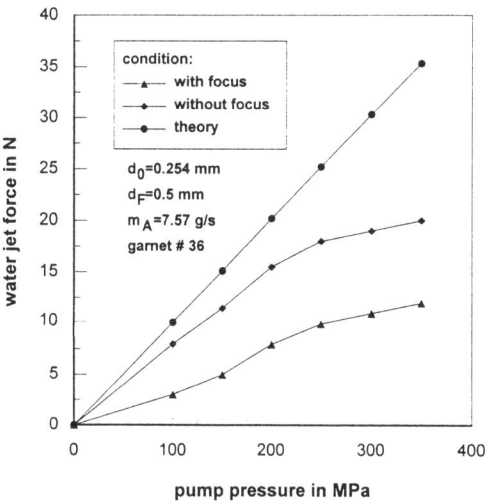

Figure 10.23b *Force signal of a plain water jet*

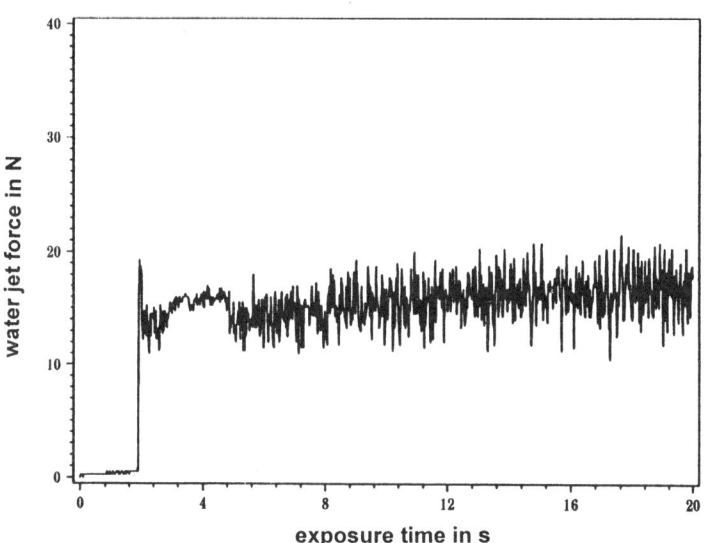

Figure 10.23c *Comparison between theoretical water jet forces and measured water jet forces [49]*

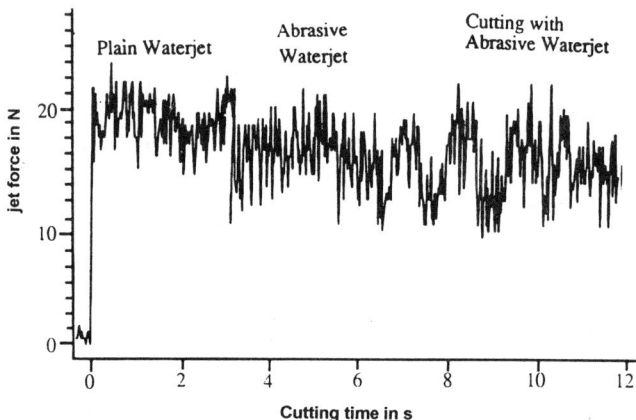

Figure 10.23d *Signals from jet impact-force measurements [49]*

Li et al. [383] give a general review on problems in estimating the impact forces of abrasive water jets. The accuracy of velocities calculated by this method is strongly influenced by the determination of the impact area. For plain water jets, Chen and Geskin [84] recommend a value equal to the area of a hole pierced in 60 s by the water jet in a 0.55 mm thick stainless-steel plate.

10.6 Measurement and Control of Abrasive Water-Jet Structure

10.6.1 Scanning-X-Ray Densitometry

Neusen et al. [145] use a Scanning-X-Ray-Densitometer to investigate the mass distribution in an injection-abrasive water jet. Figure 10.24 shows the basic components of an x-ray-densitometer-system.

The x-ray tube is a copper target x-ray-diffraction tube with a tungsten-filament cathode that is surrounded by a focusing cub. The anode consists of a copper electrode or target. Figure 10.24 shows the x-ray beam-collimator. In order to obtain a high resolution density map, a very thin x-ray beam is desireable to allow many data points to be taken across the diameter of the abrasive water jet. The x-ray scanner emits a very small beam of x-ray-energy that is passed through the jet. This beam loses intensity as it passes the mass that comprises the abrasive water jet. By measuring the decrease in the beam intensity, the amount of mass through which the beam passed is determined. The x-ray intensity is measured after it passes through the jet with the x-ray detector that sends an electrical signal to a data measuring system. This system displays a digital indication of the x-ray density. Neusen et al.

[145] describe the general acquisition procedure and the image reconstruction algorithm in detail. Section 4.2 discusses experimental results.

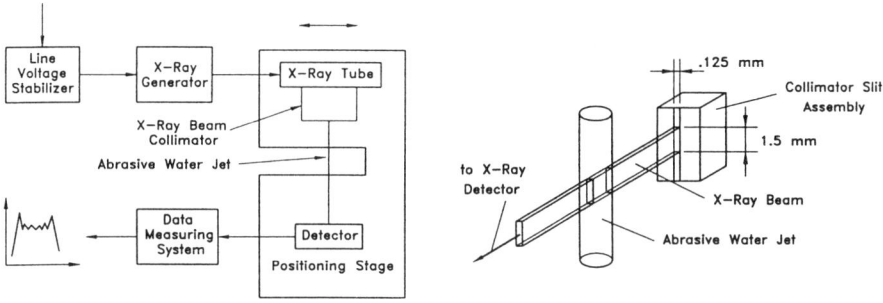

Figure 10.24 *Scanning-X-Ray-Densitometer for particle-distribution measurements [145]*

10.6.2 Flow-Separation Technique

Geskin et al. [144] and Simpson [90] use the flow-separation-technique as schematically illustrated in Figure 10.24.

Geskin's et al. [144] experiments involve the jet impingement on a plane surface holding a circular diamond washer with a hole that contains an orifice of a given cross section area. The experiments also involve the subsequent evaluation of the weight of the abrasive particles accumulated beneath the washer. By changing the diameter of the orifice, the relative mass-flow rate of the abrasives through a selected area of the jet-cross section is evaluated. The values of this ratio for several sub-regions determine the abrasive-particle distribution within an abrasive water jet. The mixtures, with and without a washer, are stored in two operate breakers, and their volumes and weights are recorded. Section 4.3 presents results of these experiments.

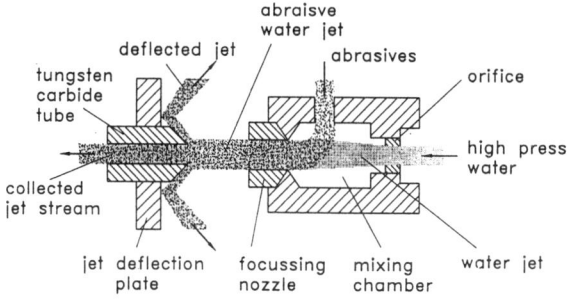

Figure 10.25 *Flow-separation-method for abrasive-particle distribution measurements [90]*

For the same purpose, Simpson [90] use a series of tungsten-carbide tubes. An external 60°-cone is ground from the internal diameter on one end of the tube length so that it is mounted axially concentric with the jet stream and deflects that portion of the jet stream that is greater than the tube internal-diameter. The portion of the jet stream that passes through the tungsten-carbide tube is collected in a horizontal tube, then dried and weighed.

10.7 Control of Material-Removal Processes

10.7.1 Acoustic-Emission Technique

10.7.1.1 Material-Removal Visualization

Momber et al. [190, 191] use acoustic-emission signals for visualization and control of material removal in brittle, multiphase materials. Figure 5.11 shows typical time-domain signals obtained from two different materials cut under identical conditions. The signal is predominantly of a continuous type for the material # 1; whereas, in the case of material # 2, the signal is primarily a burst-emission type. Figure 5.11 shows the corresponding cut surfaces. The wall structure for the material # 1 clearly demonstrates that the material removal is primarily due to inter-granular fracture mechanism that is characterized by a high degree of undamaged inclusion grains. The failure path is along the interface between the inclusion and matrix. The matrix is removed separately due to the action of the high-velocity water flow. This failure regime generates the comparatively smooth acoustic-emission signal in Figure 5.11. In contrast, the wall structure of the material # 2 indicates a failure mechanism that is characterized by a trans-granular fracture. During cutting of this material, the fracture preferably runs through the inclusion grains. Although this mechanism shorts the path of the cracks because there is no crack deflection of branching, the inclusion grains absorb a high amount of energy during the fracture. Therefore, the sudden failure of an inclusion grain is detected by the acoustic-emission technique as a burst emission with the corresponding sudden rise in the signal amplitude. This behavior is well illuminated in Figure 5.11.

Moreover, Momber et al. [384] find that acoustic emission is capable of detecting a mixed-mode material-removal mechanism. As Figure 10.26 shows, a transition from a steady-state removal mode to an unsteady removal mode is detected by changes in the structure of the frequency-domain acoustic-emission signals. Therefore, the acoustic-emission technique is sensitive to changes in the material-removal regime during abrasive water-jet cutting and offers some possibilities of on-line control of the material removal at least in composite and multiphase materials.

10.7 Control of Material-Removal Processes 359

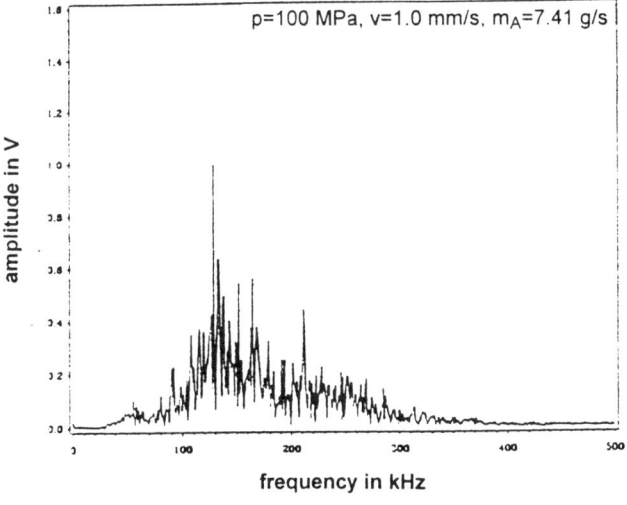

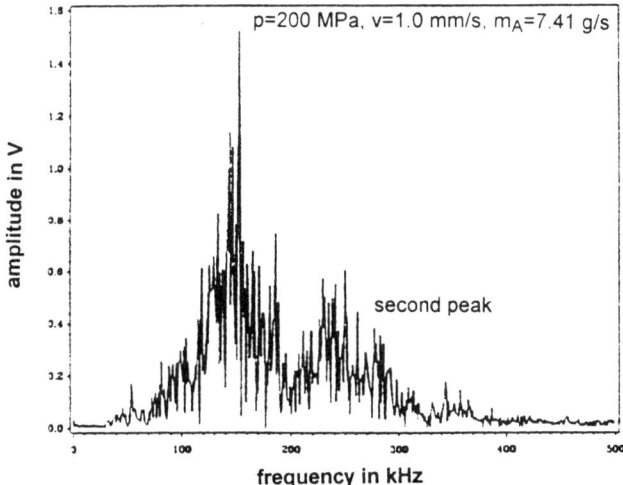

Figure 10.26 *Frequency-domain acoustic-emission signals for cutting bauxite ceramics [384]*

10.7.1.2 Cutting-Process Visualization and Cutting-Through Control

Figure 10.27 shows typical printouts of event-counts and RMS-voltage for kerfing an aluminum specimen with an abrasive water jet. Since the signals are plotted over the cutting time, the three temporal stages of the cutting process are clearly visualized from these graphs. This result agrees very well with the results of high-speed camera observations in cutting transparent materials (section 5.4).

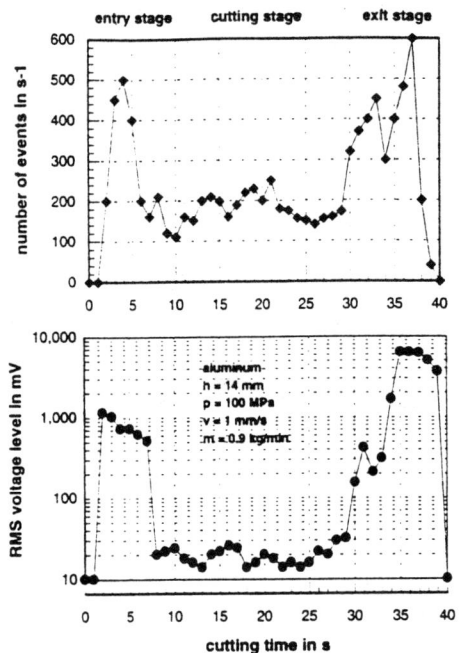

Figure 10.27 *Acoustic-emission signals acquired during abrasive water-jet cutting [209]*

Figure 10.28 shows typical FFT-plots for acoustic-emission raw signals for cutting-through and kerfing operations. The respective amplitudes for cutting-through condition are much higher than those for kerfing,. This difference indicates that the energy of the emission waves for the cutting-through-condition is significantly higher in this case. From the FFT-graph, for the cutting-through-condition, multiple frequency peaks are observed within the range of 100 kHz to 200 kHz. The primary frequency concentrates around 120 kHz and the secondary around 180 kHz. Whereas, in the case of kerfing, only one peak appears in the FFT-graph. Mohan et al. [209] and Momber et al. [190] assume that in case of kerfing, the presence of the rebounded abrasive-water mixture in the narrow kerf region has a damping effect on the acoustic-emission signal. This idea is supported by the significant changes in the acoustic-emission behavior in the entry stage and the exit stage of the cutting process compared to the cutting stage (Figure 10.27). Thus, the acoustic-emission signals reach high levels if the cutting situation allows an unrestrained outflow of the abrasive-water suspension (through-cutting, entry stage, or exit stage).

Figure 10.29 shows a plot of the PSD-peak versus depth of cut. The initial region that is characterized by a smaller slope represents the kerfing-condition, and the region with much higher gradient reflects a cutting-through-condition. If the initial region of smaller gradient is extended until the full thickness (shown by dotted line in Figure 10.29), the PSD-peak gradually increases as the depth of cut increases, and at the instant of full penetration, the peak jumps to a relatively much higher level.

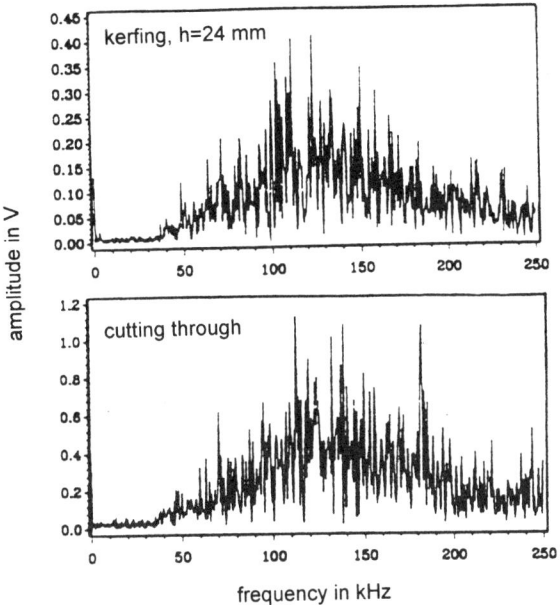

Figure 10.28 *FFT-graphs of acoustic-emission signals for different cutting conditions [209]*

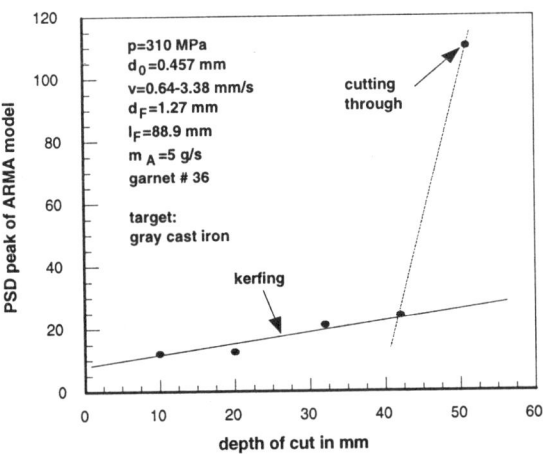

Figure 10.29 *Relation between the depth of cut and PSD-peak of acoustic-emission signals [209]*

10.7.1.3 Cutting-Efficiency Control

Mohan et al. [209] link acoustic-emission signals with the energy-dissipation parameter χ (section 5.5). For $0<\chi<1$, the abrasive water jet is able to cut through the sample, but the cutting conditions are not optimum from the energy point of view. As $\chi \to 1$, the energy conditions for through-cutting become optimum. For $\chi=1$, the maximum depth of cut is reached, and the abrasive water jet is only able to kerf the material. From Figure 10.29,

$$h = b_1 \cdot P_{PSD} + b_2. \tag{10.13}$$

Substitute h in Eq. (5.24) by Eq. (10.13), and set the resulting equation equal to zero. This yields [209]

$$\chi(PSD) = \frac{-a_2}{2 \cdot a_1} + \frac{1}{\sqrt{a_1}} \cdot \sqrt{\frac{a_2^2}{4 \cdot a_1} - a_3 + b_1 \cdot PSD + b_2} \tag{10.14}$$

(as χ should always be positive). As a particular value of the PSD is fixed for a certain demand, the condition 'optimum cutting-through' is monitored by acoustic-emission measurements.

10.7.2 Infrared-Thermography

As an alternative visualization method for the material removal in opaque materials, Kovacevic et al. [208] and Mohan et al. [246] use infrared-thermography as already addressed in sections 5.8.2 and 10.4.5. Figure 10.30 plots the temperature distribution on the workpiece surfaces for three different instances of the jet position, namely the entry stage (when the abrasive water jet starts cutting), the cutting stage (when the abrasive water jet has traversed approximately half the length of cut), and the exit stage (when the jet leaves the workpiece).

The isotherms follow the jet flow pattern in the workpiece. Thus, information regarding the behavior of the abrasive water jet during the cutting process are retrieved through the pattern of the isotherms. The inclination of the isotherms corresponds to the striation marks observed on the surfaces. The isotherms of the titanium specimen have a larger inclination than those of the aluminum specimen. This result is due to the difference in the striation angle of these two specimens. From Figure 10.30, as the cut progresses, the average workpiece temperature gradually increases. Ohadi and Cheng [245] also obtain this result by using thermocouples (section 5.8.1). This steady increase in the peak temperatures of the specimen from the entry stage to the exit stage is because of the heat conduction from the cut portion to the uncut portion.

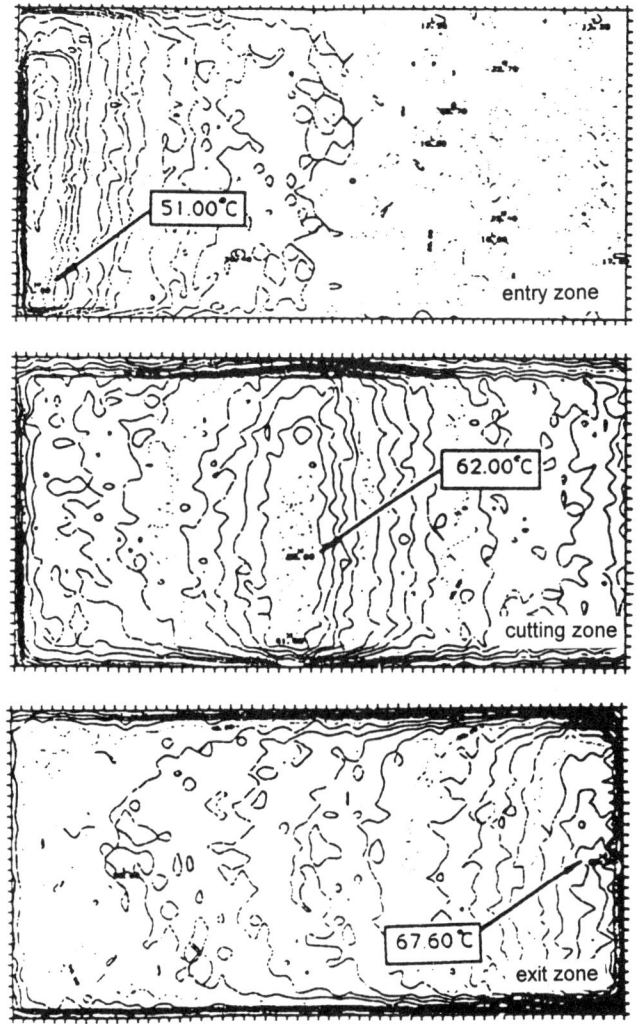

Figure 10.30 *Visualization of the abrasive water-jet cutting by infrared thermography [208]*

One of the peak temperatures observed in the titanium sample is at the middle of the smooth-cutting zone, and the other is at the middle of the rough-cutting zone. These results indicate the existence of a relatively colder region in the transition zone. The effect of the uncut triangle at the exit edge of the specimen is depicted by a hot spot in Figure 10.30. At the exit stage of the cutting process, the trailing part of the jet shifts towards the direction of cut. Thus, the top of the uncut triangle is hit by the abrasive water jet from both sides, which causes the hot spot. The temperature gradient from the entry stage to the exit stage is higher in case of aluminum compared to that of titanium, which is due to higher thermal diffusivity of aluminum. Thus, infrared-thermography is a useful tool to conduct visualization studies on opaque materials.

10.8 Control of Depth of Penetration

10.8.1 Acoustic Sensing

As chapter 7 discusses, the depth of cut generated by an abrasive water jet is a complex function of a number of process and material parameters. For automated machining, the on-line control of the depth of cut is a necessary condition. Nevertheless, investigations into the direct on-line measurement of the depth of cut by using a substitutive parameter are limited.

Louis and Meier [76] try to correlate the sound pressure that is generated during the abrasive water-jet cutting with the state of the cutting. The authors find differences in the sound-pressure signal for kerfing and cutting through, respectively. In case of cutting through, comparatively constant time-domain sound signals are acquired; whereas, the case of kerfing is characterized by a changing signal intensity over the measured time period. For the latter case, Louis and Meier [76] make an attempt to correlate the sound signal with the geometry of the generated cut. But, as Figure 10.31 illustrates, there is no noticeable correlation between the sound intensity and geometry of the bottom of the cut.

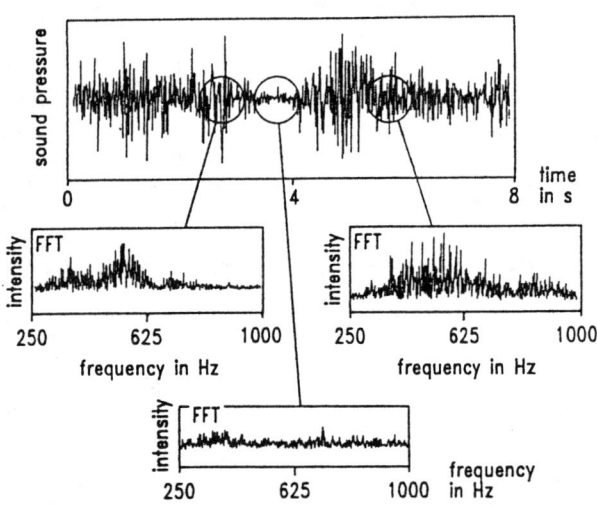

Figure 10.31 *Acoustic-sound signals acquired during submerged cutting with an abrasive water jet [76]*

10.8.2 Acoustic-Emission Technique

Section 10.7.1 already discusses the application of the acoustic-emission technique for monitoring material-removal processes. This technique is also proven for controlling the depth of cut [209]. Illustratively, Figure 10.32 plots PSD-values that represent the respective time-domain signal corresponding to different depths of cut. As the depth of cut increases, the amplitude of the acoustic-emission signal also increases. This trend is not quite evident from the FFT-plots. However, after treatment by an auto-regressive model, acoustic-emission signals are capable of reflecting the change in the depth of cut that is noted from the behavior of the PSD-curve. As the initial-gradient-region in Figure 10.29 shows, the peak of the PSD-curve exhibits a linear relationship with the depth of cut,

$$P_{PSD} = 0.423 \cdot h + 6.326, \tag{10.15}$$

This behavior of the acoustic-emission signal is the same when the different depths are generated by a changing traverse rate or pump pressure. Thus, irrespective of the abrasive water jet-parameter adopted for changing the depth of cut, the time-domain acoustic-emission signal is capable of representing the corresponding depth. Mohan et al. [209] assume that the flow regime of the water-solid suspension in the kerf strongly depends on the depth of cut, and that the acoustic-emission technique is able to detect these different flow-conditions. This argument explains the reduced sensitivity of the acoustic-emission equipment in the case of very shallow cuts. This idea is also supported by the dramatic rise of amplitude of the acoustic-emission signal in the case of cutting-through. The cutting-through-condition is characterized by very different outflow conditions in comparison to the kerfing process (Figure 10.28).

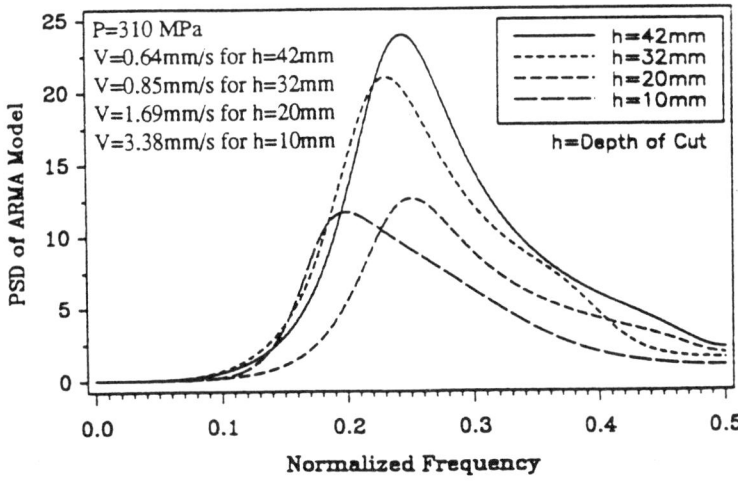

Figure 10.32 *Relation between the depth of cut and acoustic-emission signal [209]*

Based on Figure 10.29 and Eq. (10.15), a closed-loop mechanism is proposed for on-line monitoring and controlling the depth of cut [209]. The acoustic-emission signals that are generated during the cutting are acquired by a sensor that is attached to the workpiece. The time-domain signal is digitized after suitable amplification and sent to the computer where it is modeled using a stochastic modeling technique. The PSD-function of the best-fit model is computed to determine its peak. The depth of cut is estimated from the PSD-peak. Information about the predicted depth of cut is sent to the control unit where it is compared with the required depth. If the predicted depth is different from the required depth, corresponding signals are sent by the control unit to the abrasive water-jet system to change the process parameters. Thus, acoustic-emission sensing can be used to achieve a uniform depth of cut in abrasive water-jet machining. As seen previously, if the purpose of machining is cutting-through, the above mechanism can be used for optimizing the process parameters.

10.8.3 Workpiece Reaction-Force

Kovacevic [271] develops a concept of on-line monitoring of the depth of cut by using the workpiece reaction-force. Due to regression analysis, he finds

$$F_W = 0.076 \cdot d_F^{0.877} \cdot \dot{m}_A^{0.136} \cdot x^{-0.072} \cdot v^{0.06} \cdot p^{0.782} . \tag{10.16}$$

There is a strong correlation between Eq. (10.16) and Eq. (6.54). From the exponents in those equations, it is evident that all input variables for the process parameter, except the traverse rate, have the same effect on the depth of cut and workpiece reaction-force. The traverse rate only slightly influences the magnitude of the specimen reaction-force, but drastically affects the depth of cut. Thus, this process parameter, despite its ease of control, cannot be used to control the depth of cut by monitoring the workpiece reaction-force.

Figure 10.33 illustrates the concept of the control of the depth of cut by the specimen-reaction force. Thus, the depth of cut can be monitored and controlled by using a force-feedback-system and adjusting, for example, pump pressure in order to compensate for the negative effect of the focus wear on the depth of cut.

10.8.4 Supervision and Control of Piercing Processes

10.8.4.1 Monitoring by Pressure Sensors

Knaupp [367] reports preliminary results in monitoring the water jet piercing. In his study, he uses sensors for measuring the suction pressures in the cutting head as well as in the catcher. Figure 10.33 shows typical signals. In the beginning, the high-pressure valve is closed. As it opens after 10 s, the signal of the cutting head jumps up signalizing the jet flow through the head. After 25 s, the jet has pierced the workpiece and exits into the catcher. In this case, the catcher signal first drops in the negative region and then increases rapidly up to a certain level. The cutting-head signal reduces just a bit and remains at a stable level.

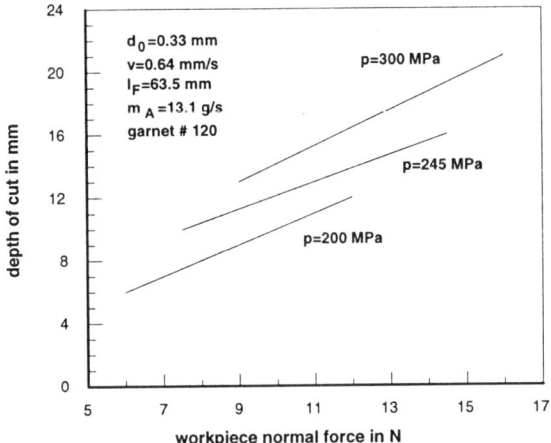

Figure 10.33 *Control of the depth of cut by measurements of the rection forces of specimens [271]*

10.8.4.2 Monitoring by Acoustic-Emission Technique

Kwak et al. [3446 and Kovacevic et al. [350] perform investigations in sensing the abrasive water-jet piercing of ceramics. Section 9.4 already presents several results. The three piercing-stages water impact, penetration, and dwelling can reliably be monitored by acoustic-emission signals. Further, strong relationships exist between the piercing time and energy parameters of the acoustic emission.

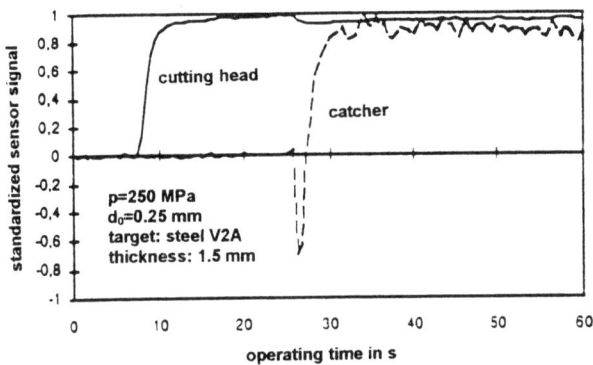

Figure 10.34 *Piercing control by monitoring the suction-pressure [367]*

The RMS-values of the raw signal linearly decrease as the piercing process continues. Also, the piercing depth and RMS-values of the raw signal are correlated by the same relationship,

$$RMS_{AE} = C_1 \cdot t_P(h_H) + C_2. \qquad (10.17)$$

In the equation, $C_1<0$ and $1.5<C_2<3$ for refractory ceramics (Figure 10.35). The cause of these effects is jet damping in the pierced hole that comes in front as the depth of the hole increases. After further treatment of the signals, separately linear regions are identified as a characterization of the piercing process. In detail, it is noted that the relation between the area under the PSD-curve and the piercing time changes at a certain drilling time, which is at about 20 % of the final piercing time.

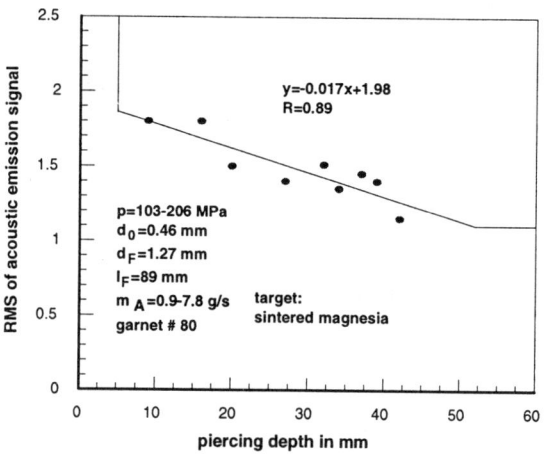

Figure 10.35 *Relation between the piercing depth and RMS-values of acoustic-emission signals [346]*

10.9 Control of the Generated Surface Topography

10.9.1 Roughness Control by Static Workpiece Reaction-Force

Hunt et al. [263, 385] first use the workpiece reaction-force to characterize the profile of surfaces that are generated by abrasive water jets. In a preliminary investigation, Hunt et al. [263] examine the influence of the traverse rate, specimen material, and abrasive type on the workpiece reaction-force. They find that the force almost linearly increases, and that the signal scattering notably increases with higher traverse rates (Figure 10.36). Also, the reaction force is very low for a hard abrasive

10.9 Control of the Generated Surface Topography

materials (aluminum oxide) compared to a softer abrasive material (garnet). The proposed functional relationship is

$$R_a = I_M \cdot f(F_W). \tag{10.18}$$

In the equation, I_M is a modified abrasive water-jet parameter, or parameter combination.

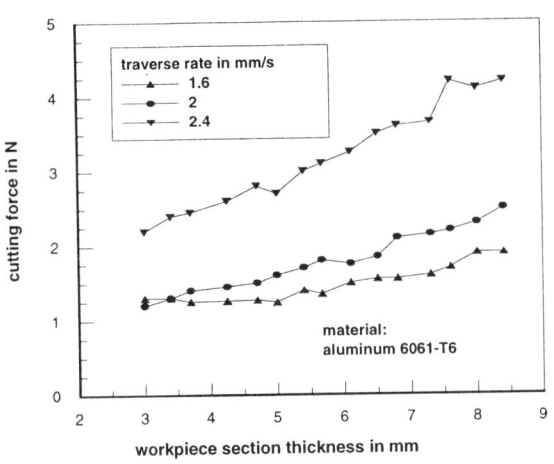

Figure 10.36 *Influence of the traverse rate on the workpiece reaction-force [263]*

Due to cutting of stepped specimens at varying cutting conditions, Hunt et al. [385] find a relation between the workpiece reaction-force, depth of cut, and traverse rate,

$$F_W = a_1 \cdot h + a_2 \cdot v - a_3. \tag{10.19}$$

A relation of the same type is found for the surface roughness,

$$R_a = b_1 \cdot h + b_2 \cdot v - b_3. \tag{10.20}$$

Eqs. (10.19) and (10.20) give,

$$R_a = c_1 \cdot v + c_2 \cdot F_W + c_3. \tag{10.21}$$

Curham et al. [324] simplify Eq. (10.21) by neglecting the small contribution from the traverse rate term (<5 %),

$$R_a = C_1 + C_2 \cdot F_W. \tag{10.22}$$

In the equation, the parameters C_1 and C_2 depend on the material properties (Figure 9.37). In a machining situation in which the sample thickness is unknown, the surface roughness can be estimated via Eq. (10.22) by measuring the workpiece reaction-force. Surface roughness can, therefore, be controlled by forming an error signal tha is based on the difference between the desired roughness and the actual roughness. The traverse rate has to be adjusted proportionally to the error signal to compensate or 'adapt' to changes in the sample thickness.

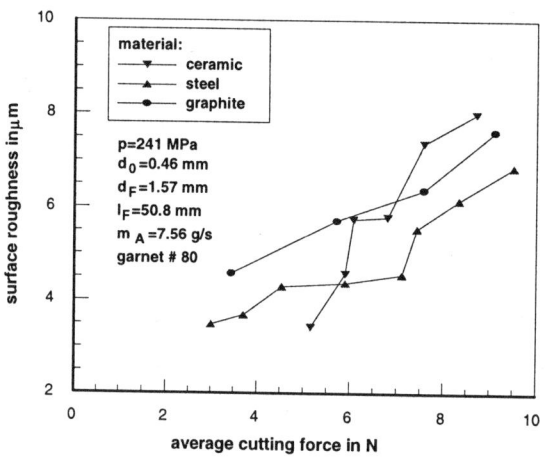

Figure 10.37 *Relation between the cutting force and surface roughness [321]*

Burnham and Kim [321] develop a multiple-regression model to solve Eq. (10.18), and to identify the structure of the parameter I_M that depends on the sample material. The linear regression-model indicates a separate trend between brittle material (ceramic) and pseudo-ductile material (aluminum, graphite composite). In order to investigate the influence of the microstructure of a family of materials, Burnham and Kim [321] test several grades of alumina ceramics, and derive a second-order relation. The results reveal a possible correlation between the surface morphology and the mechanical properties of the workpiece materials.

Curham et al. [324] use an implementation of integral force-control to control the surface roughness in real time. A single control run demonstrates the control of the surface finish for a commanded roughness value of $R_a=100$ μm ($F_W=8.0$ N) to final values of $R_a=90$ μm to $R_a=130$ μm. These values are reached in 20 s to 40 s after encountering the step change in the depth.

10.9.2 Roughness Control by Dynamic Workpiece Reaction-Force

Kovacevic et al. [206] use the dynamic component of the workpiece reaction-force to control the surface roughness. The authors model the roughness profile as well as the force signals by an auto-regressive signal processing model (ARMA) to estimate the dynamic characteristics of the measurements.

Figure 10.38a shows a plot of the PSD-peak of an ARMA-model that represents the surface profile versus the PSD-peak of an ARMA-model that represents the dynamic force with a change in the traverse rate and pump pressure. These PSD-peaks are derived from their respective models. The bottom left-hand corner of the plot indicates the smoothest surface that can be obtained. The relation in Figure 10.38a is characterized very reliably by,

$$\log PSD_{R_a} = a_1 \cdot (\log PSD_{F_W})^2 + a_2 \cdot \log PSD_{F_W} + a_3. \tag{10.23}$$

Table 10.2 gives the corresponding regression parameters.

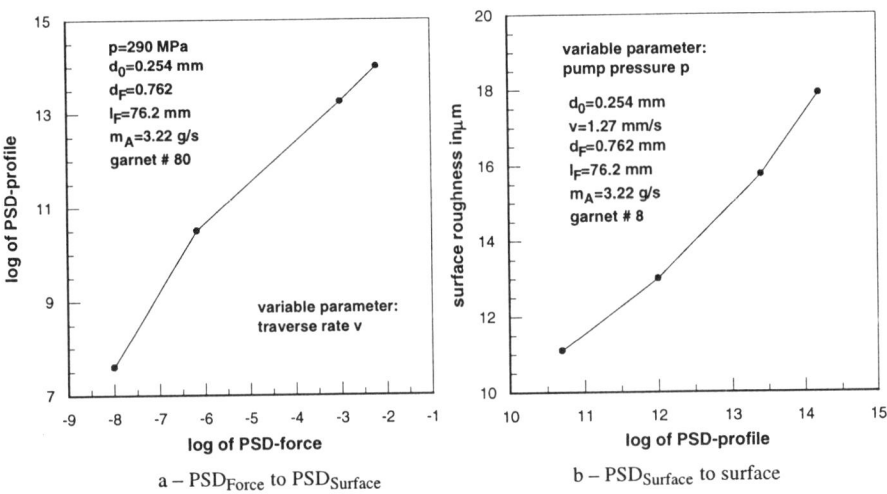

Figure 10.38 *Relations between the surface profile and dynamic reaction force [206]*

Figure 10.38b shows the corresponding plots of the surface roughness against the PSD-peak of the ARMA-model that represents the surface profile with a change in the traverse rate and pump pressure. This relationship is

$$R_a = b_1 \cdot (\log PSD_{R_a})^2 + b_2 \cdot \log PSD_{R_a} + b_3. \tag{10.24}$$

Table 10.2 lists the corresponding regression parameters.

Table 10.2 *Regression coefficients for Eqs. (10.23) and (10.24) [206]*

Equation	Coefficient	Parameter changed	
		Traverse rate	Pump pressure
(10.23)	a_1	-0.10	-0.511
	a_2	0.08	-3.252
	a_3	14.57	9.397
(10.24)	b_1	0.33	0.34
	b_2	-4.43	-6.50
	b_3	19.63	41.69

The roughness of the kerf wall can be predicted by Eqs. (10.23) and (10.24), once the control parameter is chosen. The predicted roughness are compared with the desired roughness, and if they are different, a corresponding control-signal is sent to the abrasive water-jet cutting machine to change the chosen parameter. Thus, the PSD-peak of the dynamic cutting-force signal is capable of on-line monitoring and controlling the surface roughness. The underlying assumption for this model is that the workpiece material is homogeneous and that water pressure and abrasive-mass flow rate are uniform. The relationships given in Figure 10.38 are verified through several trials conducted under identical conditions and are found to be repeatable [206].

10.9.3 Surface-Quality Monitoring by Acoustic-Emission Technique

Section 5.4.2 shows that the available jet energy and surface structure are related [217]. Thus, any detection method that has the ability of monitoring the energy dissipation in samples cut by abrasive water jetting is suitable also for on-line surface-quality monitoring. In exploring studies, Mohan et al. [221, 222] find a significant relation between the energy dissipated in a workpiece and the energy of the acquired acoustic-emission signals (Figure 10.39),

$$E_{AE} = A_1 \cdot \sqrt{E_{Diss}} . \tag{10.25}$$

The experimental points on the left-hand bottom corner in Figure 10.39 belong to shallow cuts; whereas, those on the right-hand top corner are for deeper cuts. The acoustic-emission signal is more responsive to change in the dissipated jet energy at shallow cuts. This trend verifies earlier observations from the frequency characteristics cited in section 10.7.1 in terms of the increase in the damping effect of the rebounded jet for deeper cuts. The abrasive water-jet energy that is dissipated by several mechanisms, such as, fluid-film damping, turbulence etc., which do not contribute to the material-removal process, is another cause for this trend. Thus, acoustic-emission technique is suitable for monitoring the dissipated jet energy in abrasive water-jet machining.

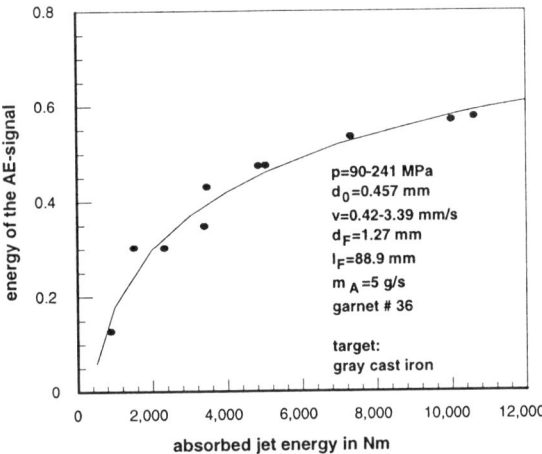

Figure 10.39 *Relation between the energy dissipated in a workpiece and the energy of the acoustic-emission signals [222]*

Figure 10.40 plots PSD-curves for various levels of dissipated energy. Any parameter change that increases the dissipated energy causes an upward shift in the PSD-curve. The hydro-dynamic effects of the abrasive water jet in the kerf region change as the dissipated energy changes. Cavitation caused by the highly turbulent flow in the kerf leads to cavitation pressure pulses that are reliably measured using the acoustic-emission technique [386]. At higher water pressures and lower traverse rates, the cuts are deeper, and the PSD-function concentrates at around a frequency of 150 kHz. The second peak becomes more insignificant. As the pump pressure reduces or the traverse speed increases, the cut becomes shallower, and the PSD-peak drops and spreads over a wider frequency range (150 kHz to 300 kHz). These promising results offer the possibility to create an on-line control system for the surface structure based on the acoustic-emission technique.

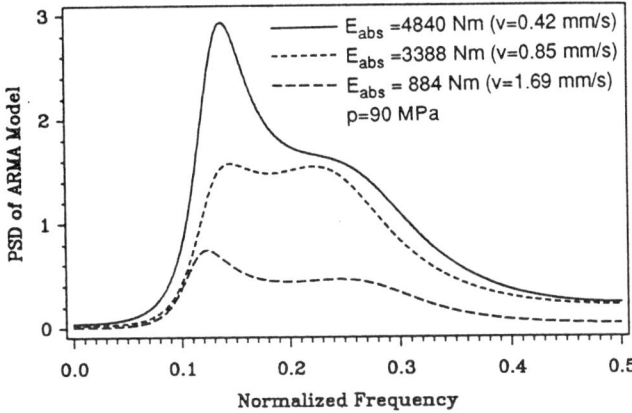

Figure 10.40 *PSD-curves of AE-signals for various levels of dissipated energy [222]*

10.10 Expert Systems for Abrasive Water-Jet Machining

Expert systems are most useful when a system cannot be managed well by rigid rules. This is the case in abrasive water-jet machining. Therefore, several approaches are made to build expert-system based software for users of the abrasive water-jet technique [352, 387]. These systems are based on semi-empirical models, including prediction equations (chapter 6).

The abrasive water-jet expert systems 'WJEXPERT™' [350] and 'Qxpert' [387] accept parameter inputs and display predicted values on a menu-driven interface (Figure 10.41). Their main features include:

- Each of the operation parameters can be entered or selected from a sub-menu, assisted by recommendation.
- Workpiece material and workpiece thickness can be entered or selected from a sub-menu. New materials can be entered.
- Surface quality is chosen from several main classes and sub-classes (see, for example, Table 8.9).
- Selection of the pump pressure and orifice diameter (water-volume flow rate) is restricted by the used intensifier capacity (if exceeded, an error message appears).
- As the orifice diameter is selected, optimum focus dimensions and abrasive-mass flow rate are recommended.
- Linear and arc-cutting traverse rate and piercing time are instantly calculated and displayed on the screen.
- The operating costs are automatically calculated.

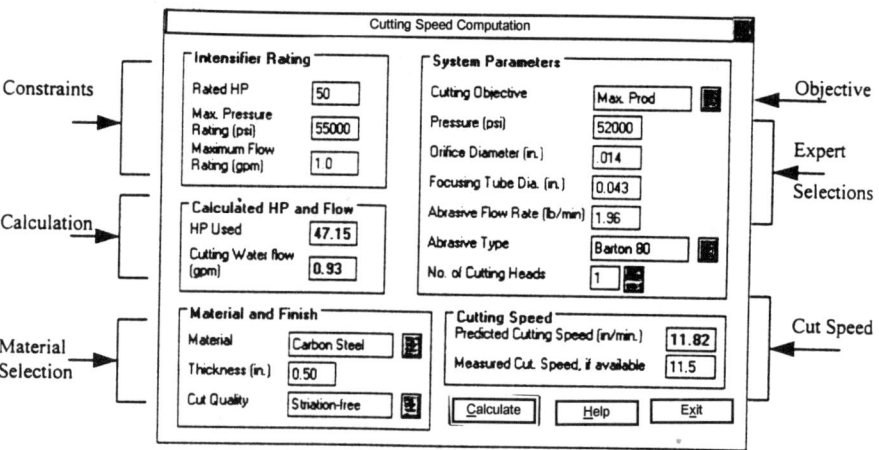

Figure 10.41 *Graphical user interface for the abrasive water-jet machining 'Qxpert' system [387]*

These expert systems can be integrated into CNC-controllers and linked to CAD/CAM-software.

References

1. Momber A W 1993 *Handbuch Druckwasserstrahl-Technik*. 1st Edn., Beton-Verlag, Düsseldorf
2. Louis H 1991 Einführung in die Wasserstrahltechnologie. In: *VDI Bildungswerk*, BW 531, pp 1-22
3. Momber A W 1998 *Water Jet Applications in Construction Engineering*. 1st ed., A.A. Balkema, Rotterdam, in print
4. Wiedemeier J 1981 Flüssigkeitsfreistrahlen hoher Relativgeschwindigkeit und Bruchkinetik spröder Werkstoffe. PhD thesis, Univ. Hannover
5. Labus T J 1991 Pulsed fluid jet technology. In: Tan J S (ed) 1991 *Proc. 1st Asian Conf. Recent Adv. Jetting Techn.*, CI Prem. Ltd., Singapore, pp 136-143
6. Vijay M M 1994 Power of pulsed jets. In: Rakowski Z (ed) 1994 *Geomechanics '93*, A.A. Balkema, Rotterdam, pp 265-274
7. Hollinger R H, Perry W D, Swanson R K 1989 Precision cutting with a low pressure, coherent abrasive suspension jet. In: Vijay M M, Savanick G A (eds) 1989 *Proc. 5th Amer. Water Jet Conf.*, Water Jet Techn. Ass., St. Louis, pp 245-252
8. Summers D A 1995 *Waterjetting Technology*. 1st ed., Chapmann & Hall, New York
9. Momber A W, Willsher J 1997 Recent developments in ultra-high pressure hydroblasting for industrial structures. In: *Proc. Int. Conf. on Protecting Industrial and Marine Structures with Coatings*, The Hague, Netherlands, 1997, pp 1-12
10. Bogaerde van de C, Momber A 1998 On-site ultra-high pressure hydroblasting for surface preparation of industrial structures. In: Momber A W (ed) 1998 *Water Jet Applications in Construction Engineering*, A.A. Balkema, Rotterdam, in print
11. Momber A W, Bogaerde van de C 1997 Entlacken mit Hochdruckwasserstrahl-Technik. *J. f. Oberflächent*. 25: 58-62
12. Momber A W, Nielsen A G 1998 Pipeline rehabilitation by water jetting. *Materials Evaluation*, in print
13. Hilmerssen S 1998 Hydrodemolition of concrete structures – basics and field experience. In: Momber A W (ed) 1998 *Water Jet Applications in Construction Engineering*, A.A. Balkema, Rotterdam, in print
14. Sondermann W 1998 Extraction and washing contaminated soils using high pressure jet grouting technique. In: Momber A W (ed) 1998 *Water Jet Applications in Construction Engineering*, A.A. Balkema, Rotterdam, in print
15. Gross H W, Wisinger F 1998 High pressure injection technique for soil conditioning. In: Momber A W (ed) 1998 *Water Jet Applications in Construction Engineering*, A.A. Balkema, Rotterdam, in print
16. Geskin E, Meng P, Tismenetskiy L, et al. 1995 Improvement of the waterjet based cleaning technology. In: Labus T J (ed) 1995 *Proc. 8th Amer. Water Jet Conf.*, Vol. 2, WJTA, St. Louis, pp 751-764
17. Momber A W 1998 On-site abrasive water jet equipment for demolition - basics and experience. In: Momber A W (ed) 1998 *Water Jet Applications in Construction Engineering*, A.A. Balkema, Rotterdam, in print
18. Momber A W 1997 Investigations into decoating and recycling of pipeline elements using the on-site high-pressure water jet technique. *Inst. Mech. Engrs., J. of Mech. Process Engng*. 211: 129-135
19. Momber A W 1995 Environmental applications of high speed water jet technology - preliminary results. *J. of Jet Flow Engng*. 12: 46-53

20. Heimhard H J 1998 Application of the water jet technology for the 'high pressure soil washing process'. In: Momber A W (ed) 1998 *Water Jet Applications in Construction Engineering*, A.A. Balkema, Rotterdam, in print
21. Schlatter M 1986 *Entgraten durch Hochdruckwasserstrahlen*. Springer–Verlag, Berlin-Heidelberg
22. Tönshoff H, Kroos F, Hartmann M 1995 Water peening - an advanced application of water jet technology. In: Labus T J (ed) 1995 *Proc. 8th Amer. Water Jet Conf.*, WJTA, St. Louis, pp 473-486
23. Kovacevic R, Cherukuthota C, Mohan R 1995 Improved milling performance with high pressure waterjet assisted cooling/lubrication. *ASME J. Engng. For Ind.* 117: 331-339
24. Kovacevic R, Cherukuthota C, Mazurkiewicz M 1996 High pressure waterjet cooling/lubrication to improve machining efficiency in milling. *Int. J. Mach. Tools Manuf.* 35: 1459-1473
25. Mohan R, Kovacevic R 1996 Performance evaluation of a hybrid machining process - high pressure waterjet assisted grinding. *MD-Vol. 74*: 131-132
26. Mort G A 1995 Results of abrasive water jet market survey. In: Labus T J (ed) 1995 *Proc. 8th Amer. Water Jet Conf.*, Vol. 1, Water Jet Techn. Ass., St. Louis, pp 259-282
27. Martinec P 1992 Mineralogical properties of abrasive materials and their role in water jet cutting process. In: Rakowski Z (ed) 1992 *Geomechanics 91*, A.A. Balkema, Rotterdam, pp 353-382
28. Agus M, Bortolussi A, Ciccu R, et al. 1995 Abrasive performance in rock cutting with AWJ and ASJ. In: Labus T J (ed) 1995 *Proc. 8th Amer. Water Jet Conf.*, Water Jet Techn. Ass., St. Louis, pp 31-48
29. Agus M, Bortolussi A, Ciccu R, et al. 1996 Abrasive-rock interaction in AWJ cutting. In: Gee C (ed) 1996 *Jetting Technol.*, Mechan. Engng. Publ., London, pp 509-520
30. Vasek J, Martinec P, Foldyna J, et al. 1993 Influence of properties of garnet on AWJ cutting process. In: Hashish M (ed) 1993 *Proc. 7th Amer. Water Jet Conf.*, Vol. 1, Water Jet Techn. Ass., St. Louis, pp 365-387
31. Dunegan H C 1961 Notes on hardness measurements of polycrystalline materials. In: Kriegel W W, Palmour H (eds) 1961 *Mechanical Properties of Engineering Ceramics*, Interscience Publ., New York, pp 521-537
32. Bowden F P, Tabor D 1964 *The Friction and Lubrication of Solids*, Part II, Clarendon Press, Oxford
33. Wadell H 1933 Sphericity and roundness of rock particles. *J. of Geology* 41: 316-331
34. Heywood H 1933 Numerical definition of particle size and shape. *Chem. Ind.* 32: 149-154
35. Bahadur S, Badruddin R 1990 Erodent particle characteristics and the effect of particle size and shape on erosion. *Wear* 158: 189-208
36. Harr M E 1977 *Mechanics of Particulate Media – A Probabilistic Approach*. McGraw-Hill, New York
37. Safanik J 1991 Garnet bohemia M.K. - a new abrasive material from Czecho-slovakia. *Int. Conf. Geomechanics*, Hradec n.M., poster presentation
38. Foldyna J, Martinec P 1992 Abrasive material in the process of AWJ cutting. In: Lichtarowicz A (ed) 1992 *Jet Cutting Technol.*, Kluwer, Dordrecht, pp 135-147
39. Martinec P 1994 A contribution to the characteristics of mineral abrasives for water jet cutting. In: Rakowski Z (ed) 1994 *Geomechanics 94*, A.A. Balkema, Rotterdam, pp 361-365
40. Cox E P 1927 A method of assigning numerical and percentage values to the degree of roundness. *J. Paleont.* 1: 179-183

41. Kelly E G, Spottiswood D J 1982 *Introduction to Mineral Processing*. John Wiley & Son, New York, pp 21-45
42. Momber A W, Kovacevic R, Schünemann R, et al. 1996 The influence of abrasive grain size distribution parameters on the abrasive water jet machining process. In: Rajukar K P (ed) 1996 *Proc. 25th North American. Manuf. Res. Conf.*, Soc. of Manuf. Engrs., Dearborn, pp 21-26
43. Schubert H 1988 *Aufbereitung fester mineralischer Rohstoffe*. Bd. 1, VEB Deutscher Verlag für Grundstoffindustrie, Leipzig, pp 21-68
44. Guo N S, Louis H, Meier G, et al. 1992 Recycling capability of abrasives in abrasive water jet cutting. In: Lichtarowicz A (1992) *Jet Cutting Technol.*, Kluwer Acad. Publ., Dordrecht, pp 503-523
45. Hashish M 1989 Pressure effects in abrasive-waterjet (AWJ) machining. *ASME J. Engng. Mat. and Techn.* 111: 221-228
46. Chen W L, Geskin E S 1991 Measurements of the velocity of abrasive waterjet by the use of Laser Transit Anemometer. In: Saunders D (ed) 1991 *Jet Cutting Technol.*, Elsevier Sci. Publ., London, pp 23-36
47. Himmelreich U 1992 Fluiddynamische Modelluntersuchungen an Wasserabrasivstrahlen. *VDI Fortschritt-Berichte*, Reihe 7, Nr. 219
48. Neusen K F, Gores T J, Labus T J 1992 Measurement of particle and drop velocities in a mixed abrasive waterjet using a forward-scatter LDV system. In: Lichtarowicz A (ed) 1992 *Jet Cutting Technol.*, Kluwer Acad. Publ., Dordrecht, pp 63-74
49. Momber A, Kovacevic R 1995 Energy dissipative processes in high speed water-solid particle erosion. In: Hoyt J W, et al. (eds) 1995 *Proc. ASME Heat Transfer and Fluids Engng. Div.*, ASME, New York, pp 243-356
50. Tazibt A Q, Parsy F, Abriak N 1996 Theoretical analysis of the particle acceleration process in abrasive water jet cutting. *Comput. Mat. Sci.* 5: 243-254
51. Shavlovsky D S 1972 Hydrodynamics of high pressure fine continuous jets. In: Brock T E, Richardson A (eds) 1972 *Proc. 1st Int. Symp. Jet Cutting Techn.*, BHRA Fluid Engng., Cranfield, pp A6-A81
52. Yanaida K, Ohashi A 1978 Flow characteristics of water jets in air. In: Clarke J, Stephens H S (eds) 1978 *Proc. 4th Int. Symp. Jet Cutting Technol.*, BHRA Fluid Engng., Cranfield, pp 39-54
53. Yanaida K, Ohashi A 1980 Flow characteristics of water jets in air. In: Stephens H S, Jarvis B (ed) 1980 *Proc. 5th Int. Symp. Jet Cutting Techn.*, BHRA Fluid Engng., Cranfield, pp 33-44
54. Whiting C E, Graham E E, Ghorashi B 1990 Evaluation of parameters in a fluid cutting equation. *ASME J. Engng. for Ind.* 112: 240-244
55. Yanaida K 1974 Flow characteristics of water jets. In: Coles N G, Barrall J S (eds) 1974 *Proc. 2nd Int. Symp. Jet Cutting Techn.*, BHRA Fluid Engng., Cranfield, pp A2/19-A2/32
56. Tikhomirov R A, Babanin V B, Pethukov E N, et al. 1992 *High-Pressure Jetcutting*. ASME Press, New York
57. Nikonov G P, Kuzmich I A, Goldin Y A 1986 *Rasruchenije Gornikh Porod Strujami Wody Wysokovo Davlenija*. Nedra Publ., Moscow
58. Neusen K F, Gores T J, Amano R S 1994 Axial variation of particle and drop velocities downstream from an abrasive water jet mixing tube. In: Allen N G (ed) 1994 *Jet Cutting Techn.*, Mechan. Engng. Publ. Ltd., London, pp 93-103
59. Davies T W, Jackson M K 1981 Optimum conditions for the hydraulic mining of China clay. In: Wang F D (ed) 1981 *Proc. 1st US Water Jet Symp.*, Colorado School of Mines Press, Golden, pp 4/1-4/16

60. Leach S J, Walker G I 1966 Some aspects of rock cutting by high speed water jets. *Phil. Trans. Roy. Soc. Lond., Ser. A* 260: 295-303
61. Yahiro T, Yoshida H 1974 Characteristics of high speed water jets in the liquid and its utilization in induction grouting. In: Coles N G, Barrall S J (eds) 1974 *Proc. 2nd Int. Symp. Jet Cutting Techn.*, BHRA Fluid Engng., Cranfield, pp G4/41-G4/63
62. Rehbinder G 1978 Erosion resistance of rock. In: Clarke J, Stephens H S (eds) 1978 *Proc. 4th Int. Symp. Jet Cutting Techn.*, BHRA Fluid Engng., Cranfield, pp E1/1-E1/11
63. Nienhaus K 1984 Ein Beitrag zur Gesteinszerstörung durch stationäre und instationäre Höchstdruckwasserstrahlen. PhD thesis, RWTH Aachen, Aachen
64. Wulf C 1986 Geometrie und zeitliche Entwicklung des Schnittspaltes beim Wasserstrahlschneiden. PhD thesis, RWTH Aachen, Aachen
65. Li Z, Geskin E S, Tismenetskiy L 1994 Improvement of water-particles mixing in the course of abrasive waterjet formation. *PED-Vol. 68-1*: 405-411
66. Galecki G, Mazurkiewicz M 1987 Hydroabrasive cutting head - energy transfer efficiency. In: Dornfeld D, Hood M (eds) 1987 *Proc 4th US Water Jet Conf.*, ASME, New York, pp 109-111
67. Mazurkiewicz M, Fincuan L, Ferguson R 1988 Investigation of abrasive cutting head internal parameters. In: Woods P A (ed) 1988 *Proc. 9th Int. Symp. Jet Cutting Techn.*, BHRA Fluid Engng., Cranfield, pp 75-84
68. Singh P J, Munoz J 1991 The alignability of jet cutting orifice and nozzle assembles. In: Saunders D (ed) 1991 *Jet Cutting Technol.*, Elsevier Sci. Publ., London, pp 207-219
69. Blickwedel H 1990, Erzeugung und Wirkung von Hochdruck-Abrasivstrahlen. *VDI Fortschritt-Berichte*, Reihe 2, Nr. 206
70. Osman A H, Buisine D, Thery B, et al. 1996 Measure of air flow rate according to the mixing chamber designs. In: Gee C (ed) 1996 *Jetting Technol.*, Mechan. Engng. Publ. Ltd., London, pp 223-236
71. Horii K, Matsumae Y, Cheng X, et al. 1991 Development of a new mixing nozzle assembly for high pressure abrasive water jet applications. In: Saunders D (ed) 1991 *Jet Cutting Technol.*, Elsevier Sci. Publ., London, pp 193-206
72. Hamada S, Kamiyama S, Tsubota M, et al. 1991 The structure of an annular jet and its characteristics. In: Saunders D (ed) 1991 *Jet Cutting Technol.*, Elsevier Sci. Publ., London, pp 69-72
73. Liu B 1991 The rotated injection abrasive jet rust cleaning system. In: Saunders D (ed) 1991 *Jet Cutting Technol.*, Elsevier Sci. Publ., London, pp 221-233
74. Yie G G 1984 Cutting hard materials with abrasive-entrained waterjet - a progress report. In: Walls I A, Stanbury J E (eds) 1984 *Proc. 7th Int. Symp. Jet Cutting Techn.*, BHRA Fluid Engng., Cranfield, pp 481-492
75. Brauer H 1971 *Grundlagen der Ein- und Mehrphasenströmungen*. Verlag Sauerländer, Aarau und Frankfurt/Main
76. Louis H, Meier G 1991 Methods of process control for abrasive water jets. In: Labus T J (ed) 1991 *Proc. 6th Amer. Water Jet Conf.*, Water Jet Techn. Ass., St. Louis, pp 427-437
77. Isobe T, Yoshida H, Nishi K 1988 Distribution of abrasive particles in abrasive water jet and acceleration mechanism. In: Woods P A (ed) 1988 *Proc. 9th Int. Symp. Jet Cutting Techn.*, BHRA Fluid Engng., Cranfield, pp 217-238
78. Wallis G B 1969 *One-Dimensional Two-Phase Flow*. McGraw-Hill, New York
79. Miller A L, Archibald J H 1991 Measurement of particle velocities in an abrasive jet cutting system. In: Labus T J (ed) 1991 *Proc. 6th Amer. Water Jet Conf.*, Water Jet Techn. Ass., St. Louis, pp 291-304

80. Ramsauer C, Beeck O, Dobke G 1927 Der Einfluß freier Oberflächen und fester Wände auf schnell bewegte Kugeln im Wasser. *Annalen der Physik*, IV. Folge, Bd. 84: 721-746
81. Abudaka M, Crofton P S 1989 Theoretical analysis and preliminary experimental results for an abrasive water jet cutting head. In: Vijay M M, Savanick G A (eds) 1989 *Proc. 5th Amer. Water Jet Conf.*, WJTA, St. Louis, pp 79-87
82. You M Q, Cui M S, Liu S L, et al. 1988 The study on the mechanism of abrasive water jet. In: Wood P (ed) 1988 *Proc. 9th Int. Symp. Jet Cutting Techn.*, BHRA Fluid Engng., Cranfield, Paper C 4
83. Meier-Wiechert G 1993 Unterwassereinsatz von Wasserabrasivstrahlen. *VDI-Fortschritt-Berichte*, Reihe 2, Nr. 389
84. Chen W L, Geskin E S 1991a Correlation between particle velocity and conditions of abrasive water jet formation. In: Labus T J (ed) 1991 *Proc. 6th Amer. Water Jet Conf.*, Water Jet Techn. Ass., St. Louis, pp 305-313
85. Buhlmann S 1970 Ein stochastisches Modell der Prallzerkleinerung. *Chemie-Ing. Technik* 42: 277-281
86. Hutchings I M 1992 Ductile-brittle transitions and wear maps for the erosion and abrasion of brittle materials. *J. of Physics, D: Appl. Phys.* 25: A/212-A/221
87. Ruppel P, Brauer H 1986 Comminution of single particles by repetitive impingement on solid surfaces. In: *Proc. 1st World Congr. Particle Technology* 1991, Nuremberg, pp 17-28
88. Yashima S, Kanda Y, Sano S 1987 Relationships between particle size and fracture energy or impact velocity required to fracture as estimated for single particle crushing. *Powder Technol.* 51: 277-282
89. Reiners E 1960 Die Prallzerkleinerung von spröden Stoffen bei sehr hohen Aufprallgeschwindigkeiten. *Chemie-Ing. Technik* 32: 136-142
90. Simpson M 1990 Abrasive particle study in high pressure water jet cutting. *Int. J. Water Jet Techn.* 1: 17-28
91. Huang H, Zhu X H, Huang Q K, et al. 1995 Weibull strength distributions and fracture characteristics of abrasive materials. *Engng. Fracture Mech.* 52: 15-24
92. Cleaver J A, Ghadiri M 1993 Impact attrition of sodium carbonate monohydrate crystals. *Powder Techn.* 76: 15-22
93. Murugesh L, Srinivasan S, Scattergood R O 1991 Models and material properties for erosion of ceramics. *J. of Mater. Engng.* 13: 55-61
94. Larsen-Basse J 1993 Effect of atmospheric humidity on the dynamic fracture strength of SiC abrasives. *Wear* 166: 93-100
95. Galecki G, Mazurkiewicz M, Jordan R. 1987 Abrasive grains disintagration effect during jet ejection. In: Wang F D (ed) 1987 *Proc. Int. Water Jet Symp.*, Beijing, pp 4/71-4/77
96. Labus T J, Neusen K F, Albers D G, et al. 1991, Factors influencing the particle size distribution in an abrasive water jet. *ASME J. Engng. for Ind.* 113: 402-411.
97. Ohlsen J 1997 Recycling von Feststoffen beim Wasserabrasivstrahlverfahren. *VDI Fortschritt-Berichte*, Reihe 15, Nr. 175
98. Kiesskalt S, Dahlhoff B 1965 Praktische Bedeutung zweier zerkleinerungsphysikalischer Effekte. *Chemie-Ing. Technik* 37: 277-283
99. Dahlhoff B 1967 Zusammenhang zwischen Zerkleinerungsgeschwindigkeit und Schallgeschwindigkeit. *Chemie-Ing. Techn.* 39: 1112-1116
100. Grady D E 1982 Local inertial effects in dynamic fragmentation. *J. Appl. Phys.* 53: 322-325
101. Glenn L A, Gommerstadt B Y, Chudnovsky A 1986 A fracture mechanics model of fragmentation. *J. Appl. Phys.* 60: 1224-1226

102. Devaswithin A, Krishnan B, Pitchumani B et al. 1987 Prediction of particle degradation during impact on a flat surface. *Wear* 118: 281-289
103. Grady D E 1990 Particle size statistics in dynamic fragmentation. *J. Appl. Phys.* 68: 6099-6105
104. Bond FC 1961 Crushing and grinding calculations, part I and II. *Br. Chem. Engr.* 6: 378-385, and 6: 543-548
105. Bond F C 1953 Work indexes tabulated. *Mining Engng.* March: 315-316
106. Mazurkiewicz M, Galecki G 1991 Energy consumed for hydro-abrasive jet formation. *Int. J. Water Jet Techn.* 1: 43-50
107. Martinec, P 1994a Changes of garnet during abrasive water jet generation and cutting of materials. In: Allen N G (ed) 1994 *Jet Cutting Technol.*, Mech. Engng. Publ. Ltd., London, pp 543-551
108. Hulsey G, Liles D, Keaton D 1989 The statistical control of abrasive waterjet nozzle wear. *MD-Vol.* 16: 17-21
109. Nakaya M, Kitagawa T, Satake S 1984 Concrete cutting with abrasive waterjet. In: Watts I A, Stanbury J F (eds) 1984 *Proc. 7th Int. Symp. Jet Cutting Techn.*, BHRA Fluid Engng., Cranfield, pp 281-292
110. Hashish M 1994 Observations on wear of abrasive-waterjet nozzle materials. *ASME J. of Tribol.* 116: 439-444
111. Mort G A 1991 Long life abrasive water jet nozzles and their effect on AWJ cutting. In: Labus T J (ed) 1991 *Proc. 6th Amer. Water Jet Conf.*, Water Jet Techn. Ass., St. Louis, pp 315-344
112. Ness E, Dubensky E, Haney C, et al. 1994 New developments in ROTOTEC composite carbides for use in abrasive waterjet applications. In: Allen N G (ed) 1994 *Jet Cutting Technol.*, Mech. Engng. Publ., London, pp 196-211
113. Schwetz K A, Greim J, Sigl L S, et al 1994 Research on design and application of industrial scale hydro-abrasive jet-cutting nozzles. In: Allen N G (ed) 1994 *Jet Cutting Technol.*, Mech. Engng. Publ., London, pp 165-175
114. Kovacevic R, Beardsley H 1989 Nozzle wear sensing in turning operation with abrasive waterjet. In: *Proc. Conf. on Nontrad. Machining*, SME MS89-809, Soc. of Manuf. Engrs., Dearborn, pp 1-11
115. Hashish M 1987 Wear in abrasive-waterjet cutting systems. In: Ludema K C (ed) 1987 *Wear of Materials*, Vol. 2, ASME, New York, pp 769-776
116. Nanduri M, Taggart D G, Kim, T J, et al. 1995 Effect of offset bores on the performance and life of abrasive waterjet mixing tubes. In: Labus T J (ed) 1995 *Proc. 8th Amer. Water Jet Conf.*, Vol. 2, Water Jet Techn. Ass., St. Louis, pp 459-470
117. Vaughan R A, Ball A 1991 The effect of hardness and toughness on the erosion of ceramics and ultrahard materials. In: Ludema K C, Bayer R G (eds) 1991 *Wear of Materials*, Vol. 1, ASME, New York, pp 71-75
118. Srinivasan S, Scattergood R O 1988 Effect of erodent hardness on erosion of brittle materials. *Wear* 128: 139-152
119. Werner M 1991 Einflußparameter und Wirkmechanismen beim Abtrag von Mörtel und Beton mit dem Hochdruckwasserstrahl. PhD thesis, RWTH Aachen
120. Holland C L 1985 Implementing abrasive waterjet cutting. *SME TP MF85-875*, pp 1-12
121. Wightman D F, Dixon M 1990 Waterjet cutting / hydroabrasive machining job shop. *SME TP MS90-410*, pp 1-12
122. Bell J F, Rogers P S 1987 Laboratory scale erosion testing of a wear resistant glass-ceramic. *Mater. Science Techn.* 3: 807-813
123. Wang D F, She J H, Ma Z M 1995 Effect of microstructure on erosive wear behaviour of SiC ceramics. *Wear* 180: 35-41

124. Ramulu M, Raju S P, Innoue H, et al. 1993 Hydro-abrasive erosion characteristics of 30vol.%SiC$_p$/6061-T6 Al composite at shallow impact angles. *Wear* 166: 55-63
125. Brandt C, Lois H, Meier G, et al. 1994 Abrasive suspension jets at working pressures up to 200 MPa. In: Allen A G (ed) 1994 *Jet Cutting Technol.*, Mech. Engng. Publ. Ltd., London, pp 489-509
126. Hashish M 1991 Cutting with high-pressure abrasive suspension jets. In: Labus T J (ed) 1991 *Proc. 6th Amer. Water Jet Conf.*, Water Jet Techn. Ass., St. Louis, pp 439-455
127. Liu B L, Cui M S 1988 Experiments in the premixed abrasive jet to cut metal plates. In: Woods P A (ed) 1988 *Proc. 9th Int. Symp. Jet Cutting Techn.*, BHRA Fluid Engng., Cranfield, pp 85-98
128. Fairhurst R M, Heron R A, Saunders D H 1986 'Diajet' - a new abrasive water jet cutting technique. In: Saunders D (ed) 1986 *Proc. 8th Int. Symp. Jet Cutting Techn.*, BHRA Fluid Engng., Cranfield, pp 395-402
129. Fair J C 1981 Development of high-pressure abrasive-jet drilling. *J. of Petroleum Technol.* Aug.: 1379-1388.
130. Andersen, C B 1992 Abrasive waterjet cutting at low water pressure. *Welding in the World* 30: 312-315
131. Bloomfield E J, Yeomans M J 1991 Diajet - a review of progress. In: Tan S J (ed) 1991 *Proc. 1st Asian Conf. Recent Adv. Jetting Techn.*, CI-Prem. Ltd., Singapore, pp 21-30
132. Guo C, Cheng D, Liu L. 1993 Mathematical modeling of the accelerating process of abrasives in diajet. In: Hashish M (ed) 1993 *Proc. 7th Amer. Water Jet Conf.*, Vol. 1, Water Jet Techn. Ass., St. Louis, pp 287-293
133. Laurinat A, Louis H, Tebbing G 1992 Premixed abrasive water jets - the influence of important parameters. In: Lichtarowicz A (ed) 1992 *Jet Cutting Technol.*, Kluwer Acad. Publ., Dordrecht, pp 577-591
134. Liu B L, Shang Y, Yao H, et al. 1992 The recent premajet advance in cutting and derusting technology. In: Lichtarowicz A (ed) 1992 *Jet Cutting Technol.*, Kluwer Acad. Publ., Dordrecht, pp 451-460.
135. Shimizu S, Wu Z L 1994 Erosion due to premixed abrasive water jet. In: Allen N G (ed) 1994 *Jet Cutting Technol.*, Mech. Engng. Publ. Ltd., London, pp 3-14
136. Walters C L, Saunders D H 1991 Diajet cutting for nuclear decommissioning. In: Saunders D (ed) 1991 *Jet Cutting Technol.*, Elsevier Sci. Publ. Ltd., London, pp 427-440
137. Yazici S, Summers D A 1989 The investigation of diajet cutting of granite. In: Vijay M M, Savanick G A (eds) 1989 *Proc. 5th Amer. Water Jet Conf.*, Water Jet Techn. Ass., St. Louis, pp 343-356
138. You M, Zhang H, Chai J P 1993 Study on a direct injection abrasive jet system. In: Hashish M (ed) 1993 *Proc. 7th Amer. Water Jet Conf.*, Vol. 1, Water Jet Techn. Ass., St. Louis, pp 295-301
139. Hollinger R H, Mannheimer M 1991 Rheological investigation of the abrasive suspension jet. In: Labus T J (ed) 1991 *Proc. 6th Amer. Water Jet Conf.*, Water Jet Techn. Ass., St. Louis, pp 515-528
140. Shimizu S, Wu Z L 1996 Acceleration of abrasive particles in premixed abrasive water jet nozzle. *JSME Int. J., Ser. B* 39: 562-567
141. Ye J, Kovacevic R 1997 Turbulent solid-liquid flow through nozzles of premixed abrasive waterjet cutting systems. *Int. J. Mach. Tool Manuf.*, submitted.
142. Shimizu S 1996 Effects of nozzle shape on structure and drilling capacity of premixed abrasive water jets. In: C Gee (ed) 1996 *Jetting Technol.*, Mech. Engng. Publ. Ltd, London, pp 13-26

143. You M, Ding, Y, Liu W, Moshen, C 1996 A study of DIAJET nozzle wear. In: Gee C (ed) 1996 *Jetting Technol.*, Mech. Engng. Publ., London, pp 45-57
144. Geskin E S, Chen W L, Chen S S, et al. 1989 Investigation of anatomy of an abrasive waterjet. In: Vijay M M, Savanick G A (eds) 1989 *Proc. 5th Amer. Water Jet Conf.*, Water Jet Techn. Ass., St. Louis, pp 217-230
145. Neusen K F, Albers A G, Gores T J, et al. 1991 Distribution of mass in a three-phase abrasive water jet using scanning X-ray densitometry. In: Saunders D (ed) 1991 *Jet Cutting Technol.*, Elsevier, London, pp 83-98
146. Öjmertz C M, Amini N 1994 A discrete approach to the abrasive waterjet milling process. In: Allen N G (ed) 1994 *Jet Cutting Technol.*, Mech. Engng. Publ. Ltd., London, pp 425-434
147. Himmelreich U, Rieß W 1991 Laser-velicometry investigations of the flow in abrasive water jets with varying cutting head geometry. In: Labus T J (ed) 1991 *Proc. 6th Amer. Water Jet Conf.*, Water Jet Techn. Ass., St. Louis, pp 355-369
148. Momber A W 1995 A generalized abrasive water jet cutting model. In: Labus T J (ed) 1995 *Proc. 8th Amer. Water Jet Conf.*, Vol. 1, Water Jet Techn. Ass., St. Louis, pp 359-371
149. Engel P A 1976 *Impact Wear of Materials.* Elsevier Sci. Publ., Amsterdam
150. Adler W F 1979 *Erosion: Prevention and Useful Application.* ASTM, New York
151. Preece C M 1979 *Erosion.* Treatise on Mat. Sci. and Technol., Vol. 16, Academic Press, New York
152. Ellermaa R R 1993 Erosion prediction of pure metals and carbon steels. *Wear* 162-164: 1114-1122
153. Meng H C, Ludema K C 1995 Wear models and prediction equations: their form and content. *Wear* 181-183: 443-457
154. Magnee A 1995 Generalized law of erosion: application to various alloys and intermetallics. *Wear* 181-183: 500-510
155. Finnie I 1958 The mechanism of erosion of ductile metals. In: Haythornthwaite R M, et al. (eds) 1958 *Proc. 3rd U.S. Nat. Congr. Appl. Mech.*, ASME, New York, pp 527-532
156. Buijs M 1994 Erosion of glass as modeled by indentation theory. *J. Amer. Ceram. Soc.* 77: 1676-1678
157. Hashish M 1984 A model study of metal cutting with abrasive water jets. *ASME J. Engng. Mat. and Techn.* 106: 88-100
158. Zeng J, Kim T J 1992 Development of an abrasive waterjet kerf cutting model for brittle materials. In: Lichtarowicz A (ed) 1992 *Jet Cutting Technol.*, Kluwer Acad. Press, Dordrecht, pp 483-501
159. Mazurkiewicz M 1991 A study of a leading edge profile for a slot formed during hydro-abrasive cutting. In: Labus T J (ed) 1991 *Proc. 6th Amer. Water Jet Conf.*, Water Jet Techn. Ass., St. Louis, pp 43-49
160. Bitter J G A 1963 A study of erosion phenomena, part I. *Wear* 6: 5-21
161. Bitter J G A 1963 A study of erosion phenomena, part II. *Wear* 6: 169-190
162. Finnie I, McFadden A 1978 On the velocity dependence of the erosion of ductile metals by solid particle at low angle of incidence. *Wear* 48: 181-190
163. Neilson J H, Gilchrist A 1968 Erosion by a stream of solid particles. *Wear* 11: 111-122
164. Hutchings I M 1979 Mechanical and metallurgical aspects of the erosion of metals. In: Levy A V (ed) 1979 *Proc. Corrosion/Erosion of Coal Convers. Syst. Mat. Conf.*, Nat Ass. Corr. Engrs., Houston, pp 393-428
165. Hutchings I M 1977 Deformation of metal surfaces by the oblique impact of square plates. *Int. J. Mechan. Sci.* 19: 45-52

166. Hutchings I M 1979 Some comments on the theoretical treatment of erosive particle impact. In: Field J E (ed) 1979 *Proc. 5th Int. Conf. Erosion by Liquid and Soil Impact*, Cavendish Lab., Cambridge, pp 36/1-36/6
167. Edington J W, Wright I G 1978 Study of particle erosion damage in haynes stellite 6B – part I+II. *Wear* 48: 131-155
168. Hutchings I M, Levy A V 1989 Thermal effects in the erosion of ductile metals. *Wear* 131: 105-121
169. Sheldon G L, Finnie I 1966 The mechanism of material removal in the erosive cutting of brittle materials. *ASME J. Engng. Ind.* 88: 393-400
170. Evans A G, Gulden M E, Rosenblatt M E 1976 Impact damage in brittle materials in the elastic-plastic response regime. *Proc. Roy. Soc. Lond., Ser. A.* 361: 343-356
171. Wiederhorn S M, Lawn B R 1979 Strength degradation of glass impacted with sharp particles. *J. Amer. Ceram Soc.* 62: 66-70
172. Ritter J E 1985 Erosion damage in structural ceramics. *Mater. Sci. and Engng.* 71: 194-201
173. Marshall D B, Lawn B R, Evens A G 1982 Elastic/plastic identation damage in ceramics: lateral crack system. *J. Amer. Ceram. Soc.* 65: 561-566
174. Kovacevic R, Liaw H H, Barrows J F 1988 Surface finish and its relationship to cutting parameters. *SME TP MR88-589*, Soc. of Manuf. Engrs., Dearborn, pp 1-5
175. Webb K E, Rajukar K P 1990 Surface characterization of inconel cut by abrasive water jet. *CSME Mechanical Engng. Forum*
176. Kovacevic R 1991 Surface texture in abrasive water jet cutting. *J. of Manuf. Systems* 10: 32-40
177. Arola D, Ramulu M 1995 Abrasive waterjet machining of titanium alloy. In: Labus T J (ed) 1995 *Proc. 8th Amer. Water Jet Conf.*, Water Jet Techn. Ass., St. Louis, pp 389-408
178. Zeng J, Kim T J 1991 Material removal of polycristalline ceramics by a high pressure abrasive water jet - a SEM study. *Int. J. Water Jet Technol.* 1: 65-71
179. Summers D A, Yao J, Wu W Z 1991 A further investigation of DIAjet cutting. In: Saunders D (ed) 1991 *Jet Cutting Technol.*, Elsevier, London, pp 181-192
180. Guo N S 1994 Schneidprozeβ und Schnittqualitat beim Wasserabrasivstrahlschneiden. *VDI-Fortschritt-Berichte*, Reihe 2, Nr. 328.
181. Arola D, Ramulu M 1993 Mechanisms of material removal in abrasive waterjet machining of common aerospace materials. In: Hashish M (ed) 1993 *Proc. 7th Amer. Water Jet Conf.*, Vol. 1. Water Jet Techn. Ass., St. Louis, pp 43-64
182. Neusen K F, Rohatgi P K, Vaidyanathan C, et al. 1987 Abrasive waterjet cutting of metal matrix composites. In: Hood M, Dornfeld D (eds) 1987 *Proc. 4th U.S. Water Jet Conf.*, ASME, New York, pp 175-182
183. Savrun E, Taya M 1988 Surface characterization of SiC whisker/2124 aluminum and Al_2O_3 composites machined by abrasive water jet. *J. Mat. Sci.* 23: 1453-1458
184. Hamatani G, Ramulu M 1990 Machinability of high temperature composites by abrasive waterjet. *ASME J. Engng. Mat. Techn.* 112: 381-386
185. Arola D, Ramulu M 1993a A study of kerf characteristics in abrasive waterjet machining of graphite/epoxy composites. *MD-Vol. 45/PED-Vol. 66*: 125-150
186. Ramulu M, Arola D 1993 Waterjet and abrasive water jet cutting of undirected graphite/epoxy composite. *Composites* 24: 299-308
187. Ramulu M, Arola D 1994 The influence of abrasive water jet cutting conditions on the surface quality of graphite/epoxy laminates. *Int. J. Mach. Tools Manuf.* 34: 295-313
188. Momber A.W, Eusch I, Kovacevic R 1995 Cutting refractory ceramics with abrasive water jets - a preliminary investigation. In: Labus T J (ed) 1995 *Proc. 8th Amer. Water Jet Conf.*, Vol. 1, Water Jet Techn. Ass., St. Louis, pp 229-244

189. Momber A W, Eusch I, Kovacevic R 1996 Machining refractory ceramics with abrasive water jet. *J. of Mater. Sci.* 31: 6485-6493
190. Momber A W, Mohan R, Kovacevic R 1995 Acoustic emission measurements on brittle materials during abrasive waterjet cutting. In: Malkin S, Zdeblick W J (eds) 1995 *Proc. 1st Internat. Machining and Grinding Conf.*, Soc. Manuf. Engrs., Dearborn, pp 441-458
191. Momber A W, Mohan R, Kovacevic R 1997 On-line analysis of hydro-abrasive erosion of pre-cracked materials by acoustic emission. *Theoret. and Appl. Fracture Mech.*, submitted
192. Ramulu M 1993 Dynamic photoelastic investigation on the mechanics of waterjet and abrasive waterjet machining. *Optics & Lasers* 19: 43-65
193. Kahlman L, Karlsson R, Carlsson R, et al. 1993 Wear and machining of engineering ceramics by abrasive waterjets. *Amer. Ceram Bull.* 72: 93-98
194. Momber A W, Kovacevic R 1994 Fundamental investigations on concrete wear by high velocity water flow. *Wear* 177: 55-62
195. Momber A W, Kovacevic 1995 Statistical character of the failure of rocklike materials due to high energy water jet impingement. *Int. J. of Fracture* 71: 1-14
196. Forman S E, Secor G A 1974 The mechanism of rock failure due to water jet impingement. *Soc. of Petrol. Engrs. J.* 14: 10-18
197. Momber A W, Kovacevic R, Ye J 1995 The fracture of concrete due to erosive wear by high velocity water flow. *Tribol. Trans.* 38: 686-692
198. Momber A W, Kovacevic R 1996 Fracture of brittle multiphase materials by high energy water jet. *J. Mat. Sci.* 31: 1081-1085
199. Momber A W, Kwak H, Kovacevic R 1997 An alternative method for the evaluation of the abrasive water jet cutting process in gray cast iron. *J. of Mater. Process. Technol.* 65: 65-72
200. Balan K P, Reddy V, Joshi V, et al. 1991 The influence of microstructure on the erosion behaviour of cast irons. *Wear* 145: 283-296
201. Okada T, Iwai Y, Yamamoto A 1983, A study of cavitation of cast iron. *Wear* 84: 297-312
202. Zhou G, Leu M, Geskin, E G, et al. 1992 Investigation of topography of waterjet generated surfaces. *PED-Vol.* 62: 191-202
203. Guo N S, Louis H, Meier G 1993a Surface structure and kerf geometry in abrasive water jet cutting: formation and optimization. In: Hashish M (ed) 1993 *Proc. 7th Amer. Water Jet Conf.*, Vol. 1, Water Jet Techn. Ass., St. Louis, pp 1-25
204. Mohan R, Kovacevic R, Zhang Y 1992 Characterization of the surface texture generated by the high-energy jets. *PED-Vol.* 62: 203-218
205. Kovacevic R, Mohan R, Zhang Y 1993 Stochastic modelling of surface texture generated by high-energy jets. *Proc. Inst. Mech. Engrs., J. of Engng. Manuf.* 207:129-140
206. Kovacevic R, Mohan R, Zhang Y M 1995a Cutting force dynamics as a tool for surface profile monitoring in AWJ. *ASME J. Engng. for Ind.* 117: 340-350
207. Hashish M 1988 Visualization of the abrasive waterjet cutting process. *Exp. Mechan.* 28: 159-169
208. Kovacevic R, Mohan R, Beardsley H 1996 Monitoring of thermal energy distribution in abrasive waterjet cutting using infrared thermography. *ASME J. Manuf. Sci. and Engng.* 118: 555-563
209. Mohan R, Momber A W, Kovacevic R 1994 On-line monitoring of the depth of AWJ penetration using acoustic emission technique. In: Allen N G (ed) 1994 *Jet Cutting Technol.*, Mech. Eng. Publ. Ltd., London, pp 649-664

210. Ohlsson L, Powell J, Magnusson C 1994 Mechanisms of striation formation in abrasive water jet cutiing. In: Allen N G (ed) 1994 *Jet Cutting Technol.*, Mech. Engng. Publ. Ltd., London, pp 151-164
211. Mohaupt U H, Burns D 1974 Machining unreinforced polymers with high velocity water jets. *Exper. Mech.* 14: 152-157
212. Chao J, Geskin E S, Chung Y 1992 Investigations of the dynamics of the surface topography formation during abrasive waterjet machining. In: Lichtarowicz A (ed) 1992 *Jet Cutting Technol.*, Kluwer Acad. Press, Dordrecht, pp 593-603
213. Chao J, Geskin E S 1993 Experimental study of the striation formation and spectral analysis of the abrasive waterjet generated surfaces. In: Hashish M (ed) 1993 *Proc. 7th Amer. Water Jet Conf.*, Vol. 1, Water Jet Techn. Ass., St. Louis, pp 27-41
214. Hashish M 1992 A modeling study of jet cutting surface finish. *PED-Vol. 58*: 151-167
215. Di Pietro P, Yao Y L 1995 A new technique to characterize and predict laser cut striations. *Int. J. Mach. Tools Manuf.* 35: 993-1002
216. Raju S P, Ramulu M 1994 Predicting hydro-abrasive erosive wear during abrasive waterjet cutting - part 1: a mechanistic formulation and its solution. *PED-Vol. 68-1*: 339-351
217. Raju S P, Ramulu M 1994a Predicting hydro-abrasive erosive wear during abrasive waterjet cutting - part 2: an experimental study and model verification. *PED-Vol. 68-1*: 381-396
218. Fukunishi Y, Kobayashi R, Uchida 1995 Numerical simulation of striation formation on water jet cutting surface. In: Labus T J (ed) 1995 *Proc. 8th Amer. Water Jet Conf.*, Vol. 2, WJTA, St. Louis, pp 657-670
219. Momber A W, Kovacevic R 1994 Calculation of exit jet energy in abrasive water jet cutting. *PED-Vol. 68-1*: 361-366
220. Momber A W, Kovacevic R 1995 Quantification of energy absorption capability in abrasive water jet machining. *Inst. Mech. Engrs., J. of Engng. Manuf.* 209: 491-498
221. Mohan R, Momber A W, Kovacevic R 1995 Detection of energy dissipation during abrasive water jet machining using acoustic emission technique. *MED-Vol. 2-1*: 243-256
222. Mohan R, Momber A W, Kovacevic R 1997 Acoustic emission-sensing as a tool for monitoring energy dissipation in abrasive water jet cutting. *ASME J. Manuf. Sci. and Engng.*, submitted
223. Momber A W 1995 A simplified mathematical energy dissipation model for water jet and abrasive water jet cutting processes. In: Labus T J (ed) 1995 *Proc. 8th Amer. Water Jet Conf.*, Vol. 2, Water Jet Techn. Ass., St. Louis, pp 829-843
224. Zeng J, Heines R, Kim T J 1991 Characterization of energy dissipation phenomena in abrasive water jet cutting. In: Labus T J (ed) 1991 *Proc. 6th Amer. Water Jet Conf.*, Water Jet Techn. Ass., St. Louis, pp 163-177
225. Capello E, Groppetti R 1993 On a simplified model for hydro-abrasive jet machining prediction, control and optimization. In: Hashish M (ed) 1993 *Proc. 7th Amer. Water Jet Conf.*, Vol. 1 Water Jet Techn. Ass., St. Louis, pp 157-174
226. Ohadi M M, Ansari A I, Hashish M 1992 Thermal energy distributiond in the workpiece during cutting with an abrasive waterjet. *ASME J. Engng. for Ind.* 114: 67-73
227. Louis H, Meier G, Ohlsen J 1995 Analysis of the process output in abrasive water jet cutting. In: Labus T J (ed) 1995 *Proc. 8th Amer. Water Jet Conf.*, Water Jet Techn. Ass., St. Louis, pp 137-152
228. Momber A W, Kwak H, Kovacevic R 1996 Investigations in abrasive water jet erosion based on wear particle analysis. *ASME J. of Tribol.* 118: 759-766
229. Johnson K L 1985 *Contact Mechnics*. Cambridge Univ. Press, Cambridge

230. Zeng J, Kim T J 1996 An erosion model for abrasive waterjet milling of polycrystalline ceramics. *Wear* 199: 275-282
231. Grady D E, Kipp M E 1987 Dynamic fragmentation. In: Atkison B K (ed) 1987 *Fracture Mechanics of Rocks*, Academic Press, London, pp 429-475
232. Momber A W 1992 Investigations on water jet processed concrete. In: Lichtarowicz A (ed) 1992 *Jet Cutting Tewchnol.*, Kluwer Acad. Publ., Dordrecht, pp 405-412
233. Ruff A W 1978 Debris analysis of erosive and abrasive wear. In: Shuh N P, Saka N (eds) 1978 *Fundamentals of Tribology*, MIT Press, Cambridge, pp 877-885
234. Rickerby R G 1983 Correlation of erosion with mechanical properties. *Wear* 84: 393-395
235. Rao P V, Buckley D H 1985 Characterization of solid particle erosion resistance of ductile materials based on their properties. *ASME J. Engn. Gas Turb. and Power* 107: 669-678
236. Matsui S, Matsumura H, Ikemoto Y, et al. 1991 Prediction equations for depth of cut made by abrasive water jet. In: Labus T J (ed) 1991 *Proc. 6th Amer. Water Jet Conf.*, WJTA, St. Louis, pp 31-41
237. Momber A W, Kovacevic R 1997 An energy balance of high-speed abrasive water jet erosion. *Inst. Mech. Engrs., J. of Engng. Tribol.* 210: in print
238. Momber A W 1998 The kinetic energy of wear particles generated by abrasive-water jet erosion. *J. Mater. Proc. Technol.*: in print
239. Zu J B, Burstein G T, Hutchings I M 1991 A comparative study on the slurry erosion and free-fall particle erosion of aluminum. *Wear* 149: 73-84
240. Clark H M, Burmeister L C 1992 The influence of the squeeze-film on particle impact velocities in erosion. *Int. J. of Impact Engng.* 12: 415-426
241. Yong Z, Kovacevic R 1997 Simulation of effects of water-mixture film on impact contact in abrasive waterjet machining. *Int. J. Mechan. Sci.* 39: 729-739
242. Kwak H 1995 The investigation of effects of abrasive sharpness on impact force. *Internal Report*, CRMS, Univ. of Kentucky, CRMS, Lexington.
243. Momber A W, Kovacevic R, Mohan, R, et al. 1995 Experimental estimation of energy dissipative processes in workpieces during abrasive water jet cutting. In: Labus T J (ed) 1995 *Proc. 8th Amer. Water Jet Conf.*, Vol. 1, Water Jet Techn. Ass., St. Louis, pp 187-206
244. Ohadi M M, Whipple R L 1991 Measurements of temperatures in a turbular workpiece during cutting with an abrasive waterjet. *Exper. Techn.* 15: 38-42
245. Ohadi M M, Cheng K L 1993 Modeling of temperature distributions in the workpiece during abrasive waterjet machining. *ASME J. of Heat Trans.* 115: 446-452
246. Mohan R, Kovacevic R, Beardsley, H E 1996 Heat flux determination at the AWJ cutting zone using IR-thermography and inverse heat conduction problem. *HTD-Vol. 332*: 245-254
247. Hunt D C, Kim T J, Sylvia J G 1986 A parametric study of abrasive waterjet by piercing experiments. In: Saunders D (ed) 1986 *Proc. 8th Int. Symp. Jet Cutting Techn.*, BHRA Fluid Engng., Cranfield, pp 287-295
248. Miranda R M, Lousa P, Miranda A, et al. 1993 Abrasive water jet cutting of portuguese marbles. In: Hashish M (ed) 1993 *Proc. 7th Amer. Water Jet Conf.*, Vol. 1, Water Jet Techn. Ass., St. Louis, pp 443-457
249. Hashish M 1991 Wear modes in abrasive water jet machining. *PED-Vol. 54*: 141-153
250. Eyre T S 1978 Wear characteristics of metals. In: Rigney D A, Glaeser W A (eds) 1978 *Source Book on Wear Control Technology*, ASM, Metals Park, pp 1-10
251. Zeng J, Kim T J, Wallace R J 1992 Quantitative evaluation of machinability in abrasive waterjet machining. *PED-Vol. 58*: 169-179

252. Momber A W, Kovacevic R, Eusch, I 1995 Resistance parameters of rocklike materials against abrasive water jet cutting. In: Daemen J J, Schultz R A (eds) 1995 *Rock Mechanics*, A.A. Balkema, Rotterdam, pp 355-359
253. Heβling M 1988 Grundlagenuntersuchungen über das Schneiden von Gestein mit abrasiven Höchstdruckwasserstrahlen. PhD thesis, RWTH Aachen.
254. Guo N S, Louis H, Meier G, et al. 1994 Abrasive water jet cutting - methods to calculate cutting performance and cutting efficiency. In: Rakowski Z (ed) 1994 *Geomechanics 93*, A.A. Balkema, Rotterdam, pp 291-299
255. Momber A W, Kovacevic R 1997 Test parameter analysis in abrasive water jet cutting of rocklike materials. *Int. J. Rock Mech. Min. Sci.* 34: 17-25
256. Momber A W, Kovacevic R 1996 Accelerated high speed water erosion test for concrete wear debris analysis. *Trib. Trans.* 39: 943-949
257. Momber A W, Kovacevic R, Pfeiffer D, et al. 1995 Relation between water jet erosion and compression testing of concrete. In: Labus T J (ed) 1995 *Proc. 8th Amer. Water Jet Conf.*, Vol. 2, Water Jet Techn. Ass., St. Louis, pp 809-828
258. Zeng J, Kim T J 1991 The machinability of porous materials by a high pressure abrasive waterjet. *PED-Vol.* 41: 37-42
259. Hashish M 1987 An improved model for erosion by solid particle impact. In: Field J E, Dear J P (eds) 1987 *Proc. 7th Int. Conf. Erosion by Liquid and Solid Impact*, Cavendish Lab., Cambridge, pp 66.1-66.9
260. Hashish M 1987 Prediction of depth of cut in abrasive waterjet (AWJ) machining. *MD-Vol. 3*: 65-82
261. White F M 1974 *Viscous Fluid Flow*. McGraw-Hill, New York
262. Zeng J, Kim T J 1993 Parameter prediction and cost analysis in abrasive waterjet cutting operations. In: Hashish M (ed) 1993 *Proc. 7th Amer. Water Jet Conf.*, Vol. 1, Water Jet Techn. Ass., St. Louis, pp 175-189
263. Blickwedel H, Guo N S, Haferkamp H et al. 1991 Prediction of abrasive jet cutting performance and quality. In: Saunders D (ed) 1991 *Jet Cutting Technol.*, Elsevier, London, pp 163-179
264. Oweinah H 1989 Leistungssteigerung des Hochdruckwasserstrahlschneidens durch Zugabe von Zusatzstoffen. PhD thesis, TU Darmstadt, Darmstadt
265. Capello E, Groppetti R 1992 On an energetic semi-empirical model of hydro-abrasive jet material removal mechanism for control and optimization. In: Lichtarowicz A (ed) 1992 *Jet Cutting Technol.*, Kluwer Acad. Publ., Dordrecht, pp 101-120
266. Hutchings I M 1979 Energy absorbed by elastic waves during plastic impact. *J. of Physics, D: Appl. Phys.* 12: 1819-1824
267. Momber A W, Kovacevic R 1997 Influence of stress-strain behaviour of quasi-brittle materials on their machinability by abrasive water jets. in preparation.
268. Iihoshi S, Nakao K, Torii K, et al. 1986 Preliminary study on abrasive waterjet assist roadheader. In: Saunders D (ed) 1986 *Proc. 8th Int. Symp. Jet Cutting Techn.*, BHRA Fluid Engng., Cranfield, pp 71-77
269. Hlavac L 1992 Physical description of high energy liquid jet interaction with material. In: Rakowski Z (ed) 1992 *Geomechanics 91*, A.A. Balkema, Rotterdam, 341-346
270. Chung Y, Geskin E S, Singh P 1992 Prediction of the geometry of the kerf created in the course of abrasive waterjet machining of ductile materials. In: Lichtarowicz A (ed) 1992 *Jet Cutting Technol.*, Kluwer Acad. Press, Dordrecht, pp 525-541
271. Kovacevic R 1992 Monitoring the depth of abrasive waterjet penetration. *Int. J. Mach. Tools and Manuf.* 32: 725-736
272. Kovacevic R, Mohan R, Hirscher J 1992 Rehabilitation of concrete pavements assiated with abrasive waterjets. In: Lichtarowicz A (ed) 1992 *Jet Cutting Technol.*, Kluwer Acad. Publ., Dordrecht, pp 425-442

273. Yazici S 1989 Abrasive jet cutting and drilling of rock. PhD thesis, Univ. of Missouri-Rolla, Rolla
274. Laidler K J 1950 *Chemical Kinetics*. McGraw-Hill, New York
275. Kovacevic R, Fang M 1994 Modeling of the influence of the abrasive waterjet cutting parameters on the depth of cut based on fuzzy rules. *Int. J. Mach. Tools Manuf.* 34: 55-72
276. Kovacevic R, Fang M, Beardsley H 1994a Development of a fuzzy logic method to select abrasive water jet parameters for notching concrete pavement slabs. *Int. J. Water Jet Techn.* 2: 28-38
277. Mazurkiewicz M 1989 Material removal by hydro-abrasive high-pressure jet mechanism study. *SEM TP MS89-811*, pp 1-15
278. Corcoran M, Mazurkiewicz M, Karlic P 1988 Computer simulation of an abrasive waterjet cutting process. In: Woods P A (ed) 1988 *Proc. 9th Int. Symp. Jet Cutting Techn.*, BHRA Fluid Engng., Cranfield, pp 49-59
279. Yong Z, Kovacevic R 1997b Simulation of chaotic particle motion in particle-laden jetflow and application to abrasive waterjet machining. *ASME J. Fluid Engng.* 119: 435-442
280. Yong Z, Kovacevic R 1996 Modeling of 3D abrasive waterjet machining, part I+II. In: Gee C (ed) 1996 *Jetting Technol.*, Mechan. Engrs. Publ., London, pp 73-89
281. Himmelreich U, Rieß W 1991 Hydrodynamic investigations on abrasive-water jet cutting tools. In: Saunders D (ed) 1991 *Jet Cutting Technol.*, Elsevier, London, pp 3-20
282. Galecki G, Summers D A 1992 Steel shot entrained ultra high pressure waterjet for cutting and drilling hard rocks. In: Lichtarowicz A (ed) 1992 *Jet Cutting Technol.*, Kluwer Acad. Publ., Dordrecht, pp 371-388
283. Hashish M 1986 Aspects of abrasive-waterjet performance optimization. In: Saunders D (ed) 1986 *Proc. 8th Int. Symp. Jet Cutting Techn.*, BHRA Fluid Engng., Cranfield, pp 297-308
284. Chalmers E J 1991 Effect of parameter selection on abrasive waterjet performance. In: Labus T J (ed) 1991 *Proc. 6th Amer. Water Jet Conf.*, WJTA, St. Louis, pp 345-354
285. Nadeau E, Stubley G D, Burns D J 1991 Prediction and role of abrasive velocity in abrasive water jet cutting. *Int. J. Water Jet Technol.* 1: 109-116
286. Laurinat A, Louis H, Meier-Wiechert G 1993 A model for milling with abrasive water jet. In: Hashish M (ed) 1993 *Proc. 7th Amer. Water Jet Conf.*, Vol. 1, Water Jet Techn. Ass., St. Louis, pp 119-139
287. Ansari A I 1990 A study on turning with abrasive waterjets. PhD thesis, Michigan Technol. Univ., Houghton
288. Dombrowski H 1974 Betonschneiden mit dem WOMA-Hochdruckwasser-Strahlsystem. *Betoninformation* 2: pp 3-6
289. Hashish M 1986 Milling with abrasive-waterjets: a preliminary investigation. In: Hood M Dornfeld D (eds) 1986 *Proc. 4th US Water Jet Conf.*, ASME, New York, pp 1-10
290. Laurinat A 1995 Abtragen mit Wasserabrasivinjektorstrahlen. *VDI-Fortschritt-Berichte*, Reihe 2, Nr. 327
291. Olsen J 1974 Jet slotting of concrete. In: Coles N G, Barrall S J (eds) 1974 *Proc. 2nd Int. Symp. Jet Cutting Techn.*, BHRA Fluid Eng., Cranfield, pp G1/1-G1/10
292. Momber A W 1991 Betonbearbeitung mit Abrasiv-Druckwasserstrahlen - Hinweise zur Prozeßoptimierung. *Bautechnik* 68: 242-249
293. Hu F, Yang Y, Geskin E S, et al. 1991 Characterization of material removal in the course of abrasive waterjet machining. In: Labus T J (ed) 1991 *Proc. 6th Amer. Water Jet Conf.*, Water Jet Techn. Ass., St. Louis, pp 17-29

294. Barton R E 1982 A safe method of cutting steel and rock. In: Stephens H S, Davies E B (eds) 1982 *Proc. 6th Int. Symp. Jet Cutting Techn.*, BHRA Fluid Engng., Cranfield, pp 503-518
295. Khan M E, Geskin E S 1994 Velocity measurements of water jet and abrasive water jet by the use of Laser Transit Anemometer. *FED-Vol. 191*: 59-63
296. Hashish M 1993 The effect of beam angle in abrasive-waterjet machining. *ASME J. Engng. for Ind.* 115:51-56
297. Wada S, Kumon Y 1993 Wear of Si_3N_4 ceramics by abrasive water jet. *J. of the Ceramic Soc. of Japan - Int. Edit.* 101: 808-811
298. Wada S, Kumon Y 1993a The effects of impact angles of jet stream in the abrasive water jet cutting of Si_3N_4 ceramics. *J. of the Ceramics Soc. of Japan - Int. Edit.* 101: 1265-1269
299. Kim T J, Sylvia J G, Posner L 1985 Piercing and cutting of ceramics by abrasive water jet. *PED-Vol. 17*: 19-24
300. Nakamura H, Narazaki T, Yanagihara S 1989 Cutting technique and system for biological shield. *Nuclear Technol.* 86: 168-178
301. Foldyna J, Fialova V 1989 Moznost vynzity nekterych tezkych mineralu pro generovani abrazivniho vodniho paprsku. In: Rakowski Z (ed) 1989 *Proc. Conf. Mining Geomech.*, Vol. 2, Dom Technika, Ostrava, pp 389-399
302. Cousens A K, Hutchings I M 1983 Influence of erodent particle shape on the erosion of mild steel. In: Field J E (ed) 1983 *Proc. 6th Int. Conf. Erosion by Liquid and Solid Impact,* Cavendish Lab., Cambridge, pp 41/1-41/7
303. Uetz H (ed) 1986 *Abrasion und Erosion*, Carl Hanser-Verlag, Muenchen
304. Echert D C, Hashish M, Marvin M 1987 Abrasive-waterjet and waterjet techniques for decontaminating and decommissioning nuclear facilities. In: Hood M, Dornfeld D (eds) 1987 *Proc. 4th US Water Jet Conf.*, ASME, New York, pp 73-81
305. Konno T, Narazaki T, Yokota M, et al. 1987 Abrasive water jet cutting technique for biological shield concrete cutting. In: *Proc. Int. Decomm. Symp.*, Pittsburgh, pp IV/270-IV/284
306. Matsumoto K, Arawasa H, Yamaguchi S 1988 A study of the effect of abrasive material on cutting with abrasive water jet. In: Woods P A (ed) 1988 *Proc. 9th Int. Symp. Jet Cutting Techn.*, BHRA Fluid Engng., Cranfield, pp 255-269
307. Kokaji C, Sakashita F, Oura S, et al. 1988 Effects of abrasives on concrete cutting. In: Woods P A (ed) 1988 *Proc. 9th Int. Symp. Jet Cutting Techn.*, BHRA Fluid Engng., Cranfield, pp 571-580
308. Labus T J, Neusen K F, Albert D G, et al. 1989 Factors influencing the abrasive mixing process. In: Vijay M M, Savanick G A (eds) 1989 *Proc. 5th Amer. Water Jet Conf.*, Water Jet Techn. Ass., St. Louis, pp 205-215
309. Guo N S, Louis H, Meier G, et al. 1994 Modeling of abrasive particle disintegration in abrasive water jet cutting in relation to the recycling capability. In: Allen N G (ed) 1994 *Jet Cutting Technol.*, Mech. Engng. Publ. Ltd., London, pp 567-587
310. Kitamura, M, Ishikawa M, Sudo K, et al. 1992 Cutting of steam turbine components using an abrasive water jet. In: Lichtarowicz A (ed) 1992 *Jet Cutting Technol.*, Kluwer Acad. Publ., Dordrecht, pp 543-554
311. Matsui S, Matsumura H, Ikemoto Y, et al. 1991 High precision cutting method for metallic materials by abrasive waterjet. In: Saunders D (ed) 1991 *Jet Cutting Technol.*, Elsevier Sci. Publ., London, pp 263-278
312. Sano T, Takahashi M, Yurakoshi Y, et al. 1991 Abrasive waterjet cutting of amorphous foil metal. In: Saunders D (ed) 1991 *Jet Cutting Technol.*, Elsevier Sci. Publ., London, pp 263-278

313. Hashish M 1989 Characteristics of surfaces machined with abrasive waterjets. *MD-Vol. 16*: 23-32
314. Hocheng H, Chang K R 1994 Material removal analysis in abrasive waterjet cutting of ceramic plates. *J. of Mat. Process. Techn.* 40: 287-304
315. Geskin E S, Chen W L, Lee W T 1988 Waterjet cutting experiments determine optimal techniques. *Glass Digest*, 66-69
316. Hashish M 1991 Advances in composite machining with abrasive-waterjets. *PED-Vol. 49/MD-Vol. 27*: 93-111
317. Yanagiuchi S, Yamagata H 1986 Cutting and drilling of glass by abrasive water jet. In: Saunders D (ed) 1986 *Proc. 8th Int. Symp. Jet Cutting Techn.*, BHRA Fluid Engng., Cranfield, pp 323-329
318. Agus M, Bortolussi A, Chiccu R 1994 Granite cutting with abrasive water jet: influence of abrasive parameters. In: Allen N G (ed) 1994 *Jet Cutting Technol.*, Mech. Engng. Publ. Ltd., London, pp 223-239
319. Pandit S M, Wu S M 1983 *Time Series and System Analysis With Applications*. Wiley & Sons, New York
320. Singh P J, Chen W L, Munoz J 1991 Comprehensive evaluation of abrasive waterjet cut surface quality. In: Labus T J (ed) 1991 *Proc. 6th Amer. Water Jet Conf.*, Water Jet Techn. Ass., St. Louis, pp 139-161
321. Burnham C D, Kim T J 1989 Statistical characterization of surface finish produced by a high pressure abrasive water jet. In: Vijay M M, Savanick G A (eds) 1989 *Proc. 5th Amer. Water Jet Conf.*, Water Jet Techn. Ass., St. Louis, pp 165-175
322. Colligan K, Ramulu M, Arola D 1993 Investigation of edge quality and ply delamination in abrasive water jet machining of graphite/epoxy. *MD-Vol. 45/PED-Vol. 66*: 167-185
323. Singh P J, Geskin E S, Li F, et al. 1994 Relative performance of abrasives in abrasive waterjet cutting. In: Allen N G (ed) 1994 *Jet Cutting Technol.*, Mech. Engng. Publ. Ltd., London, pp 521-541
324. Curham J, Reuber M, Kim T J 1989 Force control of surface finish in abrasive waterjet cutting. *PED-Vol. 41*: 31-36
325. Tan D K 1986 A model for the surface finish in abrasive-waterjet cutting. In: Saunders D (ed) 1986 *Proc. 8th Int. Symp. Jet Cutting Techn.*, BHRA Fluid Engng., Cranfield, pp 309-313
326. Singh J, Jain S C 1995 Mechanical issues in laser and abrasive water jet cutting. *JOM* 35: 28-30
327. Struck D 1990 Titan- und Aluminiumbleche im Flugzeugbau mit Abrasiv-Hochdruckwasserstrahlschneiden. *Werkstatt und Betrieb* 123: 861-864
328. Arola D, Ramulu M 1996 A residual stress analysis of metals machined with the abrasive waterjet. In: Gee C (ed) 1996 *Jetting Technol.*, Mechan. Engng. Publ., London, pp 269-290
329. Lavander C A, Smith M T 1985 Evaluation of waterjet-machined metal matrix composites tensile test specimens. Pacific Northwest Laboratory, Richland, WA, *Rep.-No. PNL-5858*
330. Kitamura M, Ishimura T, Ishikawa M, et al. 1993 Development of water jet cutting technology for steam turbine components. *Quarterly J. of the Japan Welding Soc.* 11: 461-465
331. Hashish M 1986 Turning with abrasive waterjets - a preliminary investigation. *PED-Vol. 22*: 79-100
332. Ives L K, Ruff A W 1977 Electron microscopy study of erosion damage in copper. *ASTM STP 664:* 5-35

333. Capello E, Monno M, Semeraro Q 1994 Delamination in water jet cutting of multi-layered composite materials: a predictive model. In: Allen N.G. (ed) 1994 *Jet Cutting Technol.*, Mech. Engng. Publ. Ltd., London, pp 463-476
334. Ricci W S, Colella U T, Macleod D W 1987 Abrasive water jet cutting of ceramics and organic based materials. In: *Proc. Eastern Manuf. Technol. Conf.*, Nat. Machine Tool Builder's Assoc., pp 6/145-6/156
335. Groppetti R, Monno M 1992 A contribution to the study of burr formation in hydro abrasive jet machining. In: Lichtarowicz A (ed) 1992 *Jet Cutting Technol.*, Kluwer Acad. Publ., Dordrecht, pp 621-633
336. Machaida T, Okai T, Ozaki J, et al. 1995 Potentiality of water jet method for cutting of sheet materials. In: Labus T J (ed) 1995 *Proc. 8th Amer. Water Jet Conf.*, Water Jet Techn. Ass., St. Louis, pp 343-358
337. Hashish M 1993 Prediction models for AWJ machining operations. In: Hashish M (ed) 1993 *Proc. 7th Amer. Water Jet Conf.*, Vol. 1, Water Jet Techn. Ass., St. Louis, pp 205-216
338. Hashish M 1994 Controlled-depth milling techniques using abrasive-waterjets. In: Allen N G (ed) 1994 *Jet Cutting Technol.*, Mech. Eng. Publ. Ltd., London, pp 449-461
339. Laurinat A, Louis H, Haferkamp H 1993 Werkstücke dreidimensional bearbeiten mit Hilfe des Wasserabrasivstrahlverfahrens. *Maschinenmarkt* 99: 22-27
340. Hashish M 1994 Controlled-depth milling of isogrid structures with abrasive-waterjets. *PED-Vol. 68-1*: 413-419
341. Öjmertz C M 1993 Abrasive waterjet milling: an experimental investigation. In: Hashish M (ed) 1993 *Proc. 7th Amer. Water Jet Conf.*, Vol. 2, Water Jet Techn. Ass., St. Louis, pp 777-791
342. Hocheng H, Tsai H Y, Shiue J J, et al. 1997 Feasibility study of abrasive-waterjet milling of fiber-reinforced plastics. *ASME J. Manuf. Sci. and Engng.* 119: 133-142
343. Freist B, Haferkamp H, Laurinat A, et al. 1989 Abrasive jet machining of ceramic products. In: Vijay M M, and Savanick G A. (eds) 1989 *Proc. 5th Amer. Water Jet Conf.*, ed., Water Jet Techn. Ass., St. Louis, pp 191-204
344. Yong Z, Kovacevic R 1997d 3D-simulation of macro and micro characteristics for AWJ machining. In: *Proc. 9th Amer. Water Jet Conf.*, Water Jet Techn. Ass., St. Louis, in print
345. Zeng J, Wu S, Kim, T J 1994 Development of a parameter prediction model for abrasive waterjet turning. In: Allen N G (ed) 1994 *Jet Cutting Technol.*, Mech. Engng. Publ. Ltd., London, pp 601-617
346. Kwak H, Kovacevic R, Mohan R 1996 Monitoring of AWJ drilling of ceramics using AE sensing technique. In: Gee C (ed) 1996 *Jetting Technol.*, Mechan. Engng. Publ. Ltd, London, pp 137-152
347. Hashish M, Whalen J 1993 Precision drilling of ceramic-coated components with abrasive-waterjets. *ASME J. Energy Gas. Turb. Power* 115: 148-154
348. Hashish M 1988 Turning, milling, and drilling with abrasive water jets. In: Woods P A (ed) 1988 *Proc. 9th Int. Symp. Jet Cutting Techn.*, BHRA Fluid Engn., Cranfield, pp 113-131
349. Hashish M 1989 Machining of advanced composites with abrasive-water jets. *Manufact. Rev.* 2: 142-150
350. Kovacevic R, Kwak, H S, Mohan R 1997 Acoustic emission sensing as a tool for understanding the mechanism of abrasive water jet drilling of difficult-to-machine materials. *Inst. Mech. Engrs., J. Engng. Manuf.* 211: in print
351. Raju S P, Ramulu M 1993 A transient model for material removal in the abrasive-waterjet machining process. In: Hashish M (ed) 1993 *Proc. 7th Amer. Water Jet Conf.*, Vol. 1, Water Jet Techn. Ass., St. Louis, pp 141-155

352. Zeng J, Munoz J 1994 Intelligent automation of abrasive water jet cutting for efficient production. In: Allen N G (ed) 1994 *Jet Cutting Technol.*, Mech. Engng. Publ. Ltd., London, pp 401-408
353. Yong Z, Kovacevic R 1997 Modeling jetflow drilling with consideration of chaotic erosion histories of particles. *Wear*, in print
354. Yong Z, Kovacevic R 1997a Fundamentals of constructing particle-laden jetflow by fractal point sets and predicting 3D solid-erosion rates. *Chaos, Solitons & Fractals* 8: 207-220
355. Ohlsson L, Powell J, Ivarson A, et al. 1992 Optimization of the piercing or drilling mechanism of abrasive water jets. In: Lichtarowicz A (ed) 1992 *Jet Cutting Technol.*, Kluwer Acad. Publ., Dordrecht, pp 359-370
356. Hashish M 1996 Deep hole drilling in metals using abrasive-waterjets. In: Gee C (ed) 1996, *Jetting Technol.*, Mechan. Engng. Publ., London, pp 691-708
357. Hashish M 1995 Investigation of new fluidjet polishing techniques. in review.
358. Hashish M 1992 Diamond film polishing with abrasive-liquid jets: an exploratory study. *PED-Vol. 58*: 29-41
359. Li F, Geskin E S, Tismenetskiy L 1996 Feasibility study of abrasive waterjet polishing. In: Gee C (ed) 1996 *Jetting Technol.*, Mechan. Engng. Publ., London, pp 709-723
360. Sheridan M D, Taggart D G, Kim T J 1994 Screw thread machining of composite materials using abrasive waterjet cutting. *PED-Vol. 68-1*: 421-432
361. Sheridan M D, Taggart D G, Kim T J, et al. 1995 Microstructural characterization of threaded composite tubes machined using AWJ cutting. In: Labus T J (ed) 1995 *Proc. 8th Amer. Water Jet Conf.*, Vol. 1, Water Jet Techn. Ass., St. Louis, pp 245-258
362. Kovacevic R, Mohan R, Ramulu M, et al. 1997a State of the art of research and development in abrasive waterjet machining. *ASME J. Manuf. Sci. and Engng.* 119: to be printed
363. Mazurkiewicz M, Karlic P 1997 Material response during hydroabrasive jet machining (HAJM). In: Hood M, Dornfeld D (eds) 1987 *Proc. 4th US Water Jet Conf.*, ASME, New York, pp 159-167
364. Hashish M, Monserud D O, Bondurant P D, et al. 1993 A new abrasive-waterjet nozzle for automated and intelligent machining. In: Hashish M (ed) 1993 *Proc. 7th Amer. Water Jet Conf.*, Vol. 2, Water Jet Techn. Ass., St. Louis, pp 829-842
365. Hunt D C, Burnham C D, Kim T J 1987 Surface characterization in machining advanced ceramics by abrasive water jet. In: Hood M, Dornfeld D (eds) 1987 *Proc. 4th Amer. Water Jet Conf.*, ASME, New York, pp 169-174
366. Zeng J, Munoz J 1994 Feasibility of monitoring abrasive water jet conditions by means of a vacuum sensor. In: Allen N G (ed) 1994 *Jet Cutting Technol.*, Mech. Engng. Publ. Ltd., London, pp 553-565
367. Knaupp M 1993 The application of sensors for process monitoring in high pressure water jet technology. In: Hashish M (ed) 1993 *Proc. 7th Amer. Water Jet Conf.*, Vol. 2, Water Jet Techn. Ass., St. Louis, pp 935-949
368. Hashish M 1995 Advances in fluidjet beam processing. In: Labus T J (ed) 1995 *Proc. 8th Amer. Water Jet Conf.*, Vol. 2, Water Jet Techn. Ass., St. Louis, pp 487-503
369. Kovacevic R 1994 Sensing the abrasive water jet nozzle wear. *Int. J. Water Jet Technol.* 2: 1-10
370. Kovacevic R 1991 A new sensing system to monitor abrasive waterjet nozzle wear. In: Venkatesh V C, McGeough J A (eds) 1991 *Computer-Aided Prod. Engng.*, Elsevier, London, pp 117-125
371. Kovacevic R 1991 Development of an opto-electric sensing system to monitor the abrasive waterjet nozzle wear. *PED-Vol. 44*: 9-16

372. Barker C R, Cummings A, Andersen M 1982 Jet noise measurements on hand-held cleaning equipment. In: Stephens H S, Davies E B (eds) 1982 *Proc. 6th Int. Symp. Jet Cutting Techn.*, BHRA Fluid Eng., Cranfield, pp 161-178
373. Kovacevic R, Evizi M 1990 Nozzle wear detection in abrasive waterjet cutting systems. *Mater. Evaluation* 48: 348-353
374. Kovacevic R, Wang L, Zhang Y M 1993 Detection of abrasive waterjet nozzle wear using acoustic signature analysis. In: Hashish M (ed) *Proc. 8th Amer. Water Jet Conf.*, Water Jet Techn. Ass., St. Louis, pp 217-231
375. Kovacevic R, Wang L, Zhang Y M 1994 Identification of abrasive waterjet nozzle wear based on parametric spectrum estimation of acoustic signals. *Proc. Inst. Mech. Engrs., J. of Engng. Manuf.* 208: 173-181
376. Mohan R, Kovacevic R, Damarla T R 1994a Real-time monitoring of AWJ nozzle wear using artificial neural network. *Trans. NAMRI* 22: 253-358
377. Kovacevic R, Chen G 1989 A workpiece reactive force as a parameter for monitoring the nozzle wear in turning operation with abrasive waterjet. *PED-Vol. 41*: 43-49
378. Kovacevic R 1990 Estimation of nozzle wear under varying cutting conditions. *PED-Vol. 55*: 73-81
379. Kovacevic R 1991 Detection of abrasive waterjet nozzle wear by using time series modeling technique. *Trans. of NAMRI/SME* 19: 96-100
380. Kovacevic R, Zhang Y M 1992 On-line fuzzy recognition of abrasive waterjet nozzle wear. In: Lichtarowicz A (ed) 1992 *Jet Cutting Technol.*, Kluwer Acad. Publ., Dordrecht, pp 329-345
381. Swanson R K, Kilman M, Cerwin S, et al. 1987 Study of particle velocities in water driven abrasive jet cutting. In: Hood M, Dornfeld D (eds) 1978 *Proc. 4th US Water Jet Conf.*, ASME, New York, pp 103-107
382. Himmelreich U, Rieβ, W 1991 Hydrodynamic investigations on abrasive water jet cutting tools. In: Schafstall H G, et al. (eds) 1991 *Proc. 3rd Int. Symp. Underwater Technol.*, GKSS GmbH, Geesthacht, pp 17.01-17.09
383. Li H Y, Geskin E S, Chen W L 1989 Investigation of forces exerted by an abrasive water jet on a workpiece. In: Vijay M M, Savanick G A (eds) 1989 *Proc. 5th Amer. Water Jet Conf.*, Water Jet Techn. Ass., St. Louis, pp 393-402
384. Momber, A W, Kovacevic R, Mohan, R 1997 An acoustic emission study on bauxite ceramics eroded by abrasive water jetting. *J. of Phys. D: Appl. Phys.*, submitted
385. Hunt D C, Kim T J, Reuber M 1988 Surface finish optimization for abrasive waterjet cutting. In: Woods P (ed) 1988 *Proc. 9th Int. Symp. Jet Cutting Techn.*, BHRA Fluid Eng., Cranfield, pp 99-112
386. Derakhshan O, Houghton J R, Jones R K et al. 1991 Cavitation monitoring of hydro-turbines with RMS acoustic emission measurement. *ASTM STP 1077*, ASTM, New York, pp 305-315
387. Singh P J 1995 Development of a window-based expert system for abrasive water jet cutting. In: Labus T J (ed) 1995 *Proc. 8th Amer. Water Jet Conf.*, Vol. 2, Water Jet Techn. Ass., St. Louis, pp 717-726